TOPICS IN FLAVONOID CHEMISTRY AND BIOCHEMISTRY

PROCEEDINGS OF THE FOURTH HUNGARIAN BIOFLAVONOID SYMPOSIUM
KESZTHELY, 1973

TOPICS IN FLAVONOID CHEMISTRY AND BIOCHEMISTRY

Proceedings of the Fourth Hungarian Bioflavonoid Symposium Keszthely, 1973

Edited by

L. FARKAS

M. GÁBOR

F. KÁLLAY

ELSEVIER SCIENTIFIC PUBLISHING COMPANY
AMSTERDAM—OXFORD—NEW YORK 1975

The distribution of this book is being handled by the following publishers:
for the U.S.A. and Canada
American Elsevier Publishig Company, Inc.
52 Vanderbilt Avenue
New York, New York 10017

for the East European countries, China, Korean People's Republic, Cuba, People's Republic of Viet-nam and Mongolia
Akadémiai Kiadó, The Publishing House of the Hungarian Academy of Sciences, Budapest

for all remaining areas
Elsevier Scientific Publishing Company
335 Jan van Galenstraat
P.O. Box 211, Amsterdam, The Netherlands

Library of Congress Card Number 75—2768

ISBN 0-444-99861-6

Joint edition published by
Elsevier Scientific Publishing Company, Amsterdam
The Netherlands and
Akadémiai Kiadó, The Publishing House of the Hungarian Academy of Sciences,
Budapest, Hungary

Printed in Hungary

CONTENTS

LIST OF CONTRIBUTORS

ACHARD, M., Laboratoire de Physiologie Pathologique, Faculté de Médecine Necker, Paris, France
ANTAL, E., Faculty of Applied Chemistry, Kossuth Lajos University, Debrecen, Hungary
ANTUS, S., Research Group for Alkaloid Chemistry of the Hungarian Academy of Sciences, Budapest, Hungary
BAKAY, M., Institute of Microbiology, Medical University, Szeged, Hungary
BÁLINT, J., BIOGAL Pharmaceutical Factory, Debrecen, Hungary
BÉLÁDI, I., Institute of Microbiology, Medical University, Szeged, Hungary
BOGNÁR, R., Institute of Organic Chemistry, Kossuth Lajos University, Debrecen, Hungary
BORDA, J., Institute of Applied Chemistry, Kossuth Lajos University, Debrecen, Hungary
CHARI, V. M., Institut für pharmazeutische Arzneimittellehre der Universität München, FRG
CHOPIN, J., Laboratoire de Chimie biologique, UER Chimie-Biochimie, Université de Lyon, Villeurbanne, France
DELLE MONACHE, F., Facoltà di Scienze Matematiche Fisiche e Naturali, Università di Roma e Centro Chimica dei Recettori del CNR, Università Cattolica del S. Cuore, Roma, Italy
DINYA, Z., Institute of Organic Chemistry, Kossuth Lajos University, Debrecen, Hungary
DONNELLY, D. M. X., Department of Chemistry, University College, Dublin, Ireland
FARKAS, L., Institute of Organic Chemistry, Technical University, Budapest, Hungary
FÖRSTER, H., Zentrum der Biologischen Chemie der Universität, Frankfurt, FRG
GAZAVE, J.-M., Laboratoire de Physiologie Pathologique, Faculté de Médicine Necker, Paris, France
(Mrs.) GÁBOR, E.-Sz., School of Food Industry, Chemical Department, Szeged, Hungary
GÁBOR, M., Department of Pharmacodynamics, Medical University, Szeged, Hungary
GOTTSEGEN, Á., Research Group for Alkaloid Chemistry of the Hungarian Academy of Sciences, Budapest, Hungary
GRIFFITHS, L. A., Department of Biochemistry, University of Birmingham, UK
GUNNING, P. J. M., Department of Chemistry, University College, Dublin, Ireland
HASLAM, E., Department of Chemistry, University of Sheffield, UK
HETÉNYI, E., BIOGAL Pharmaceutical Factory, Debrecen, Hungary
JANZSÓ, G., Research Institute for Organic Chemical Industry, Budapest, Hungary
KAVANAGH, Ph. J., Department of Chemistry, University College, Dublin, Ireland
KÁLLAY, F., Research Institute for Organic Chemical Industry, Budapest, Hungary
KEIL, D., Institut für Tierernährung, Tierärztliche Hochschule, Hannover, FRG
KRÜGER, H., Institut für Tierernährung, Tierärztliche Hochschule, Hannover, FRG
LÉVAI, A., Institute of Organic Chemistry, Kossuth Lajos University, Debrecen, Hungary
L.-TŐKÉS, A., Institute of Organic Chemistry, Kossuth Lajos University, Debrecen, Hungary

Litkei, Gy., Institute of Organic Chemistry, Kossuth Lajos University, Debrecen, Hungary
Marini-Bettòlo, G. B., Facoltà di Scienze Matematiche Fisiche e Naturali, Università di Roma e Centro Chimica dei Recettori del CNR, Università Cattolica del S. Cuore, Roma, Italy
Nógrádi, M., Research Group for Alkaloid Chemistry of the Hungarian Academy of Sciences, Budapest, Hungary
Parrot, J.-L., Laboratoire de Physiologie Pathologique, Faculté de Médecine Necker, Paris, France
Pusztai, R., Institute of Microbiology, Medical University, Szeged, Hungary
Rákosi, M., Institute of Organic Chemistry, Kossuth Lajos University, Debrecen, Hungary
R.-Dávid, É., Institute of Organic Chemistry, Kossuth Lajos University, Debrecen, Hungary
Roger, C., Laboratoire de Physiologie Pathologique, Faculté de Médecine Necker, Paris, France
Schultz, G., Institut für Tierernährung, Tierärztliche Hochschule, Hannover, FRG
Seligmann, O., Institut für pharmazeutische Arzneimittellehre der Universität München, FRG
Strelisky, J., Institute of Organic Chemistry, Technical University, Budapest, Hungary
Szabó, V., Institute of Applied Chemistry, Kossuth Lajos University, Debrecen, Hungary
Takács, Ö., Department of Biochemistry, Medical University, Szeged, Hungary
Vermes, B., Institute of Organic Chemistry, Technical University, Budapest, Hungary
Wagner, H., Institut für pharmazeutische Arzneimittellehre der Universität München, FRG
Weinges, K., Organisch-Chemisches Institut der Universität, Heidelberg, FRG

OPENING ADDRESS

by

PROF. L. FARKAS

Ladies and Gentlemen,

Allow me to extend you a very hearty welcome on behalf of the Hungarian Academy of Sciences and the Hungarian Chemical Society, the organizers of this Symposium.

It was eight years ago that our Academy decided to establish a Work-Committee for Flavonoid Research with the task of promoting the research activity and coordinating the work of chemists, biochemists, pharmacologists and botanists in this interesting field, which has – as you know – great traditions in Hungary ever since the time of Géza Zemplén.

In our Work-Committee, our plans of research and problems of common interest are discussed at sessions held twice or three times a year, whereas new results are reported at yearly meetings, which we call "colloquia". On the other hand, we present features of our recent research work to foreign scientists and get acquainted in exchange with their current activity in the flavonoid field at Symposia with international participation, of which the one beginning today is the fourth.

It is a very great pleasure to see here many scientists from abroad and from Hungary, who have been our usual participants at these Symposia; it is just as great a satisfaction to find an increase in the number of those present, signalling a growing interest in our work, and ever-improving contacts.

We thank you all for coming and we hope you will have much pleasure in listening to the interesting lectures of a successful Symposium; we wish you a very pleasant stay in Hungary.

Herewith I open the Symposium.

NATURAL PRENYLATED FLAVONOIDS

by
G. B. MARINI-BETTÒLO and F. DELLE MONACHE
Facoltà di Scienze Matematiche Fisiche e Naturali, Università di Roma e Centro Chimica dei Recettori del CNR, Università Cattolica del S. Cuore
Rome, Italy

The occurrence in nature of a number of prenylated phenols, coumarins and xanthones has been described in the last years.

Until recently only very few examples were known of prenylated flavonoids.

Among the isoflavones, osajin (Ia) and pomiferin (Ib), the pigments of osage orange (*Maclura pomifera* Raf.) studied by Wolfrom *et al.* [1] contain two isoprenyl groups in their molecule.

Piscidone (II) is also an isoflavone prenylated in ring B [2].

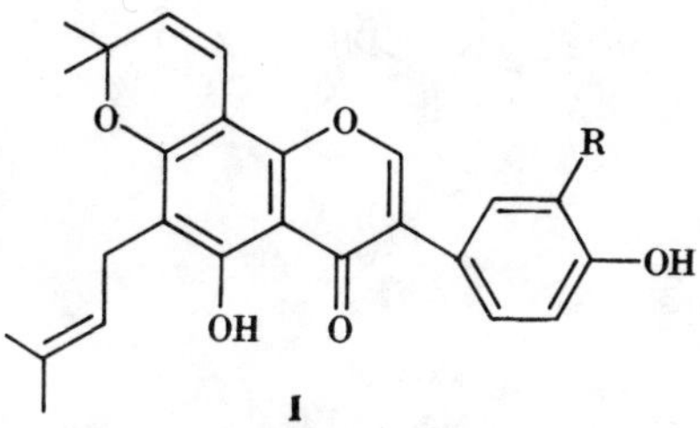

I
a. R = H Osajin
b. R = OH Pomiferin

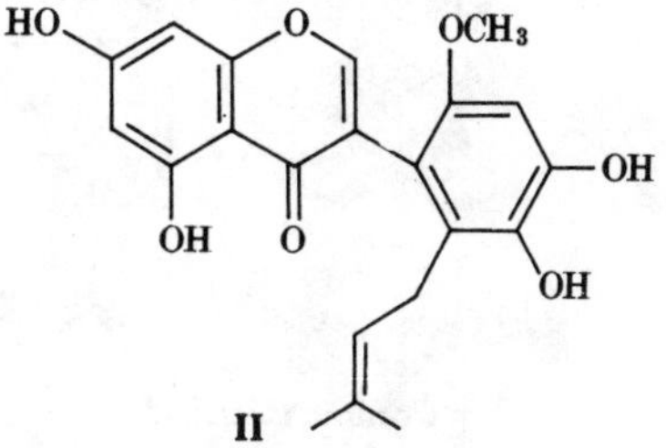

II
Piscidone

In 1957 Verzele, Stockx, Fontijn and Anteunis [3] reported the isolation of a C-prenylated chalcone, xanthohumol (III), whereas Geissman [4] described an O-prenylated flavanone (IV) obtained from *Melicope sarcococca,* a Rutacea from New Guinea.

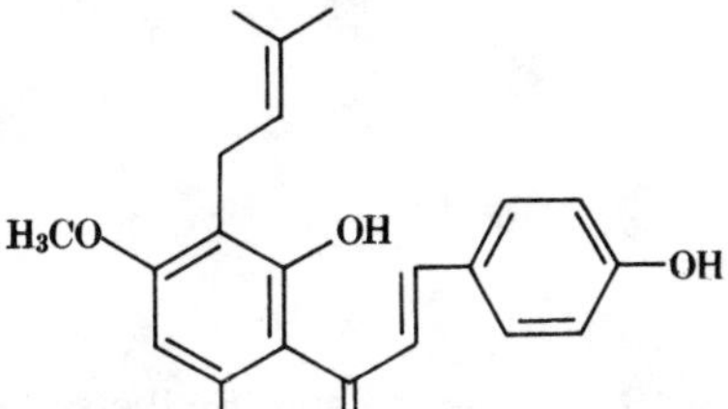

III

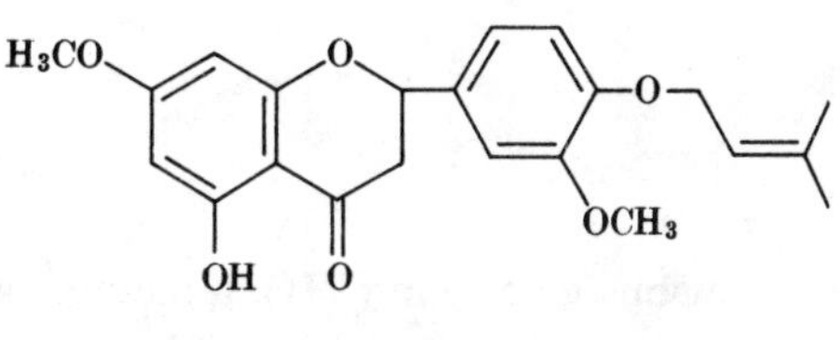

IV

Selinone, described later by Seshadri and Sood [5], is closely related to the latter compound, and is (–)-naringenin 4′-(γ,γ-dimethylallyl ether). Syntheses of these compounds have been achieved by Aurnhammer, Chari and Wagner [6].

Recent investigations on the flavonoids of the *Lonchocarpus* genus, mainly *Derris sericea* and *Cordoa piaca (Derris* sp.*)*, have shown the presence of several prenylated chalcones and flavanones and thus permit us to have a more complete picture of the relationship between these substances.

Baudrenghien *et al.* [7] isolated a chromeno-chalcone, named lonchocarpin (XXXII), from *Derris sericea* of Western Africa.

Later Gonçalves de Lima *et al.* [8] obtained an O-prenylated chalcone, cordoin (V), and a C-prenylated chalcone, isocordoin (VI) from another *Lonchocarpus* species from North East Brasil, called also *Cordoa piaca.* Mors and Nascimiento [9] reported simultaneously in Rio de Janeiro the isolation of the same O-prenylated chalcone, which they called derricidin.

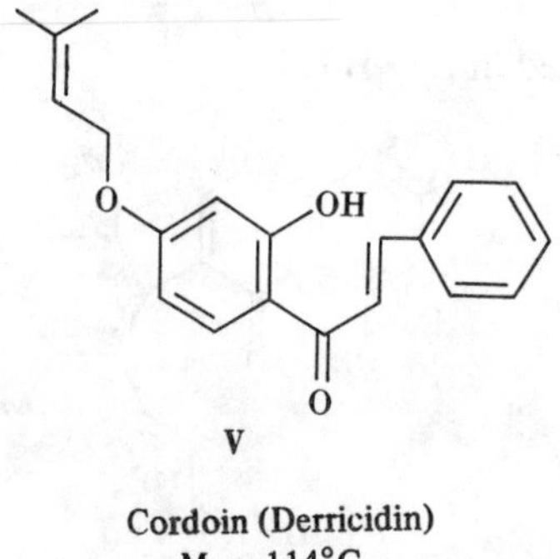

O OH O

V

Cordoin (Derricidin)
M. p. 114°C

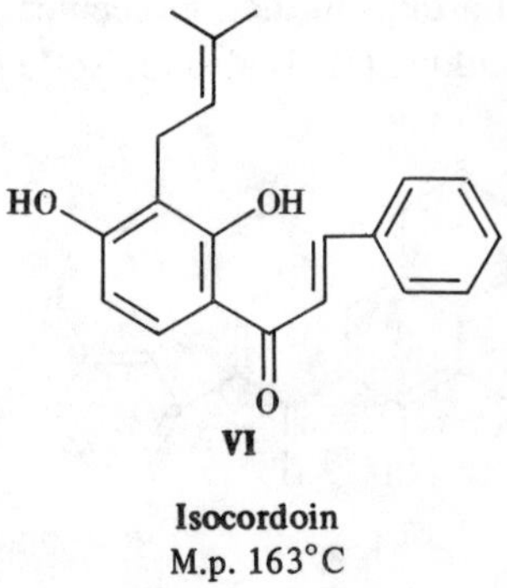

HO OH O

VI

Isocordoin
M.p. 163°C

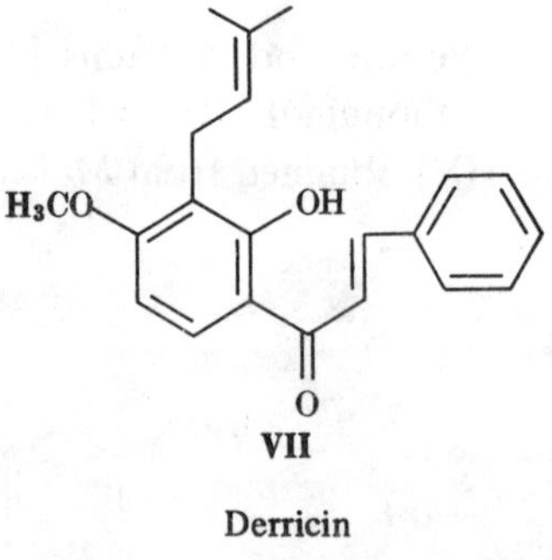

H_3CO OH O

VII

Derricin

Previously derricin (VII), a methoxyisocordoin, had been isolated by the same authors from *Derris sericea* [10].

Further investigations on *Cordoa piaca* have shown also the presence of derricin and five new substances: dihydrocordoin (XV), 2′,4′-dihydroxy-3′-(α,α-dimethylallyl)chalcone (ψ-isocordoin) (XVI), 4-hydroxyderricin (XVIII), 4-hydroxyisocordoin (XIX) and 4-hydroxylonchocarpin (XVII). The occurrence in the same plant of both prenylated chalcones and chromeno-chalcones (lonchocarpin and 4-hydroxylonchocarpin) is rather interesting, and the former may be considered the biogenetical precursors of the latter.

We shall now report the physico-chemical behaviour and reactions of the prenylated chalcones of *Lonchocarpus*, which have recently formed the object of joint investigations by research groups in Recife (Brazil) and Rome [11].

The present results support the hypothesis that the prenylated chalcones may be considered the precursors of the chromeno derivatives which have been found in many polyheterocyclic compounds, like rotenoids, etc., present in plants [12].

Both cordoin (derricidin) and isocordoin show the characteristic reactions of chalcones; yellow crystals, becoming red with NaOH solution; the UV spectra have maxima at 258 (log ε 3.70), 320 (log ε 4.30) and 340 nm (log ε 4.38), with shoulder at 310 nm (log ε 4.34).

In the case of cordoin (derricidin) the NMR spectrum indicates the presence of the system

$$-O-CH_2-CH=C(CH_3)_2$$

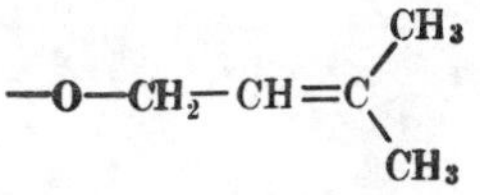

(2H at δ = 4.50 (d) (J = 7 Hz); 3H at δ = 1.75 (s) and 1.70 (s)).

Cordoin is easily hydrogenated to a tetrahydro derivative which has the NMR spectrum of the saturated system:

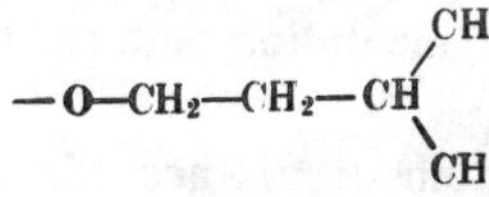

$$-O-CH_2-CH_2-CH(CH_3)_2$$

(2H at δ = 4 (t) (J = 5 Hz); 3H between δ = 1.4 and 1.9 (broad multiplet); 6H at δ = 0.95 (d)).

On the other hand, isocordoin shows the presence of the system

$$-\overset{|}{\underset{|}{C}}-CH_2-CH=C(CH_3)_2$$

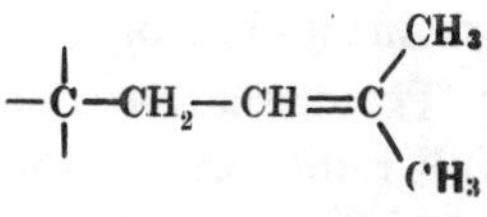

(2H at δ = 3.4 (d) (J = 8 Hz); 1H at δ = 5.32 (broad triplet) (J = 8 Hz); 3H each at δ = 1.8 and 1.66 (two broad singlets)).

Both cordoin (derricidin) and isocordoin undergo alkaline cleavage to give 2-hydroxy-4-γ,γ-dimethylallyloxyacetophenone (VIII) and 2,4-dihydroxy-3-γ-γ-dimethylallylacetophenone (IX), respectively.

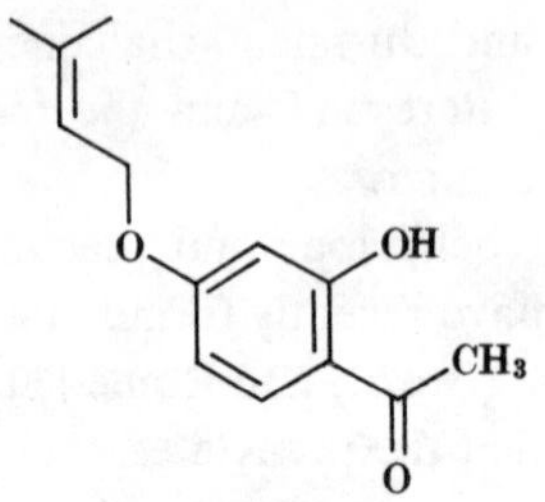

VIII 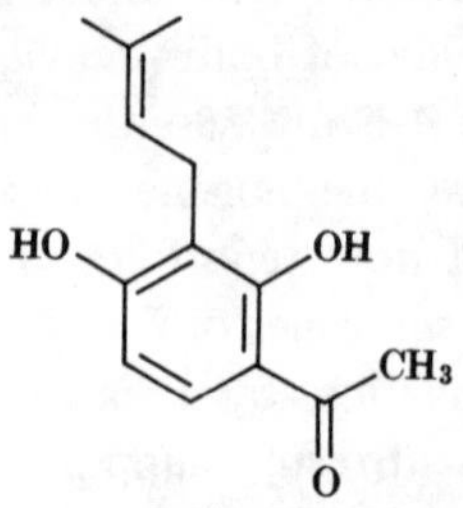 IX

The proposed structures of both these acetophenones have been confirmed by UV and IR spectroscopy as well as by NMR and MS data.

Moreover cordoin (derricidin) in methanolic HCl gives 2′,4′-dihydroxy-chalcone (X) and (±)-7-hydroxyflavanone (XI) in equal amounts.

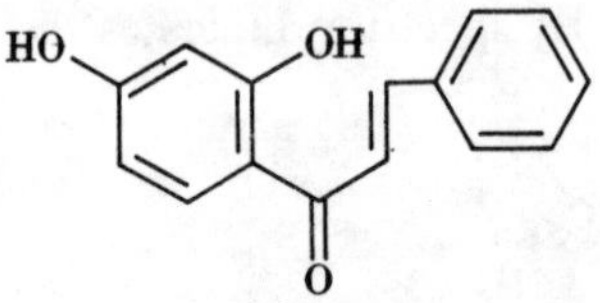

X
M. p. 135°C

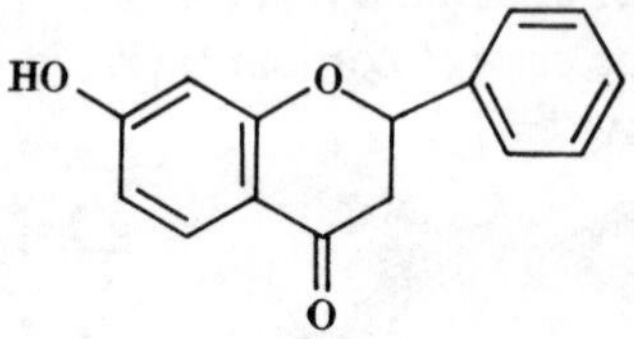

XI
M.p. 182°C

Both substances were identified through direct comparison with synthetic specimens.

The mass spectroscopic fragmentation patterns of cordoin and isocordoin are shown in Figs 1 and 2.

The fragmentation of cordoin occurs according to the pattern of chalcones and flavanones. In effect, the compound loses first of all 68 m.u. from the O-prenylated group to give 2′,4′-dihydroxychalcone (m/e 240). The chalcone successively gives the ions m/e 239, 163 and 137 (136) which correspond to the loss of one proton, of the ring B and of the group C_6H_5–CH=CH, respectively.

On the other hand, the fragmentation of isocordoin is complicated by the presence of a C-prenyl group. The direct losses of the molecular ion as chalcone (M—1, M—77 and M—104) is rather scarce (ions m/e 307, 231 and 204). The preferred way is the loss of 15 m.u. from the side chain and then 28 m.u., in order to give ions m/e 293 and 265, respectively.

Another possible way is the loss of 55 m.u. to give the ion m/e 253.

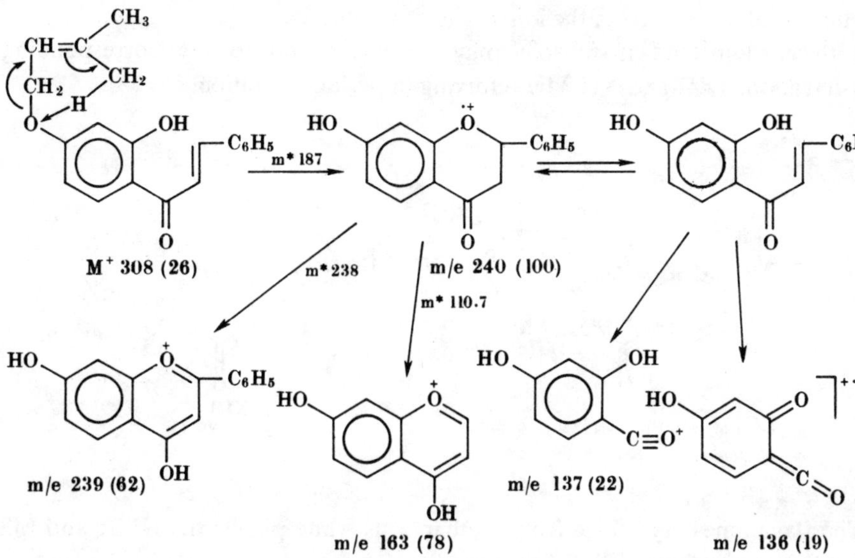

Fig. 1. Scheme of fragmentation of cordoin

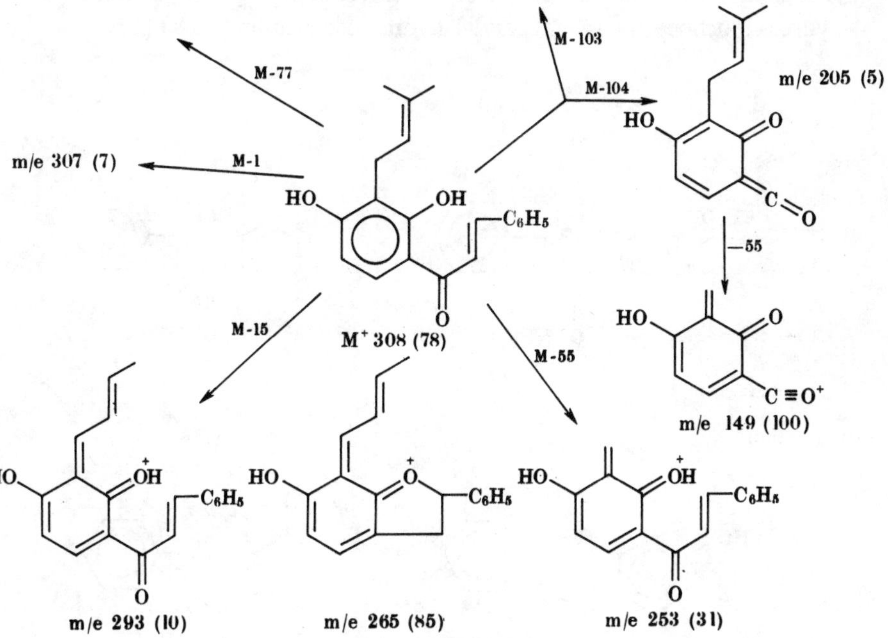

Fig. 2. Scheme of fragmentation of isocordoin

The loss of 55 m.u. from the ion 204 is considerable.

Both cordoin and isocordoin may be isomerized to the corresponding (±)-flavanone (XII and XIII) by refluxing in pyridine solution.

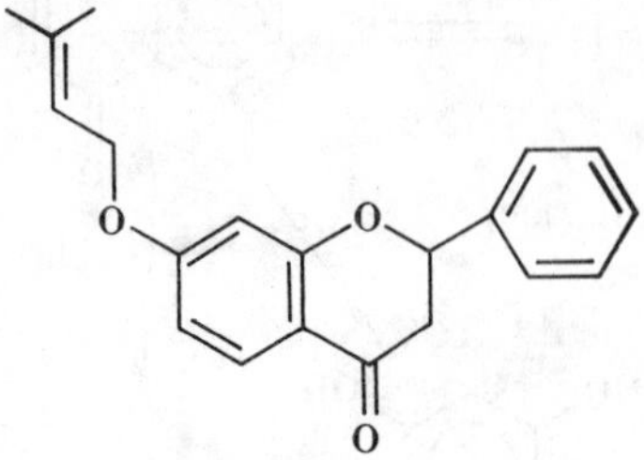

XII
M.p. 62 °C

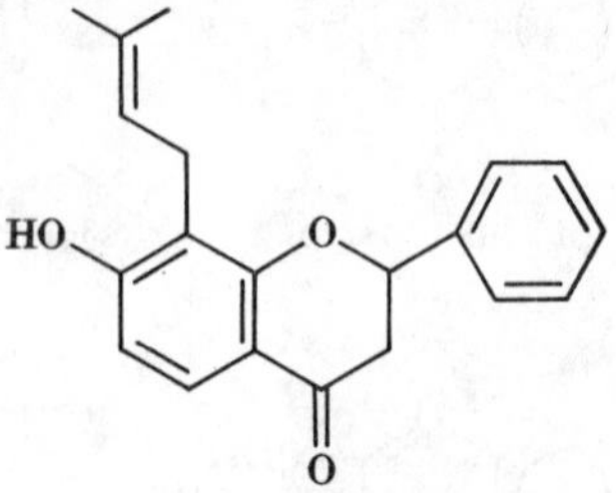

XIII
M.p. 129 °C

The flavanones crystallize from methanol as white plates; the NMR and MS data are in accordance with the proposed structures.

As it has been mentioned, from *Cordoa piaca* there were also isolated derricin (XIV) and five new compounds: dihydroxycordoin (XV), 2′,4′-dihydroxy-3′-(α,α-dimethylallyl)chalcone or ψ-isocordoin (XVI), 4-hydroxyderricin (XVIII), 4-hydroxylonchocarpin (XVII) and 4-hydroxyisocordoin (XIX) [13].

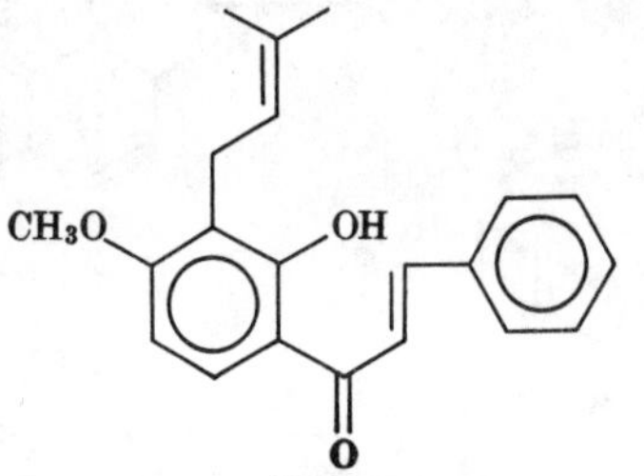

XIV

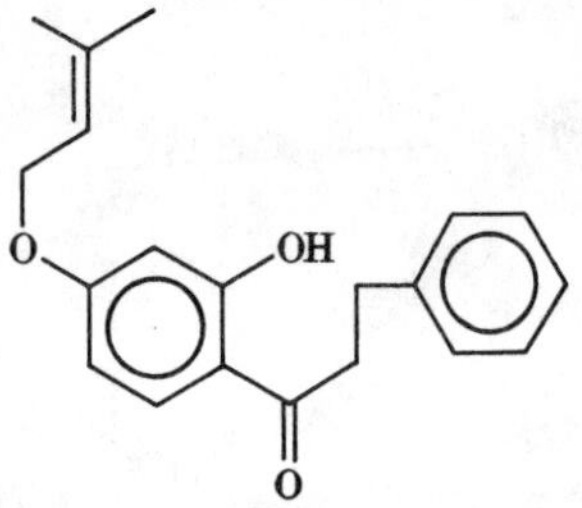

XV

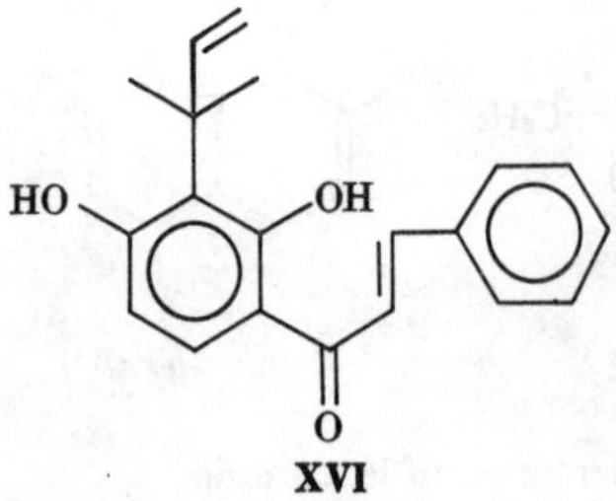

XVI

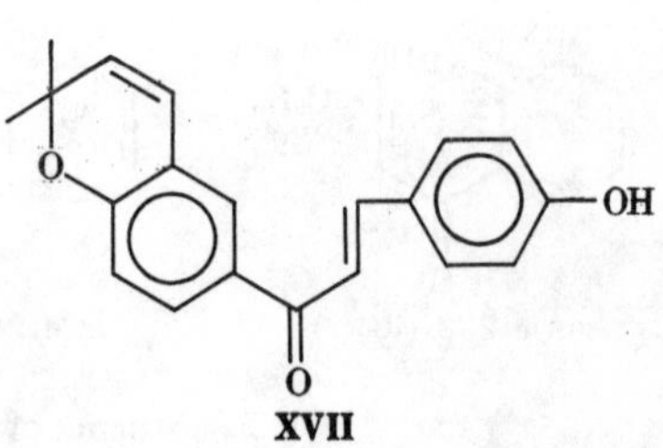

XVII

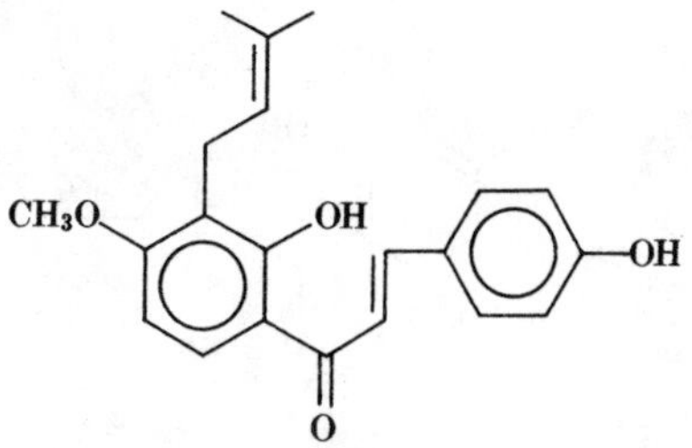

CH_3O OH OH HO OH OH

O O

XVIII 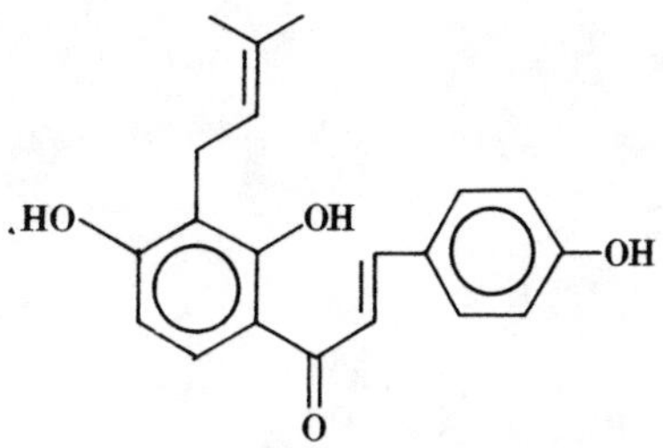 XIX

All the structures have been confirmed by the NMR and mass spectra, which also allow to establish unambiguously the presence of an OH group in position 4.

Special attention must be given to the structure of ψ-isocordoin (XVI), an isomer of isocordoin, where the C_5 unit is attached to the chalcone moiety according to a head-to-tail inversion:

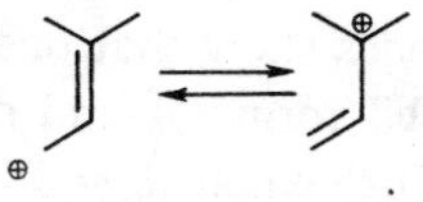

Of particular biological significance are the derivatives containing an OH group in ring B.

As mentioned above, the presence of chromeno-chalcones, such as lonchocarpin, in the *Lonchocarpus* genus may represent a stage of further cyclization and dehydrogenation of the related prenylated chalcones, which all derive from resacetophenone.

Delle Monache, Marini-Bettòlo, de Mello *et al.* [14] have therefore studied the possible isomerization of isocordoin and cordoin into the corresponding chromanes.

In the case of an *ortho*-prenylated phenol, it is known that it will cyclize in strong acid medium to give the corresponding chroman:

HO HO O

$+H^+$ $-H^+$

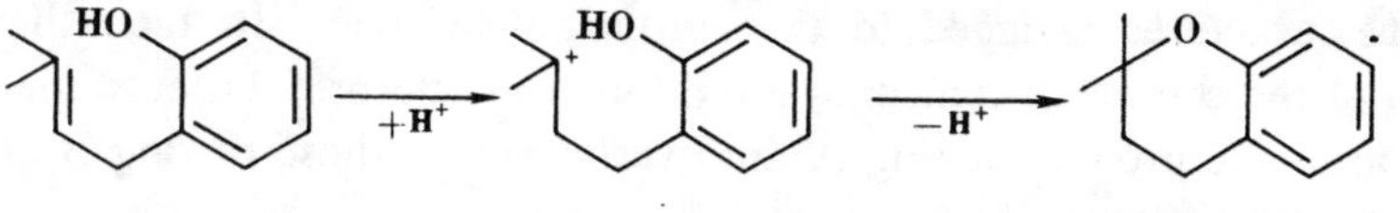

When isocordoin is dissolved in acetic acid, and a small quantity of sulfuric acid is added, the prenylated compound is quantitatively transformed into a mixture of the two chromano-chalcones (XX and XXI) formed by cyclization of the C-prenyl in 3′ with the hydroxyl group at C-2′ and C-4′, respectively.

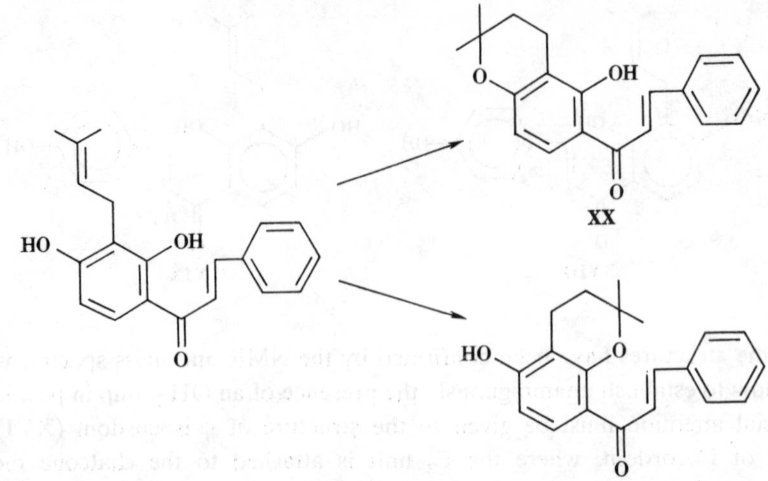

HO OH O

OH O XX

HO O O XXI

Both these chromano-chalcones show chalcone properties. The NMR spectra are consistent with the assigned formulas and differ only in the signal of the hydroxyl group, this being CO-chelated (δ = 13.9) in XX, but falling into the aromatic and olefinic proton envelope in XXI.

By alkaline cleavage both XX and XXI give the substituted acetophenones XXII and XXIII, respectively.

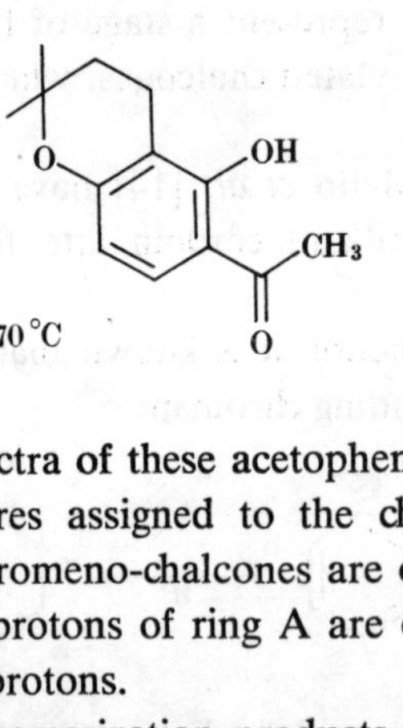

O OH CH_3 O
XXII
M.p. 70 °C

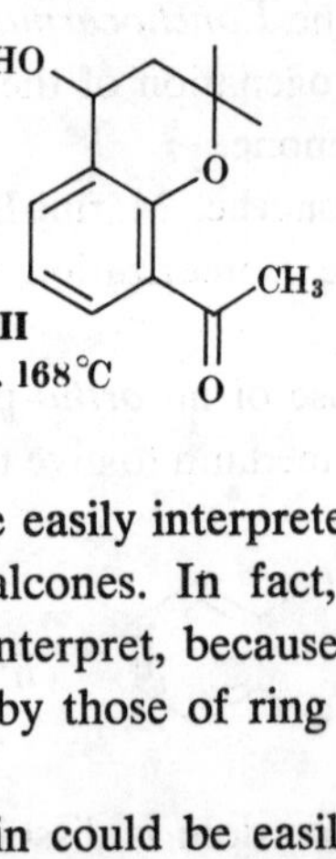

HO O CH_3 O
XXIII
M.p. 168 °C

The NMR spectra of these acetophenones can be easily interpreted and confirm the structures assigned to the chromano-chalcones. In fact, the NMR spectra of the chromeno-chalcones are difficult to interpret, because the signals of the aromatic protons of ring A are overlapped by those of ring B and also by the ethylenic protons.

Whereas the isomerization products of isocordoin could be easily predicted on the basis of the behaviour of C-prenylated phenols, the reaction of cordoin (derricidin), under the same experimental conditions, was completely unpredictable.

Indeed, a very complex mixture is formed when cordoin is treated with acetic acid and traces of concentrated sulfuric acid at room temperature. From this reaction mixture unaltered cordoin (V), the corresponding hydrolysis product, i.e. 2′,4′-dihydroxychalcone (X), three chromano-chalcones and one dichromano-chalcone have been isolated.

Of the chromano-chalcones, two were identified with those obtained from isocordoin, i.e. XX, XXI.

The third is 6-cinnamoyl-7-hydroxy-2,2-dimethylchroman (XXIV).

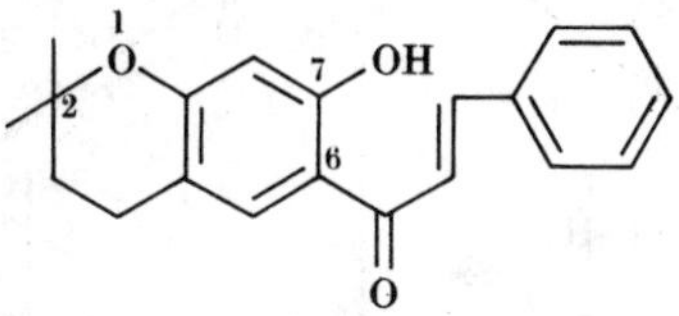

1 O 7 OH 2 6 O

XXIV
M.p. 195°C

The structure of this substance was established on the basis of the analytical and physico-chemical data, and mainly by NMR and mass spectroscopy.

The dichromano-chalcone, $C_{25}H_{28}O_3$ (M^+ 376) does not contain free OH groups (the UV spectrum remains unalterated on the addition of sodium methoxide) and in the NMR spectrum the peaks due to the protons of the two pyrane rings are partially split because they are not perfectly equivalent. The integration figures are consistent with the presence of two pyrane rings (XXV).

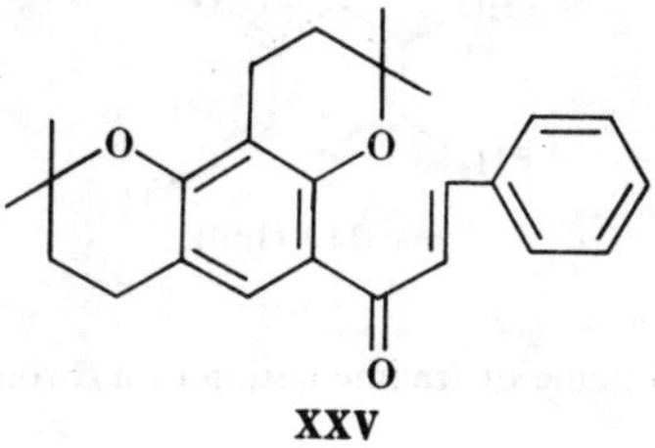

O O O

XXV

The mass spectra of the chromano-chalcones XX, XXI, XXIV are very similar and show only differences in the intensities of the peaks. Chromano-chalcones behave in fragmentation both as chalcones and chromans (Fig. 3). For example the chromano-chalcone (XXIV) as a chroman loses 15 and 55 m.u. to give the ions m/e 293 (M—15) and m/e 253 (M—55), while as a chalcone it gives the ions m/e 307 (M—1), 231 (M—Ph) and 204 (M—Ph-CH=CH).

The latter can undergo further fragmentation as a chroman to give the ions m/e 189 (204–15) and 149 (204–55).

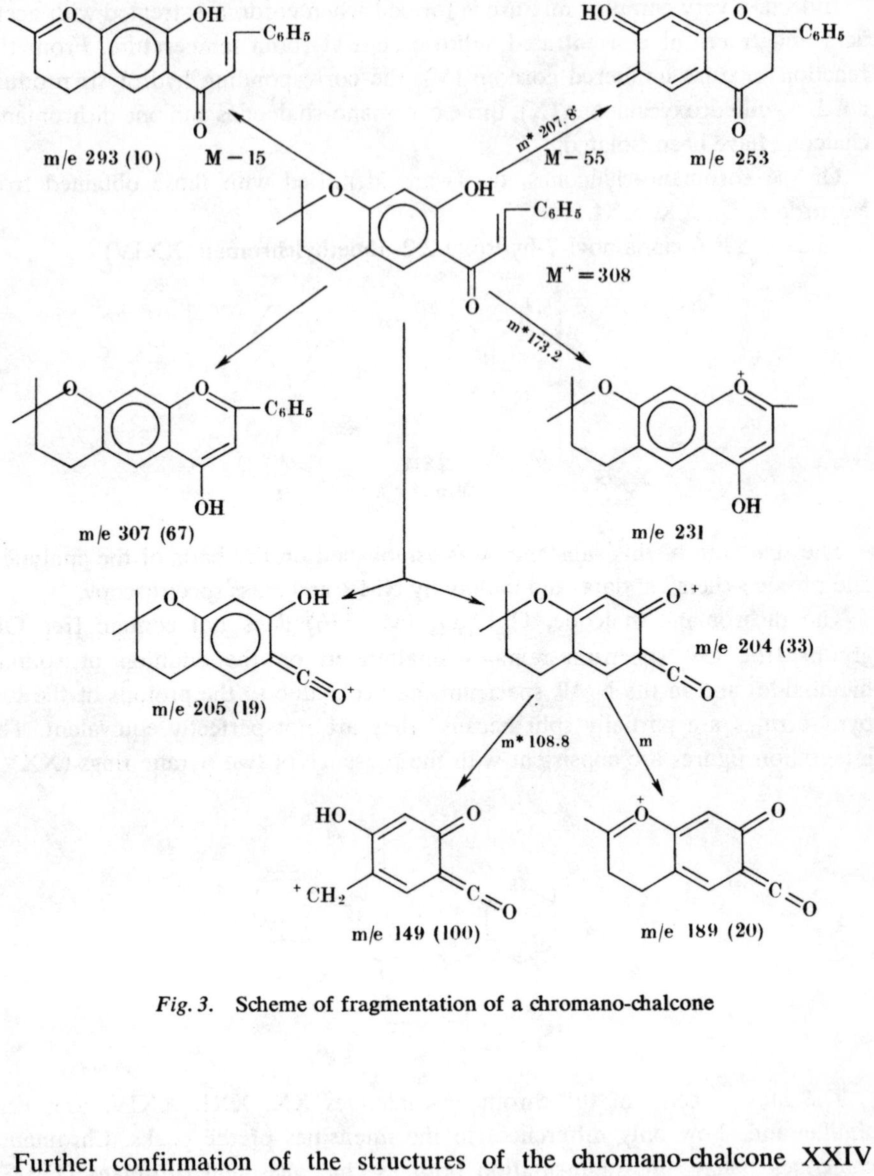

Fig. 3. Scheme of fragmentation of a chromano-chalcone

Further confirmation of the structures of the chromano-chalcone XXIV and the dichromano-chalcone XXV is given by the alkaline cleavage, which yields a chromano-acetophenone (XXVI) and a dichromano-acetophenone (XXVII), respectively.

Both were easily identified through their chemical behaviour and physico-chemical properties (MS and NMR spectra).

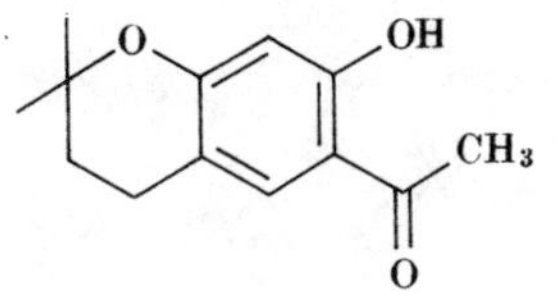

XXVI
M.p. 118 °C

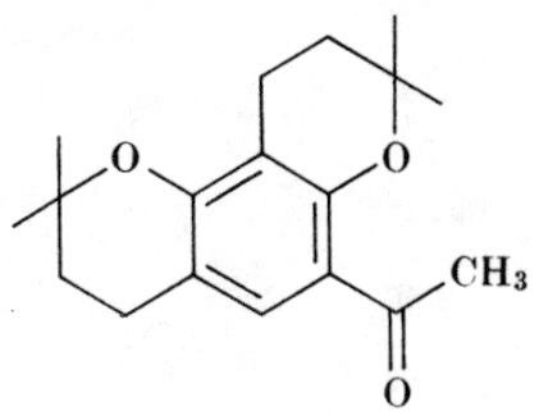

XXVII
M.p. 128 °C

Whereas the isomerization of isocordoin into the chromano derivatives XX and XXI can be easily interpreted according to the general mechanism of cyclization of prenylated phenols, a completely different mechanism must be taken into account for the isomerization of cordoin (derricidin).

On the basis of the experimental data it is evident that the reaction must proceed from cordoin in three steps:

(1) hydrolysis of the prenyl ether;
(2) C-prenylation (which may occur on C-3′ or C-5′ of the 2′,4′-dihydroxychalcone;
(3) cyclization of the intermediate C-prenylated chalcone.

These steps are shown in Fig. 4. The hydrolysis of cordoin yields 2′,4′-dihydroxychalcone and activated isopentyl; the latter reacts with the former by nucleophilic attack on the two relatively most reactive positions, C-3′ and C-5′, of the aromatic ring A to yield the intermediate compounds. Of these, isocordoin (VI) is directly cyclized under strongly acidic conditions to yield the known chromano-chalcones, XX and XXI. Isocordoin as well as the other intermediate, the hypothetical 5′-prenyl-2′,4′-hydroxychalcone (XXVIII), may be successively prenylated at positions 5′ and 3′, respectively, to yield the compound XXIX, which can cyclize under the given conditions to yield the bichromano-chalcone XXV.

This hypothesis has been supported by showing recently that C-prenylation may occur even under much milder conditions than those employed above.

The classic mechanism of the Fries randomized transposition of phenol ethers is also in accordance with the proposed mechanism. In fact, the formation of the unsubstituted 2′,4′-dihydroxychalcone and of the C-3′, C-5′ and C-3′–C-5′ prenylated chalcones is in complete agreement with the Fries transposition rule.

Owing to the strong acidity of the medium it is evident that the intermediate prenylated chalcones are not stable and thus cannot be isolated, because they isomerize to the corresponding chromano derivatives, whereas 2′,4′-dihydroxychalcone, which is stable under these conditions, can be readily separated.

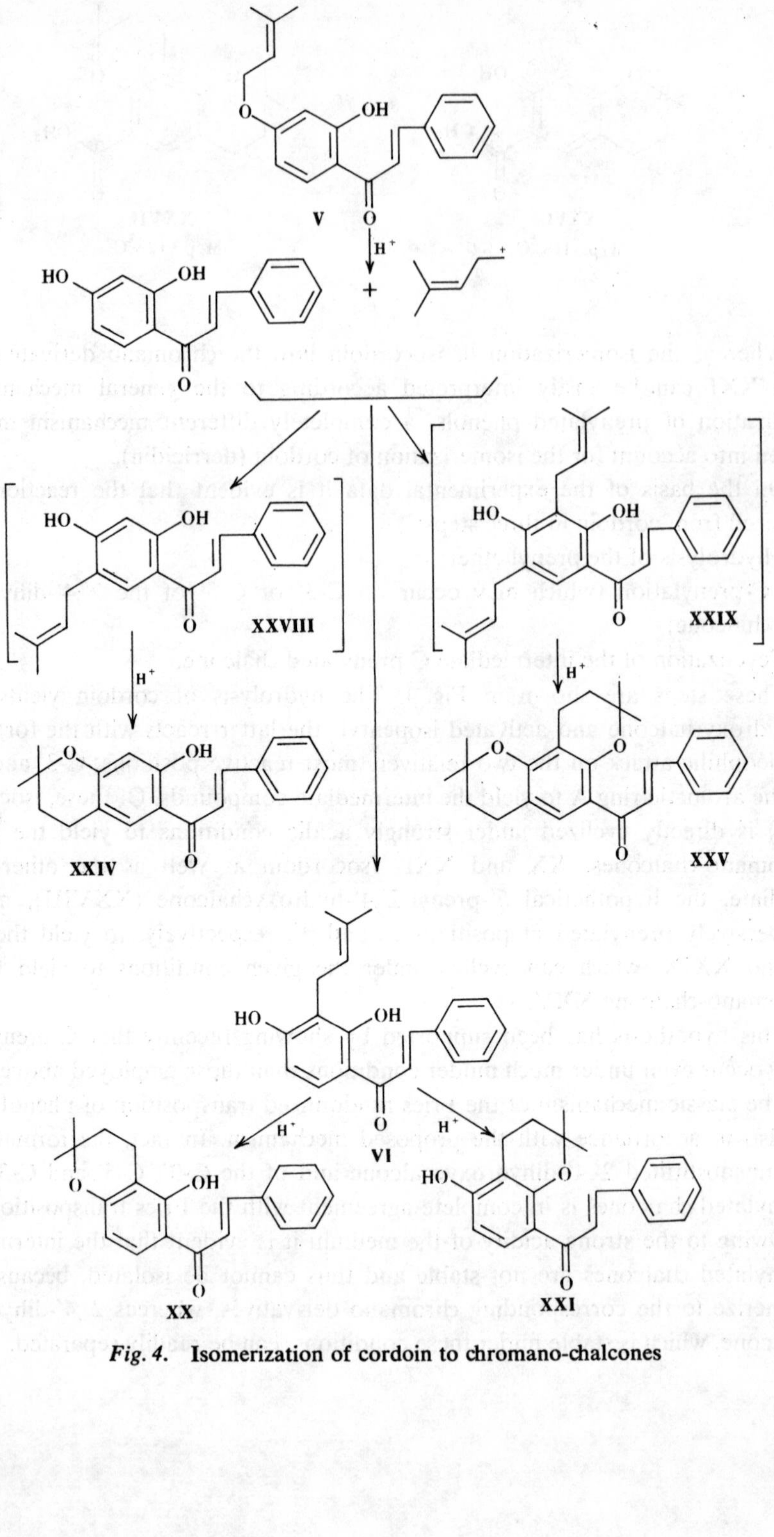

V

H^+

XXVIII

XXIX

H^+

H^+

XXIV

XXV

VI

H^+

H^+

XX

XXI

Fig. 4. **Isomerization of cordoin to chromano-chalcones**

The chromano-chalcones XX and XXIV are isomerized by refluxing in aqueous pyridine into the corresponding (±)-flavanones, XXX and XXXI.

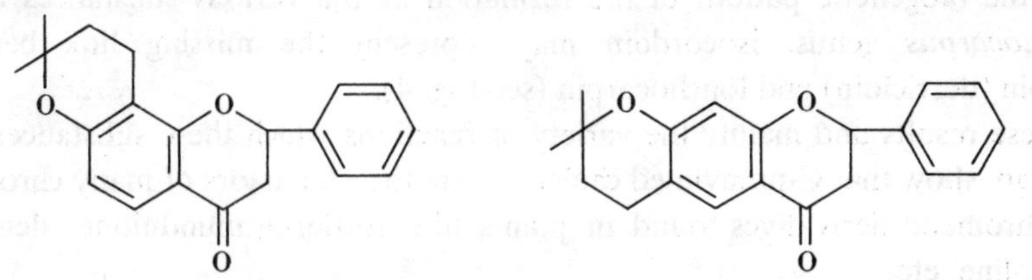

XXX XXXI

Dehydrogenation of the chromano-chalcones XX and XXIV with 2,3-dichloro-5,6-dicyano-1,4-benzoquinone (DDQ) gives, in low yields, the corresponding chromeno-chalcones XXXII and XXXIII.

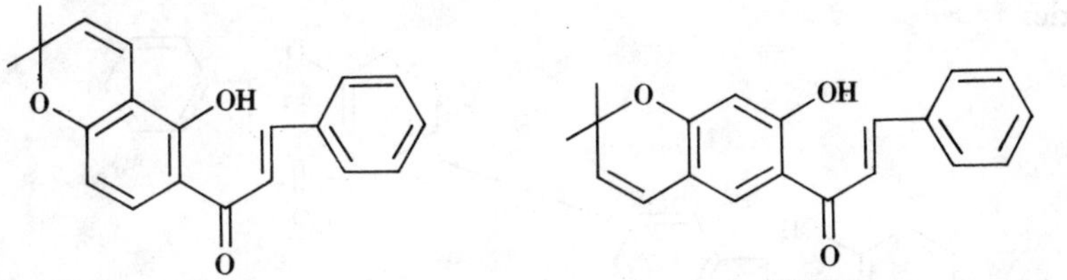

XXXII XXXIII

Compound XXXII is identical with lonchocarpin. Lonchocarpin was also obtained, but now in high yields, in a one-step cyclodehydrogenation with DDQ from isocordoin:

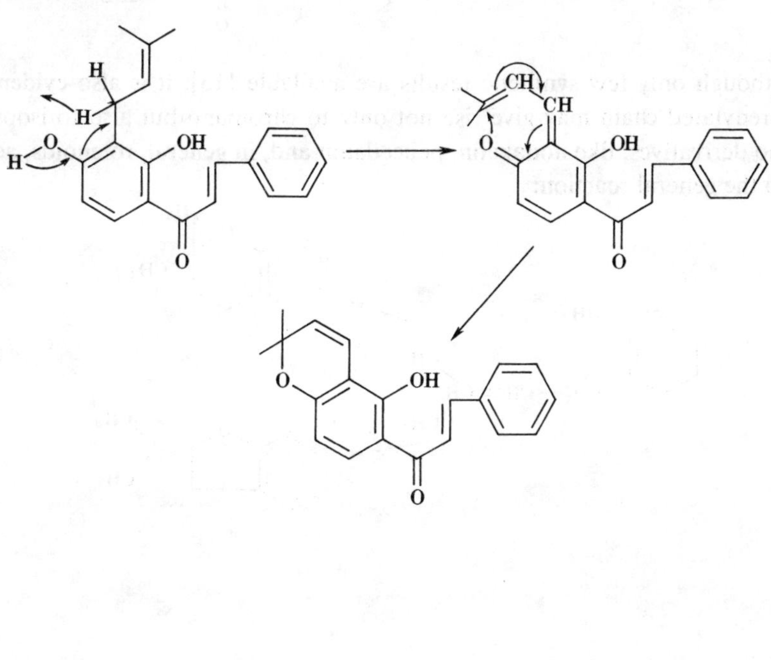

This result may have a remarkable biogenetic significance and may justify the reason why chromenes are more common in nature than the chromans.

In the biogenetic pattern of the formation of the various substances in the *Lonchocarpus* genus, isocordoin may represent the missing link between cordoin (derricidin) and lonchocarpin (see Fig. 4).

These results and mainly the variety of reactions which these substances may undergo, show that C-prenylated chalcones are the precursors of many chromano and chromeno derivatives found in plants like rottlerin, mundulone, deguelin, phaseoline, etc.

Although chromeno-isoflavones are more abundant than other chromano-flavonoids, it must be emphasized that according to recent results 2′-hydroxy-chalcones do not only give rise to flavanone and flavone derivatives through ring closure involving the C-2′ OH group, but also to isoflavanones by means of a transposition reaction, probably due to the formation of an intermediate epoxide:

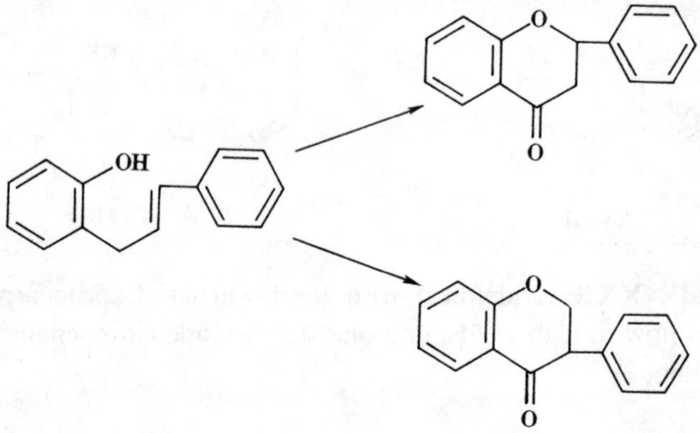

OH

O

O

O

O

Although only few synthetic results are available [15], it is also evident that the prenylated chain may give rise not only to chromano but also to isopropyl-furano derivatives, like nodaketin, peucedanin and, in general, rotenoids, according to the general reaction:

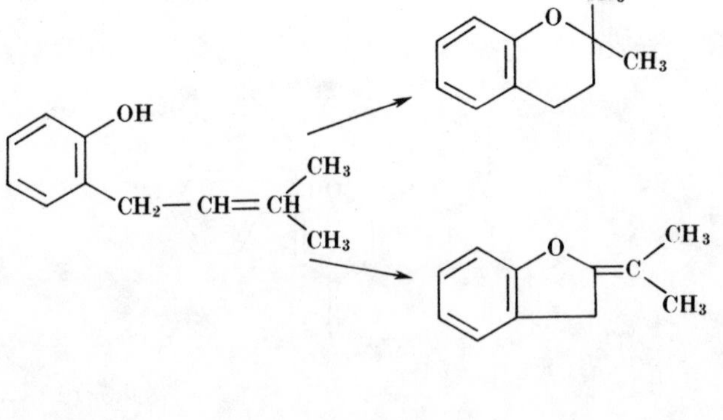

CH_3 O CH_3

OH

CH_3

$CH_2—CH=CH$

CH_3

O CH_3 $=C$ CH_3

These facts should direct the interest of chemists and biochemists to the role of prenylated flavonoids in plants as the potential key products of the biogenesis of many natural substances of biological interest.

REFERENCES

1. Wolfrom, M. L. and Wildi, B. S., *J. Am. Chem. Soc., 73,* 235 (1951).
2. Falshaw, C. P., Lane, S. A. and Ollis, W. D., *JCS, Chem. Comm., 1973,* 491.
3. Verzele, M., Stockx, J., Fontijn, F. and Anteunis, M., *Bull. Soc. Chim. Belg., 66,* 452 (1957).
4. Geissman, T. A., *Aust. J. Chem., 11,* 376 (1958).
5. Seshadri, T. R. and Sood, M. S., *Tetrahedron Lett., 1967,* 853.
6. Aurnhammer, G., Chari, V. M. and Wagner, H., *Chem. Ber., 105,* 3511 (1972).
7. Baudrenghien, J., Jadot, J. and Huls, R.: *Bull. classe sci., Acad. roy. Belg., 39,* 105 (1953).
8. Gonçalves de Lima, O., Marini-Bettòlo, G. B., de Mello, J. F., Delle Monache, F., de Andrade Lyra, F. D. and Machado de Albuquerque, M., *Atti. Acad. Naz. Lincei, 8,* 53, 433 (1972).
9. de Nascimiento, M. C. and Mors, W. B., *Phytochemistry, 11,* 3023 (1972).
10. de Nascimiento, M. C. and Mors, W. B., *An. Acad. Bras. Cienc., 42* (Sup.), 87 (1970).
11. Gonçalves de Lima, O., Marini-Bettòlo, G. B., de Mello, J. F., Delle Monache, F., de Barros Coelho, J. S. and Machado de Albuquerque, M., *Gazz. Chim. Ital., 103,* 771 (1973).
12. Dean, F. M., "Naturally Occurring Oxygen Ring Compounds", Butterworth, London, 1963.
13. Delle Monache, G., de Mello, J. F., Delle Monache, F., Marini-Bettòlo, G. B., Gonçalves de Lima, O., and de Barros Coecho, J. S., *Gazz. Chim. Ital., 104,* 861 (1975).
14. Delle Monache, F., Gonçalves de Lima, O., de Mello, J. F., Delle Monache, G. and Marini-Bettòlo, G. B., *Gazz. Chim. Ital., 103,* 779 (1973).
15. King, F. E., Housley, J. R. and King, T. J., *J. Chem. Soc., 1954,* 1392.

SYNTHESIS OF NATURAL ISOFLAVANOIDS BY THE OXIDATIVE REARRANGEMENT OF CHALCONES BY THALLIUM(III) NITRATE*

by

M. NÓGRÁDI, S. ANTUS, Á. GOTTSEGEN and L. FARKAS
Research Group for Alkaloid Chemistry, Hungarian Academy of Sciences
Budapest, Hungary

As a continuation of the published [1] syntheses of the natural isoflavanones (±)-dalbergioidin, (±)-ferreirin and (±)-ougenin, the synthesis of sophorol has been undertaken (Fig. 1).

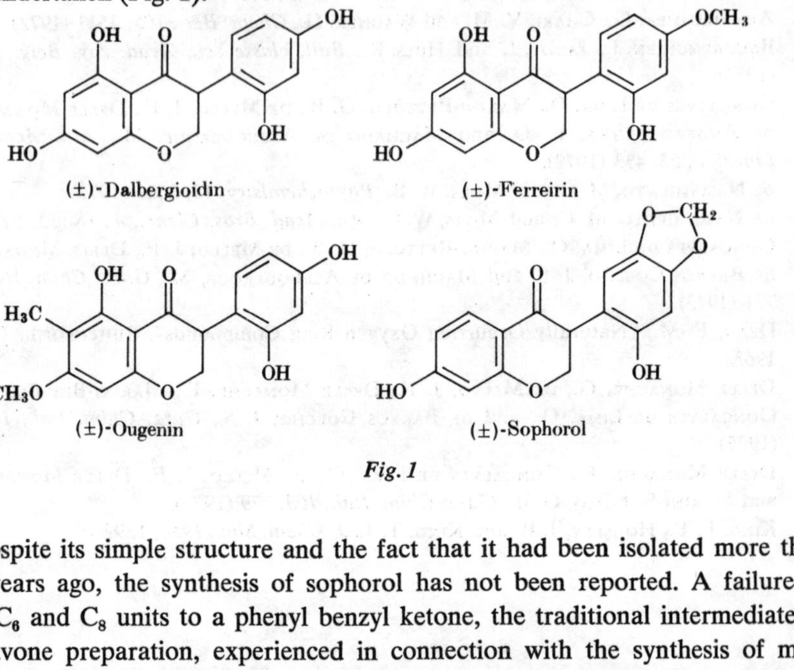

(±)-Dalbergioidin (±)-Ferreirin

(±)-Ougenin (±)-Sophorol

Fig. 1

Despite its simple structure and the fact that it had been isolated more than ten years ago, the synthesis of sophorol has not been reported. A failure to join C_6 and C_8 units to a phenyl benzyl ketone, the traditional intermediate of isoflavone preparation, experienced in connection with the synthesis of mill-durone [2], suggested that some other approach should be attempted.

Oxidative rearrangement of fully protected chalcones by thallium(III) acetate discovered in 1968 by Ollis [3] (Fig. 2) seemed to be attractive, but no appreciable amount of an isoflavone could be isolated when the rearrangement of the chalcone which would lead ultimately to sophorol was attempted. This

* Published in detail in *J. Chem. Soc. Perkin I, 1974,* 304 and *Acta Chim. (Budapest)* *82,* 225 (1974).

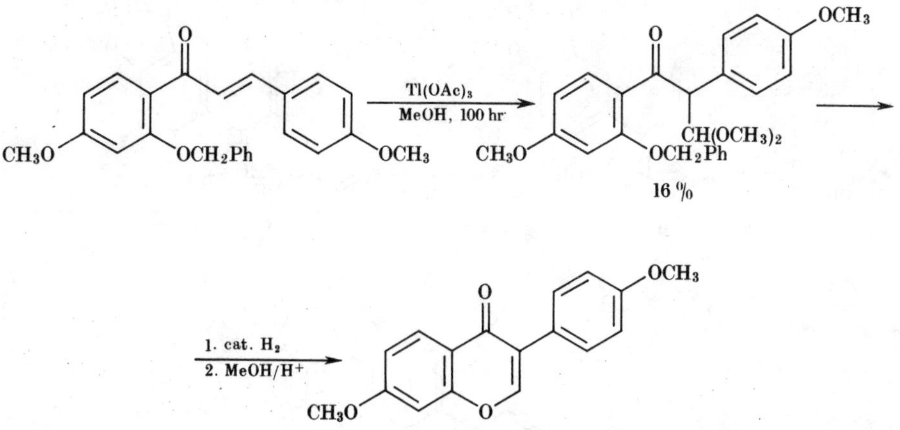
Tl(OAc)$_3$
MeOH, 100 hr
CH$_3$O
OCH$_2$Ph
OCH$_3$
CH(OCH$_3$)$_2$
16 %
1. cat. H$_2$
2. MeOH/H$^+$

Fig. 2

route was not further pursued since we became aware of the results of McKillop, Taylor and their co-workers [4], who found that for the rearrangement of simple chalcones the nitrate of thallium is a much more efficient reagent than the acetate. Whereas with the acetate the reaction requires refluxing in methanol for up to 100 hours giving moderate to low yields of the intermediate acetal, reaction with the nitrate is usually complete within a few minutes at room temperature and affords the acetal in high yield. The authors did not explore the application of the reaction to the synthesis of isoflavones.

After having successfully tested the thallium(III) nitrate method in the synthesis of dehydrosophorol dimethyl ether (Fig. 3), we proceeded to prepare 2′-hydroxy-7-methoxy-4′,5′-methylenedioxyisoflavone (dehydrosophorol 7-methyl ether).

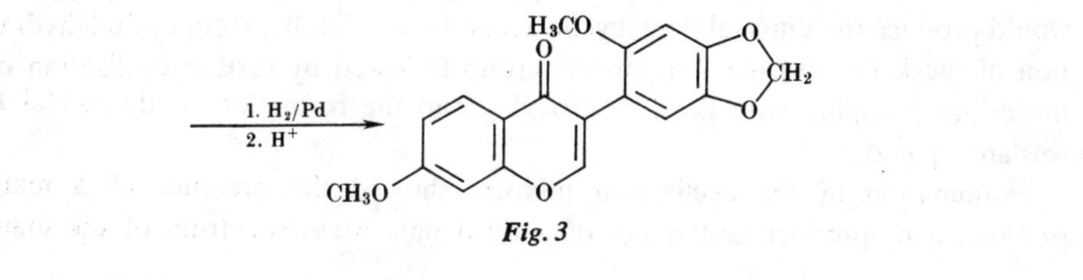
Tl(NO$_3$)$_3$
MeOH
CH$_3$O
OCH$_2$Ph
OCH$_3$
H$_3$CO
CH$_2$
CH(OCH$_3$)$_2$
1. H$_2$/Pd
2. H$^+$

Fig. 3

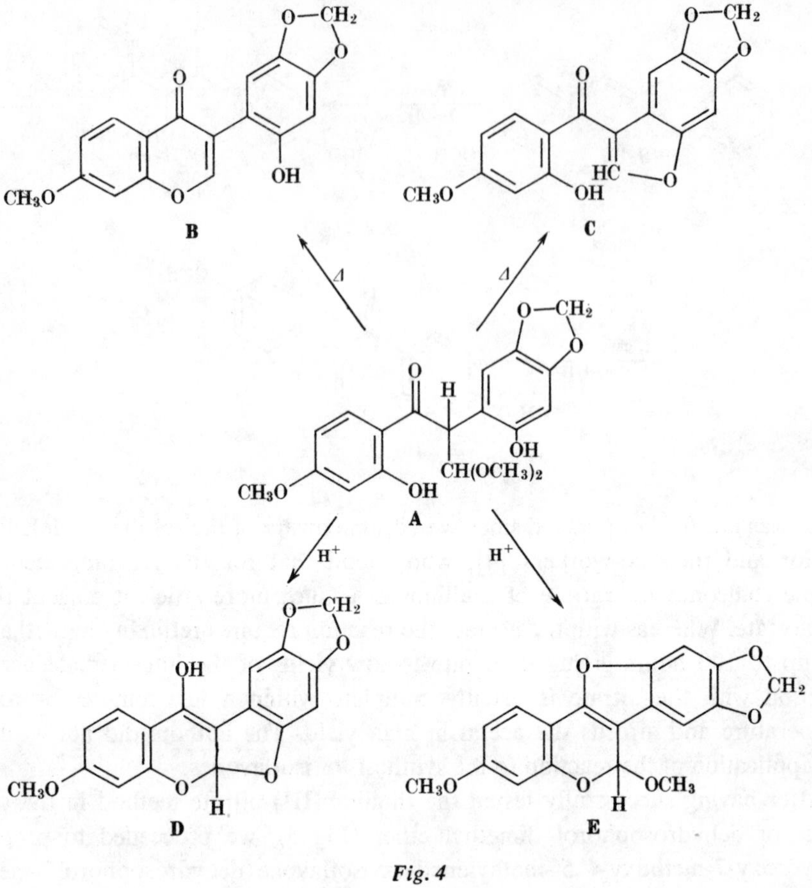

Fig. 4

Preparation of the dihydroxyacetal intermediate (Fig. 4, **A**) did not pose any difficulties. Formation of at least four products (**B**, **C**, **D**, and **E**) could be envisaged on acid-catalyzed ring closure of **A**. Cyclization involving the 2-hydroxyl with double elimination of the elements of methanol would give the desired isoflavone **B**; a similar procedure with the 2′-hydroxyl group was expected to yield the acylfuran **C**. Ring closure involving both hydroxyl groups would produce the unusual pentacyclic acetal **D**, and finally, from cyclodehydration of the keto- and the 2′-hydroxyl group followed by further cyclization of the acetal grouping with the 2-hydroxyl group the formation of the acetal **E** was anticipated.

Examination of the cyclization mixture showed the presence of a main product, a by-product and traces of a third one. Mass spectrum of the main product gave $C_{17}H_{12}O_6$ as the molecular formula. This would have been in accord with the isoflavone structure B, but there was no carbonyl absorption in the IR, and the characteristic low-field peak of isoflavone around $\delta = 7.9$ ppm in the NMR was clearly absent. Instead of the latter, a one-proton singlet at $\delta = 7.37$ ppm was observed, which we assigned to an acetal proton. This and the fragmentation pattern of the mass spectrum suggested the pentacyclic structure D for the main product. Except for the natural isoflavanoid lisetin [5], D is the only known example of the benzofuro[2,3-*b*][1]benzopyran ring system.

The by-product contained two methoxyl groups and an acetal-type proton as shown by the NMR; also, there was no carbonyl absorption in the IR. These facts and the molecular formula ($C_{18}H_{14}O_6$) supported the mixed acetal structure E for the by-product.

As it will be shown later the third product was the acylfuran C. Not even traces of the required isoflavone (B) could be detected in the acid-catalyzed ring closure of A, which is all the more remarkable since later we found that, though only as a by-product, this isoflavone was formed in the thermolysis of A, the main product being the acylfuran C. The structure of C was based on a positive ferric chloride reaction, the molecular formula ($C_{17}H_{12}O_6$) and bands in the IR spectrum characteristic of the benzofuran system.

Since results of the thermolysis became only available at a later stage of our investigations, we looked for a scheme which would ensure cyclization into the desired direction. Before embarking upon investigations aiming at the differential blocking of the free hydroxyls which would enable us to liberate the one in position 2 first, we tried the oxidation of a 2′-hydroxychalcone. Since thallium-(III) acetate oxidation of 2′-hydroxychalcones failed to give an isoflavone [3], the prospects of this reaction were unpromising. Nevertheless, 2′-hydroxy-4,4′-dimethoxychalcone (Fig. 5, a) could be converted to the corresponding acetal and ultimately to 4′,7-dimethoxyisoflavone in a good yield. The feasibility of the transformation of 2′-hydroxychalcones to isoflavones was further confirmed by the synthesis of a number of model compounds (Fig. 5 b, c, and d).

Unfortunately, oxidation of the 2′-hydroxychalcone precursor of sophorol (F) (Fig. 6) gave the corresponding isoflavone G in only very low yields. This was ascribed to its insolubility in methanol, which required long reaction times and/or higher temperatures for the completion of the oxidation. Since it was suspected that side reactions diminishing the yield were due to the presence of a free phenolic hydroxyl, the transformation was carried out with the corresponding acetate. This was successful and gave, after saponification and treatment with acid, the key isoflavone (G) in a fair yield.

Transformation of the isoflavone to sophorol was best carried out by hydrogenation of the corresponding acetate in acetone (Fig. 7). This gave the acetate

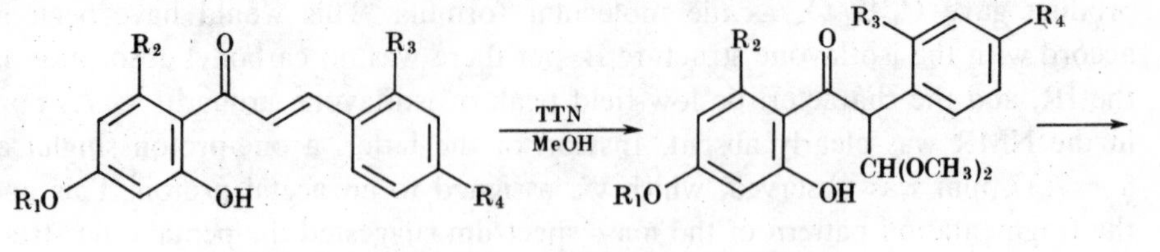

	R_1	R_2	R_3	R_4
a.	CH_3	H	H	OCH_3
b.	$PhCH_2$	H	OCH_2Ph	OCH_3
c.	CH_3	H	OCH_2Ph	H
d.	CH_3	OCH_3	OCH_3	OCH_3

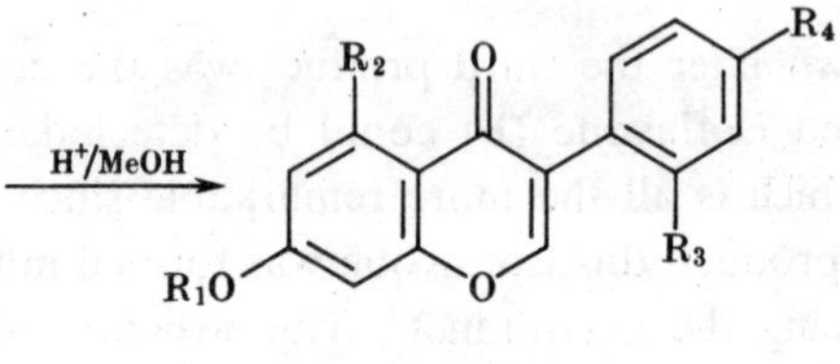

Fig. 5

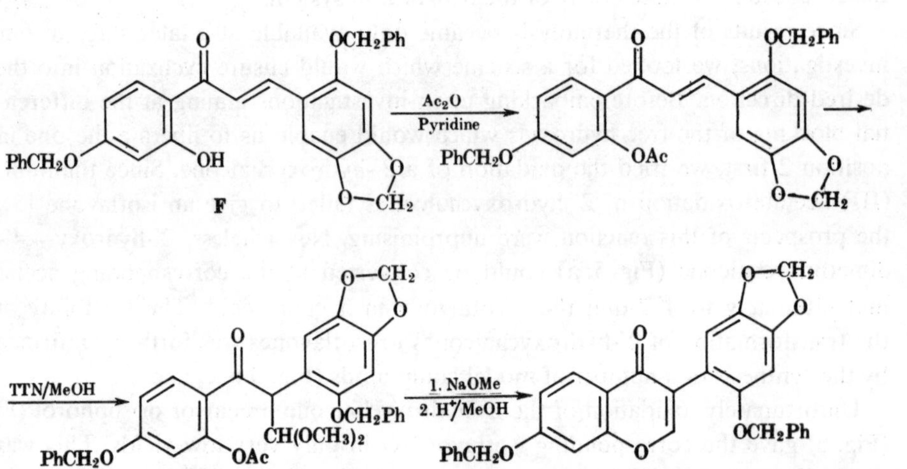

Fig. 6

of (±)-sophorol, a substance that cannot be prepared by acetylation of sophorol, because of dehydration of the latter.

The possession of sophorol made possible a new synthesis of (±)-maackiain [6] (Fig. 8).

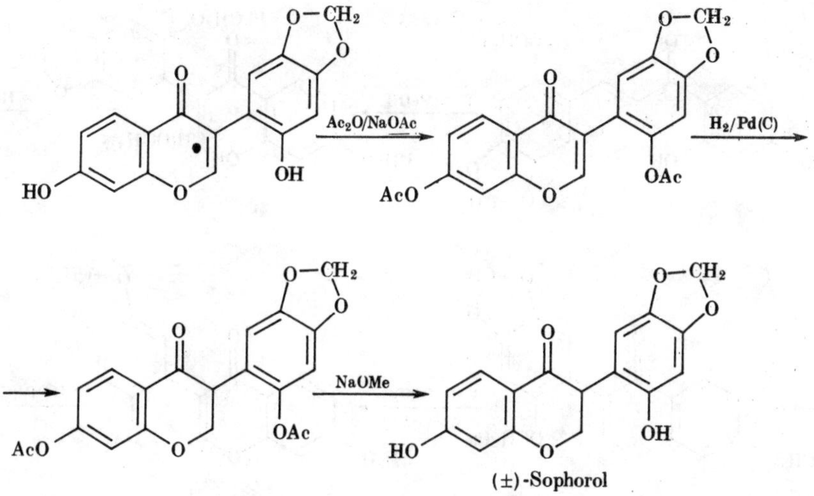

(±)-Sophorol

Fig. 7

(±)-Maackiain

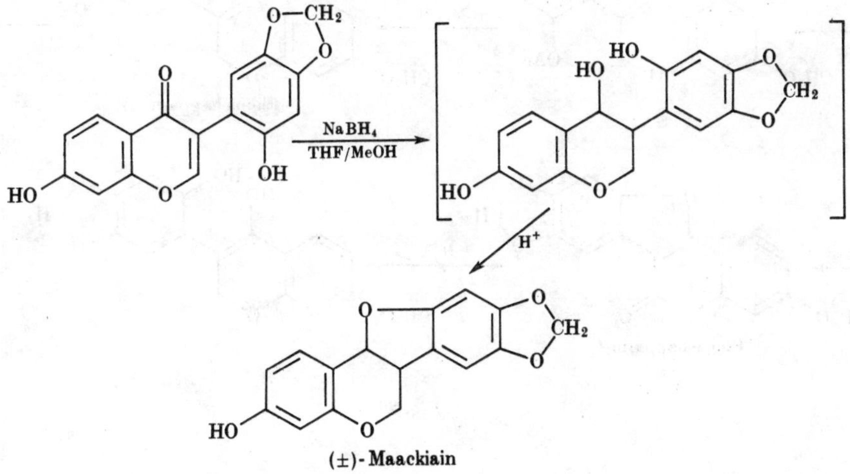

Fig. 8

The possibilities of this convenient method for the preparation of isoflavones were further explored in the synthesis of flemichapparin-B and C, two isoflavonoids recently recognized as the constituents of *Flemingia chappar* [7].

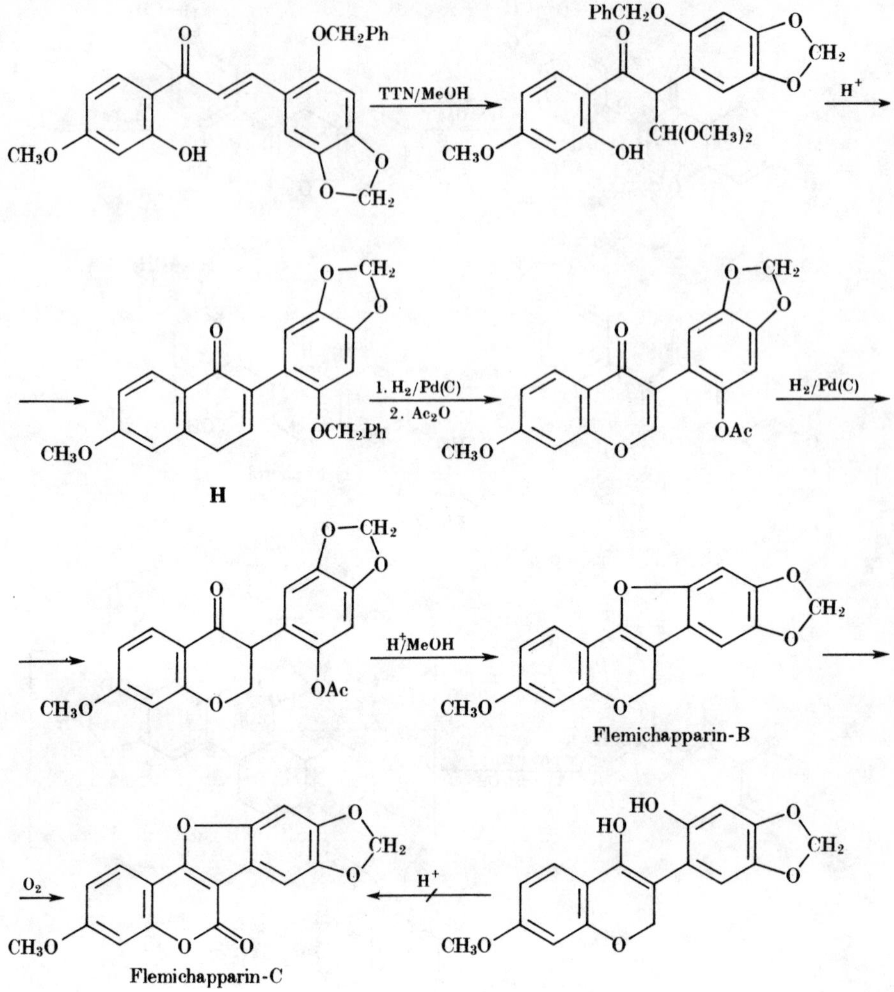

O OCH₂Ph TTN/MeOH PhCH₂O O O CH₂ H⁺
CH₃O OH O O—CH₂ CH₃O OH CH(OCH₃)₂

O—CH₂ O O O—CH₂ O O
1. H₂/Pd(C) 2. Ac₂O H₂/Pd(C)
CH₃O OCH₂Ph CH₃O O OAc
H

O—CH₂ O O O O CH₂
H⁺/MeOH
CH₃O O OAc CH₃O O O
Flemichapparin-B

O O CH₂ O HO HO O CH₂ O
O₂ H⁺
CH₃O O O CH₃O O
Flemichapparin-C

Fig. 9

For the synthesis of flemichapparin-B (Fig. 9) we used the isoflavone H, prepared by the $Tl(NO_3)_3$ oxidation of the corresponding 2′-hydroxychalcone. This was transformed *via* the corresponding isoflavone acetate to an isoflavanone acetate, which on treatment with hydrochloric acid in methanol directly gave flemichapparin-B. In the course of the purification of flemichapparin-B we noticed that when its solution was exposed to air, sometimes an oxidation product appeared which we identified with flemichapparin-C.

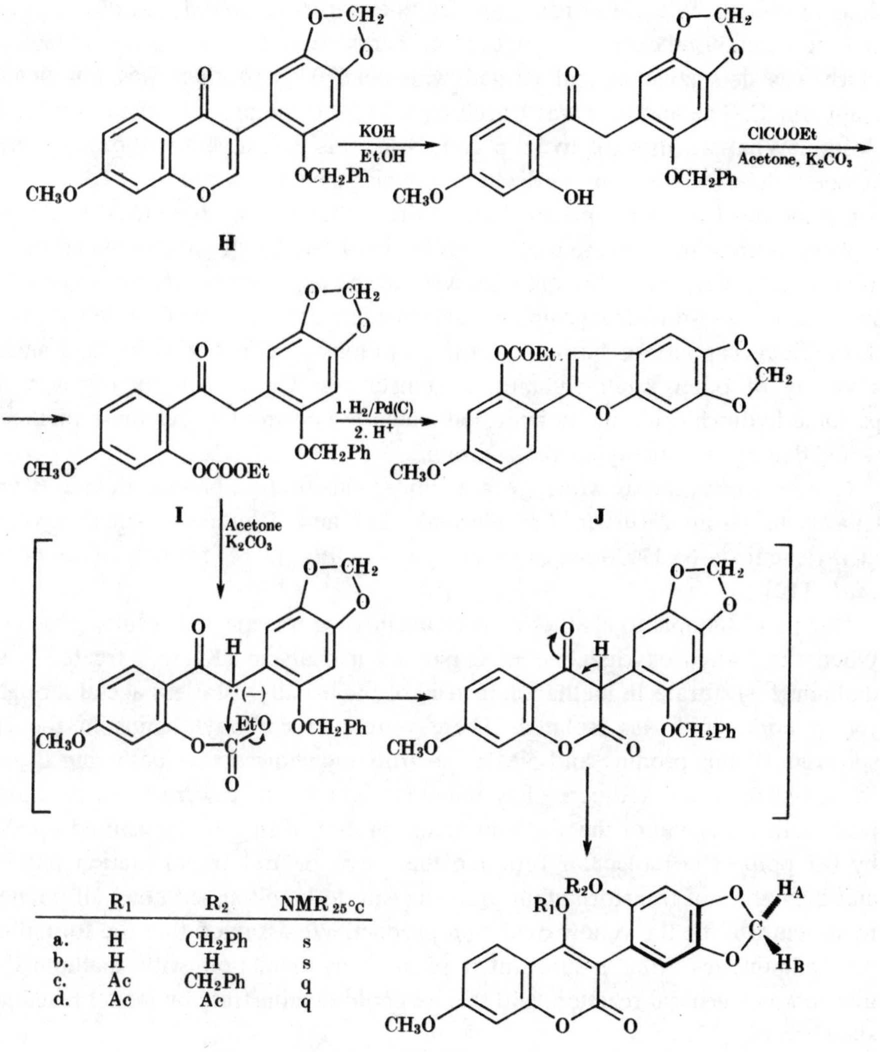

	R_1	R_2	NMR $_{25°C}$
a.	H	CH_2Ph	s
b.	H	H	s
c.	Ac	CH_2Ph	q
d.	Ac	Ac	q

Fig. 10

An intended but eventually miscarried synthesis of flemichapparin-C (Fig. 10) was based on the observation of Fukui *et al.* [8] that 2′,4,7-trihydroxy-3′,4′-dimethoxy-3-phenylcoumarin could be dehydrated under mild conditions to the corresponding coumestone. In order to prepare the necessary 2′,4-dihydroxy-coumarin, the isoflavone H was degraded by alkali to give a ketone which was treated, according to Robertson's method [9] for the synthesis of 4-hydroxy-3-phenylcoumarins, with ethyl chloroformate in acetone in the presence of potas-

sium carbonate. When according to TLC the starting material had disappeared and the reaction mixture was worked up, a crystalline compound was obtained, which was debenzylated and treated with acid. This product was not flemichapparin-C. The anomaly was traced back soon to the product of the first ring closure. What we thought to be a coumarin, was actually the ethoxycarbonyl ketone **I**, debenzylation and cyclization of which gave the benzofuran **J**.

Compound **I** is a true intermediate in the Robertson synthesis of 4-hydroxy-3-phenylcoumarins, because when **I** was further boiled with potassium carbonate in acetone, it was converted quantitatively to the expected coumarin. Debenzylation gave a 2′,4-dihydroxycoumarin intermediate, but we failed in dehydrating this to flemichapparin-C both under the conditions specified by Fukui and under a variety of other relatively mild circumstances. Cyclization by fusion with pyridine hydrochloride or by hydroiodic acid is known [10] but these methods would damage the methylenedioxy group.

Natural isoflavanoids which have a similar substitution pattern in ring B are leiocalycin (from *Swartzia leiocalycina*) [11] and 2-hydroxy-3-methoxy-8,9-methylenedioxy-6a,11a-dihydropterocarpan (see Fig. 13, **M**) from *Neorautanenia edulis* [12].

Our first attempts to synthesize leiocalycin gave unexpected results (Fig. 11). When the 2′-hydroxychalcone precursor of leiocalycin (**K**) was treated with thallium(III) nitrate in methanol, instead of the usually colorless acetal a bright yellow compound was isolated. There were two carbonyl bands in the IR spectrum of this product and NMR spectroscopy showed that both ring B and the α,β-unsaturated ketone moiety remained intact. There were three methoxyl peaks and the signal of the single aromatic proton of ring A was shifted upfield by 0.7 ppm. The molecular formula, the mass spectral fragmentation pattern and the nature of transformation products (not to be discussed now) all pointed to structure **L** for the yellow oxidation product. We assumed that the formation of semiquinones from *p*-substituted phenols by oxidation with thallium(III) nitrate was a general reaction, and this we could demonstrate by model reactions shown in Fig. 12.

These observations indicated that for the successful oxidative rearrangement of **K** the 2′-hydroxyl group had to be blocked temporarily. Acetylation served this purpose well; the rearrangement proceeded smoothly and a sequence of transformations, analogous to those leading to flemichapparin-B, gave leiocalycin.

In the synthesis of the *Neorautanenia* isoflavone **M** we made use of our previous observations and rearranged the 2′-acetoxychalcone. The full sequence yielding ultimately racemic **M** is shown in Fig. 13.

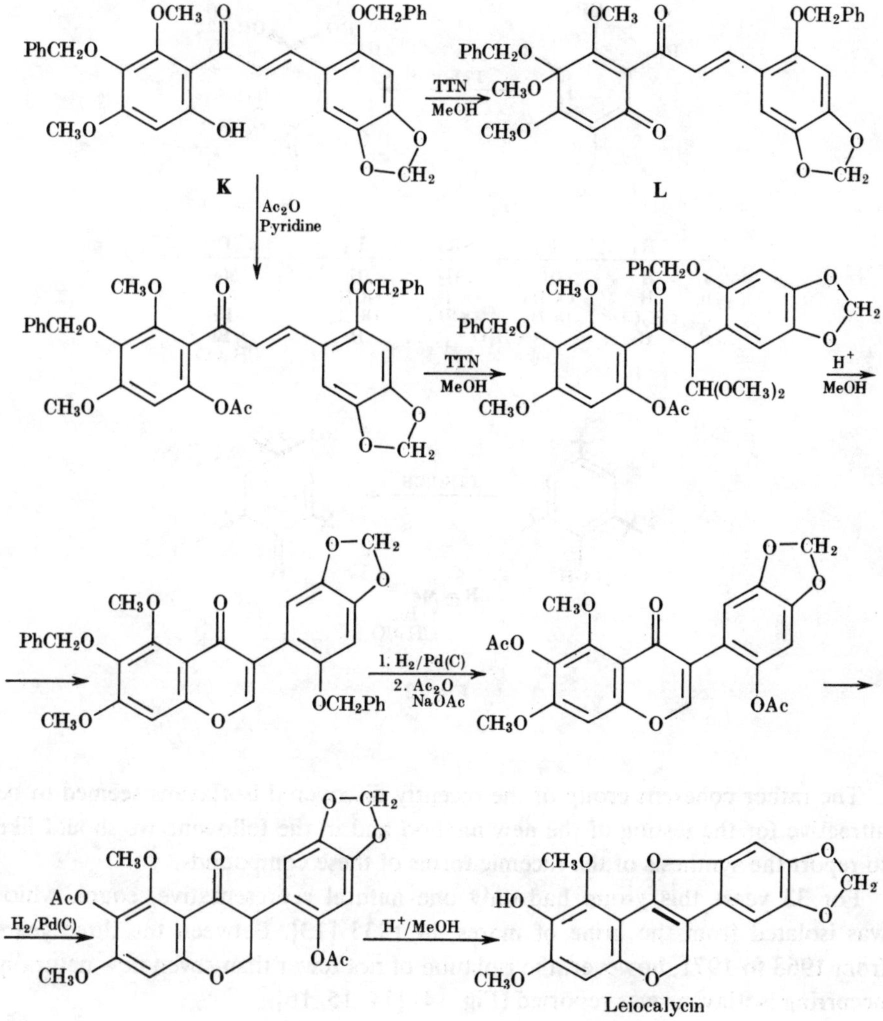

OCH_3 O OCH_2Ph
$PhCH_2O$
CH_3O OH
O
O—CH_2
K

TTN / MeOH →

OCH_3 O OCH_2Ph
$PhCH_2O$
CH_3O
CH_3O O
O
O—CH_2
L

Ac_2O / Pyridine ↓

CH_3O O OCH_2Ph
$PhCH_2O$
CH_3O OAc
O
O—CH_2

TTN / MeOH →

$PhCH_2O$
CH_3O O
O
CH_2
O
$PhCH_2O$
$CH(OCH_3)_2$
CH_3O OAc

H^+ / MeOH →

O—CH_2
O
CH_3O O
$PhCH_2O$
CH_3O O
OCH_2Ph

1. H_2/Pd(C) 2. Ac_2O NaOAc →

O—CH_2
O
CH_3O O
AcO
CH_3O O
OAc
→

O—CH_2
O
CH_3O O
AcO
H_2/Pd(C) →
CH_3O O
OAc

H^+/MeOH →

CH_3O O
O
CH_2
O
HO
CH_3O O
Leiocalycin

Fig. 11

So far an account has been given above describing how we arrived at the development of a rather general and practical method for the synthesis of isoflavones.

In the following the utilization of this method for the synthesis of a number of other natural isoflavanoids will be presented.

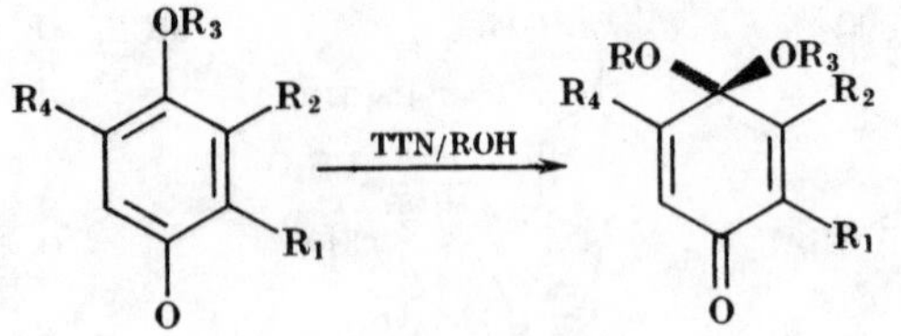

TTN/ROH

	R_1	R_2	R_3	R_4
a.	H	H	CH_3	H
b.	H	OCH_3	OCH_3	OCH_3
c.	CH_3CO	OCH_3	OCH_3	OCH_3
d.	H	OCH_2O		H

R
Me
Et
i-Pr
t-Bu
CH_3CO

TTN/ROH

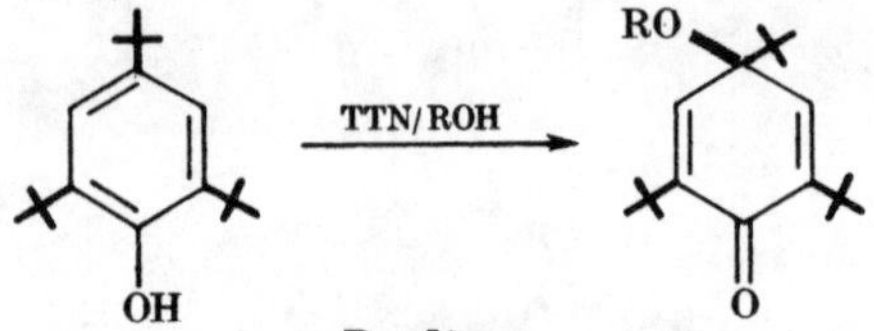

R = Me
t-Bu
CH_3CO

Fig. 12

The rather coherent group of the recently discovered isoflavans seemed to be attractive for the testing of the new method and in the following we should like to report the synthesis of the racemic forms of these compounds.

For 33 years this group had only one natural representative, equol, which was isolated from the urine of mares, in 1935 [13]. Between the three years from 1968 to 1971, however, the isolation of not fewer than seven new naturally occurring isoflavans was reported (Fig. 14) [14, 15, 16].

Although the difficulties incident to the synthesis of sophorol were not anticipated with these isoflavans, the application of the new method promised a more convenient and significantly shorter way to the end-products.

The synthesis of vestitol which we carried out in both ways, served as a test case (Fig. 15).

The classical route (Fig. 15) required as starting material an appropriately substituted phenylacetonitrile (N). Starting from 2-benzyloxy-4-methoxybenzaldehyde, the preparation of this nitrile required seven steps. The Hoesch reaction of the nitrile with resorcinol gave the phenyl benzyl ketone intermediate. Ring closure to the dibenzyloxyisoflavone, catalytic hydrogenation first in acetone to

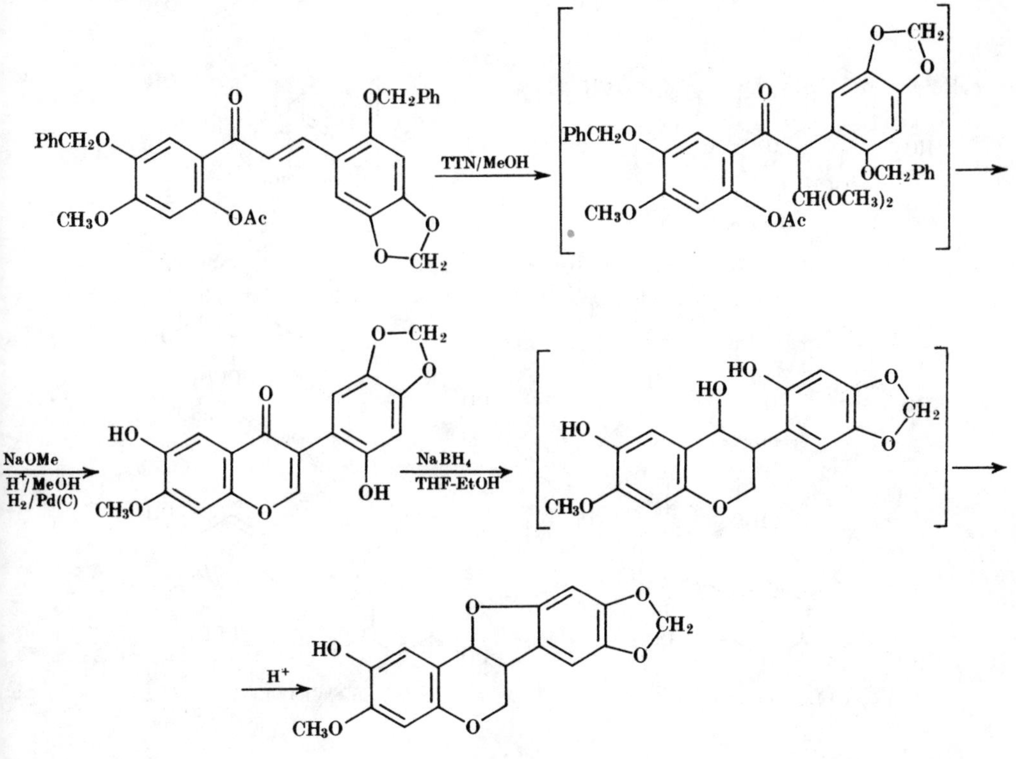

Fig. 13

the dihydroxyisoflavone and then in acetic acid to the dihydroxyisoflavan completed the synthesis of racemic vestitol.

Applying the new method, i.e. using thallium trinitrate for the oxidative rearrangement of the chalcone (Fig. 16), 2-benzyloxy-4-methoxybenzaldehyde could be directly used for assembling a C_{15}-unit by chalcone condensation with an equally simple acetophenone. The primary product of the oxidative rearrangement need not be isolated, since it can be transformed directly to isoflavone by boiling a few hours with dilute hydrochloric acid. Catalytic hydrogenation of the latter on palladium-charcoal catalyst in acetic acid afforded racemic vestitol.

Comparing the two routes, the advantages of the second one over the classical method are obvious. Thus instead of the phenylacetic acid derivatives the more easily available benzaldehydes are used for construction of ring B of the

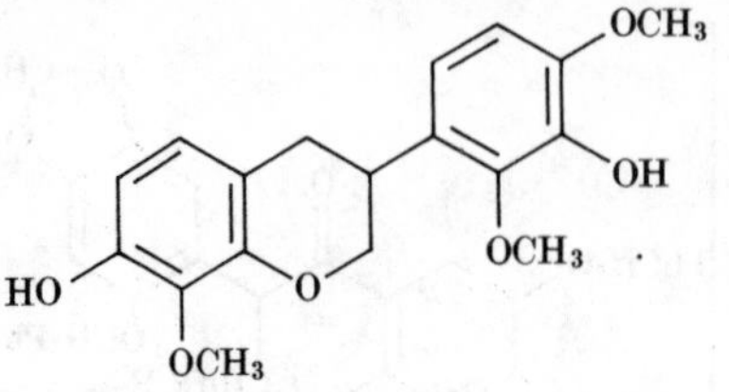

Duartin

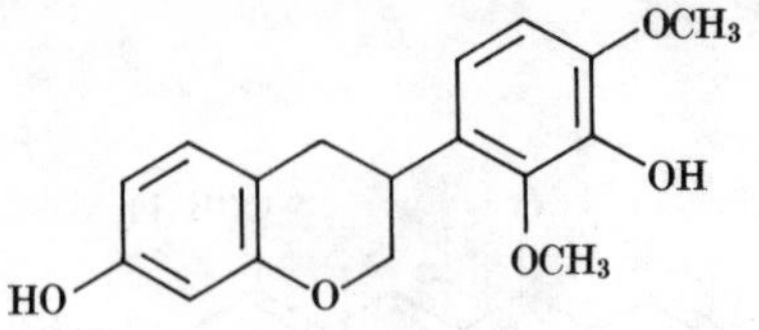

Mucronulatol

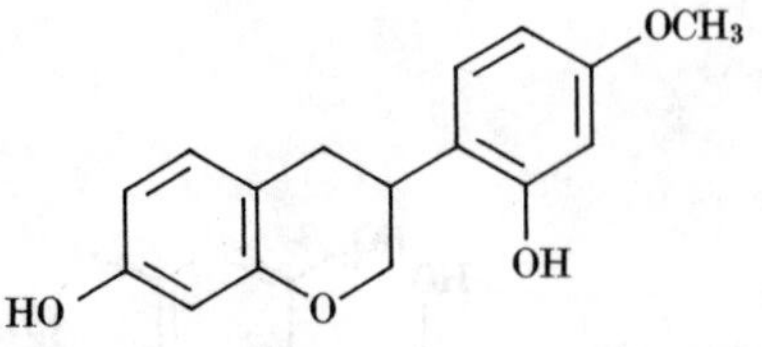

Vestitol

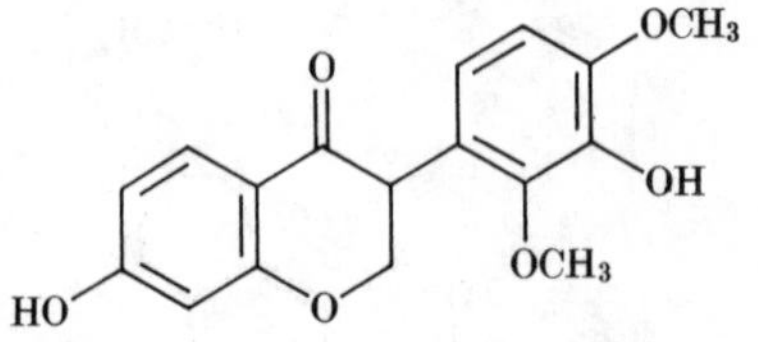

Violanon

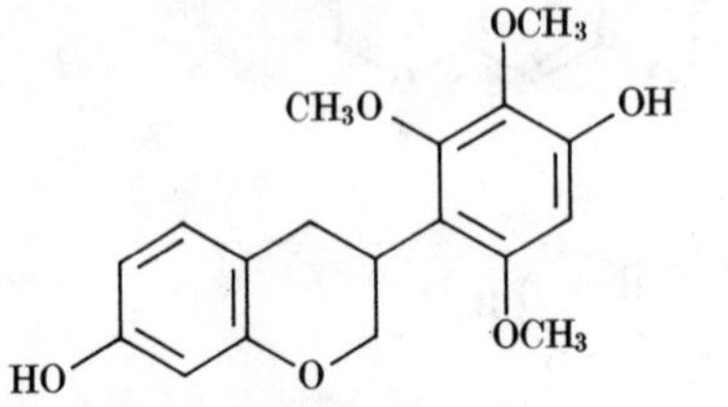

Lonchocarpan

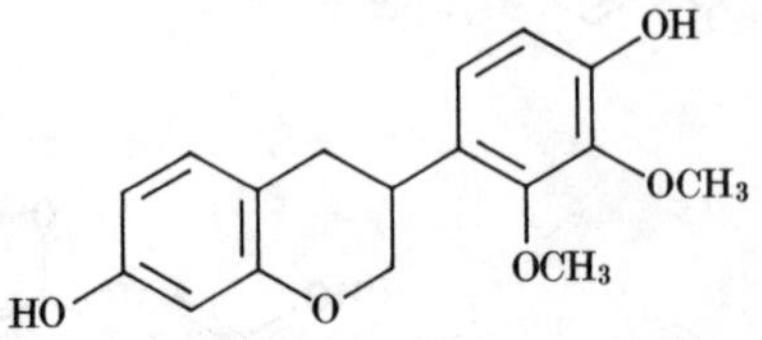

Laxifloran (?)

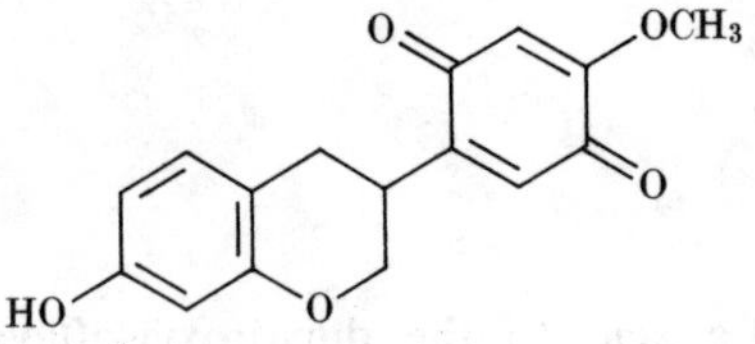

Claussequinone

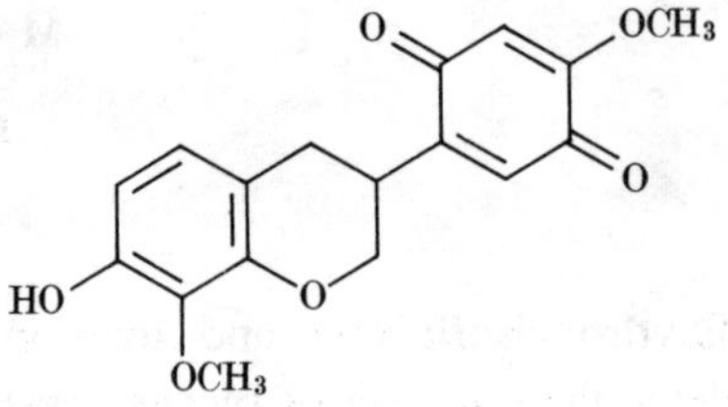

Mucroquinone

Fig. 14

isoflavones, and strongly acidic media as required by the Hoesch- or Friedel-Crafts acylation are avoided, permitting a wider range of usable protecting groups. These advantages were already given in the thallium triacetate method of chalcone oxidation discovered by Ollis and his co-workers [3]. With the thallium acetate method, however, the simplicity of the pathway is largely offset by the usually very low yields in the oxidation step, and the good yields (30–80 %) secured by the use of thallium trinitrate clearly turn the scale in favour of the oxidative approach.

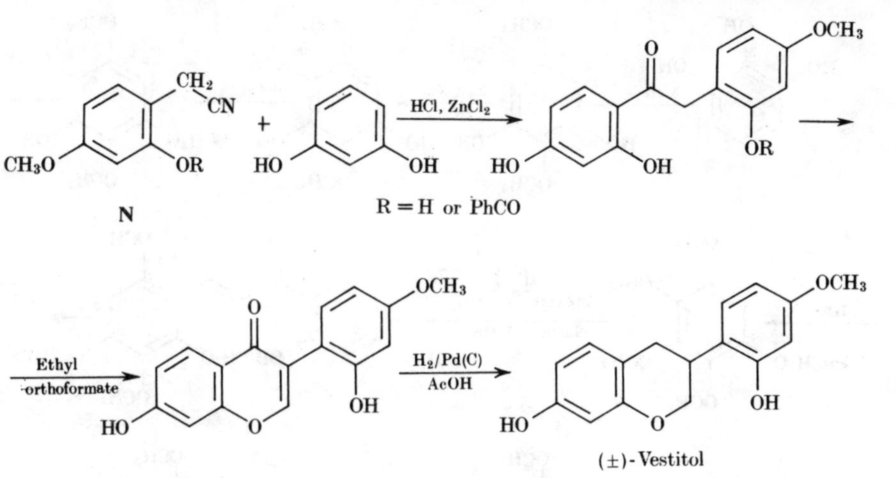

CH_2CN CH_3O OR **N** + HO OH $\xrightarrow{HCl,\ ZnCl_2}$ O HO OH OCH_3 OR →

$R = H$ or PhCO

$\xrightarrow{Ethyl\ orthoformate}$ O HO O OCH_3 OH $\xrightarrow{H_2/Pd(C),\ AcOH}$ HO O OCH_3 OH

(±)-Vestitol

Fig. 15

O CH_3 $PhCH_2O$ OH + CHO OCH_2Ph $PhCH_2O$ OCH_3 $\xrightarrow{NaOH,\ MeOH}$ O $PhCH_2O$ OH OCH_2Ph OCH_3 →

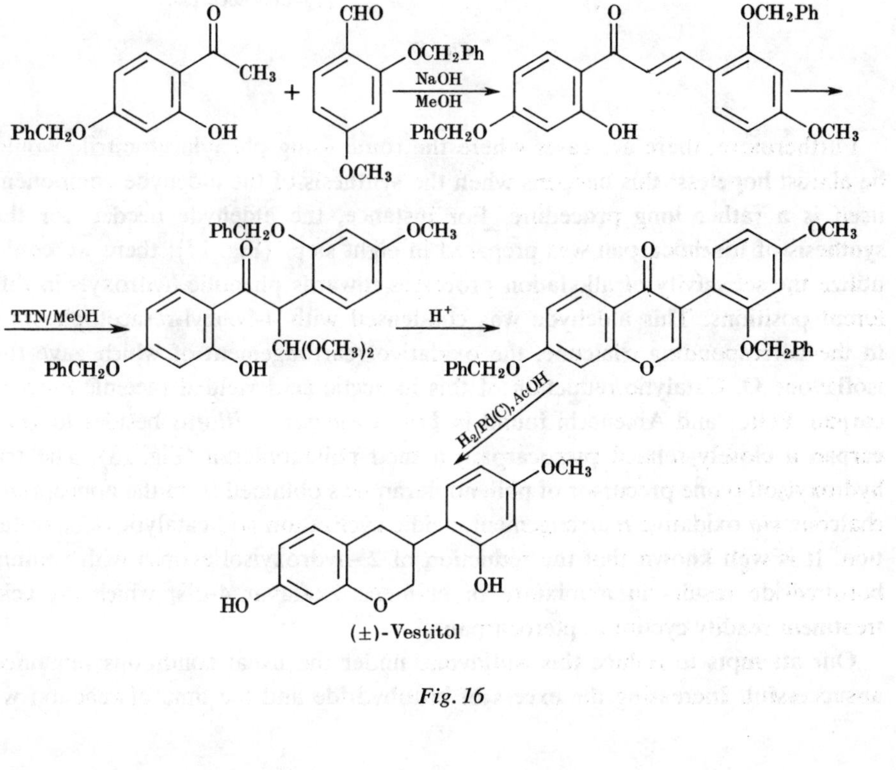

$\xrightarrow{TTN/MeOH}$ $PhCH_2O$ O $PhCH_2O$ OH OCH_3 $CH(OCH_3)_2$ $\xrightarrow{H^+}$ O $PhCH_2O$ O OCH_3 OCH_2Ph

$H_2/Pd(C), AcOH$

HO O OCH_3 OH

(±)-Vestitol

Fig. 16

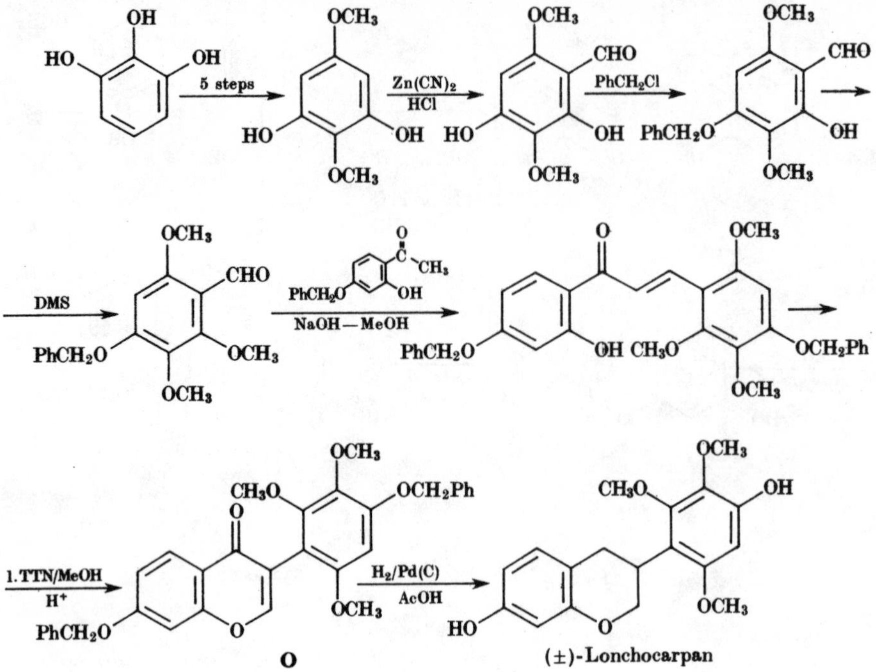

O

(±)-Lonchocarpan

Fig. 17

Furthermore, there are cases where the route using phenylacetonitrile would be almost hopeless; this happens when the synthesis of the aldehyde component itself is a rather long procedure. For instance, the aldehyde needed for the synthesis of lonchocarpan was prepared in eight steps (Fig. 17); there we could utilize the selectivity of alkylation processes towards phenolic hydroxyls in different positions. This aldehyde was condensed with 4-benzyl-resacetophenone to the corresponding chalcone, the oxidative rearrangement of which gave the isoflavone O. Catalytic reduction of this in acetic acid yielded racemic lonchocarpan. Pelter and Amenechi found in *Lonchocarpus laxiflorus* besides lonchocarpan a closely related pterocarpan, named philenopteran (Fig. 18). The trihydroxyisoflavone precursor of philenopteran was obtained from the appropriate chalcone *via* oxidative rearrangement, acidic cyclization and catalytic debenzylation. It is well known that the reduction of 2′-hydroxyisoflavones with sodium borohydride results in a mixture of epimeric isoflavan-4-ols, which on acid treatment readily cyclize to pterocarpans.

Our attempts to reduce this isoflavone under the usual conditions remained unsuccessful. Increasing the excess of borohydride and the time of reaction we

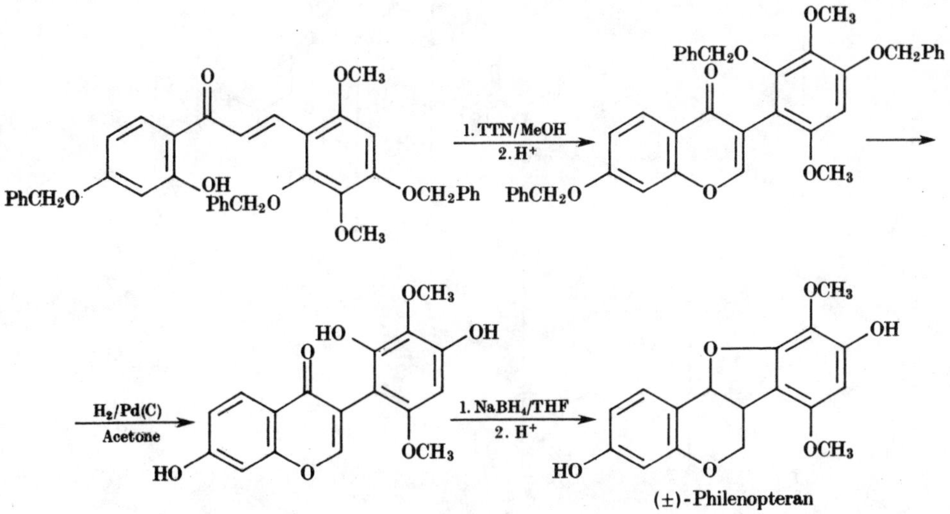

Fig. 18

obtained a new product, but its infrared- and mass spectra showed unambiguously that it was merely an isoflavanone. On further increase of the excess of the reagent and the reaction time, the isoflavan-4-ols were obtainable, in low yields, only after a 20-hour boiling in tetrahydrofuran. These difficulties may be ascribed to the strong steric hindrance caused by the *o*-substituents of the phenyl ring. The epimeric mixture of alcohols gave on acidification racemic philenopteran.

The plant, *Lonchocarpus laxiflorus* seems to be rich in isoflavonoids. Besides lonchocarpan and philenopteran, Pelter and Amenechi [15] found still another isoflavan, laxifloran, which could only be isolated as its dimethyl ether.

According to the NMR spectra of the dimethyl ether in different solvents and the mass spectrum of an extract rich in laxifloran — which gave the fragments shown in Fig. 19 — the presence of two free hydroxyls, one of them being at C_7, was reliably established. There was, however, no information available about the position of the other hydroxyl in ring B. That it might be in position 4′ was only postulated because of the co-occurrence of laxifloran and lonchocarpan in the same plant.

Considering the position of the free hydroxyls in ring B, there are three possible structures for laxifloran and we wished to locate the second free hydroxyl of laxifloran by the synthesis of all of the isomers.

One of them was identical with mucronulatol, isolated by Ollis and his coworkers [14]. For the synthesis of mucronulatol (Fig. 20) the appropriate

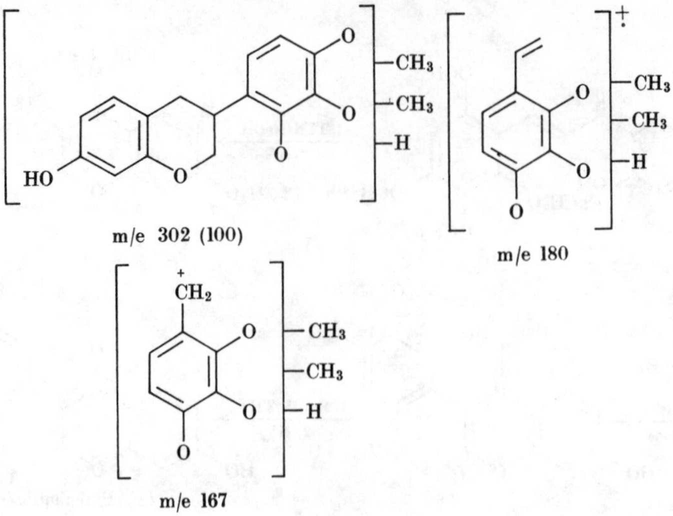

m/e 302 (100)

m/e 180

m/e 167

Fig. 19

1. TTN/MeOH
2. H^+

H_2/Pd(C) Acetone

H_2/Pd(C) AcOH

(±)-Violanone

(±)-Mucronulatol

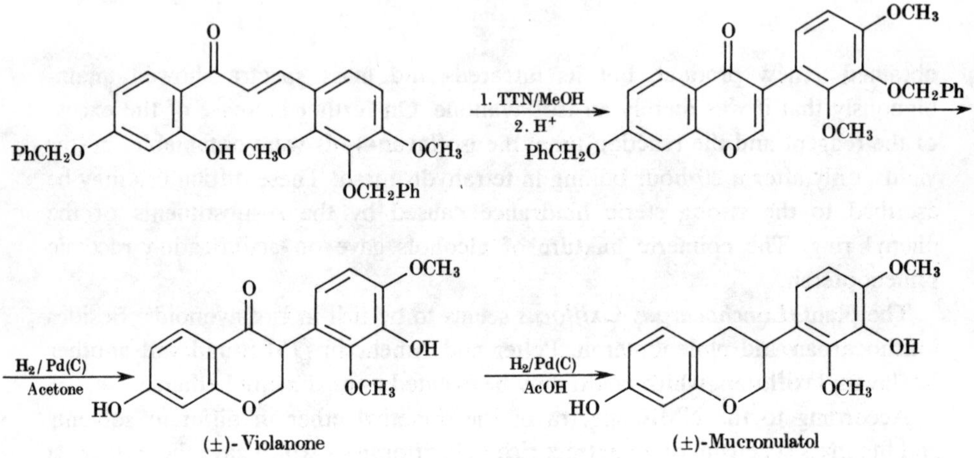

Fig. 20

2′-hydroxychalcone was converted in methanol with thallium(III) nitrate to the corresponding dibenzyloxyisoflavone.

As the isoflavanone of the very same substitution pattern is also a natural product, the catalytic hydrogenation was carried out first in acetone until the uptake of 3 equivalents of hydrogen, resulting in racemic violanone. This was hydrogenated further in acetic acid to give racemic mucronulatol.

For the synthesis of laxifloran of the structure given by Pelter, our starting material had to be 4-benzyloxy-2,3-dimethoxybenzaldehyde (Fig. 21, P;

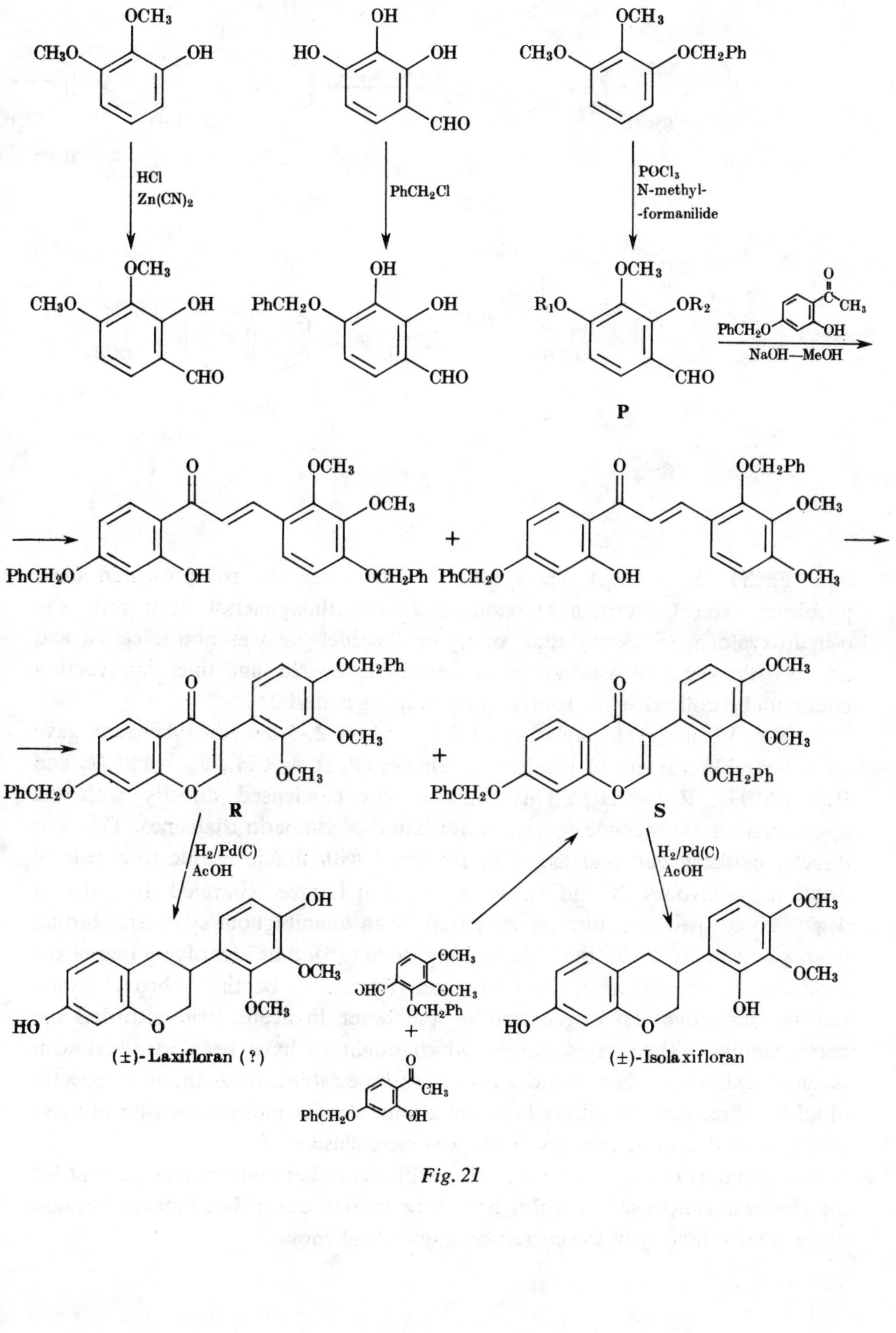

OCH3
CH3O
OH
HCl
Zn(CN)2
OH
HO
OH
CHO
PhCH2Cl
OCH3
CH3O
OCH2Ph
POCl3
N-methyl-
-formanilide
OCH3
CH3O
OH
CHO
OH
PhCH2O
OH
CHO
OCH3
R1O
OR2
CHO
P
PhCH2O
OH
CH3
NaOH—MeOH
OCH3
OCH3
PhCH2O
OH
OCH2Ph
OCH2Ph
OCH3
PhCH2O
OH
OCH3
OCH2Ph
OCH3
PhCH2O
OCH3
R
OCH3
OCH3
PhCH2O
OCH2Ph
S
H2/Pd(C)
AcOH
H2/Pd(C)
AcOH
OH
OCH3
HO
OCH3
(±)-Laxifloran (?)
OCH3
OHC
OCH3
OCH2Ph
CH3
PhCH2O
OH
OCH3
OCH3
HO
OH
(±)-Isolaxifloran

Fig. 21

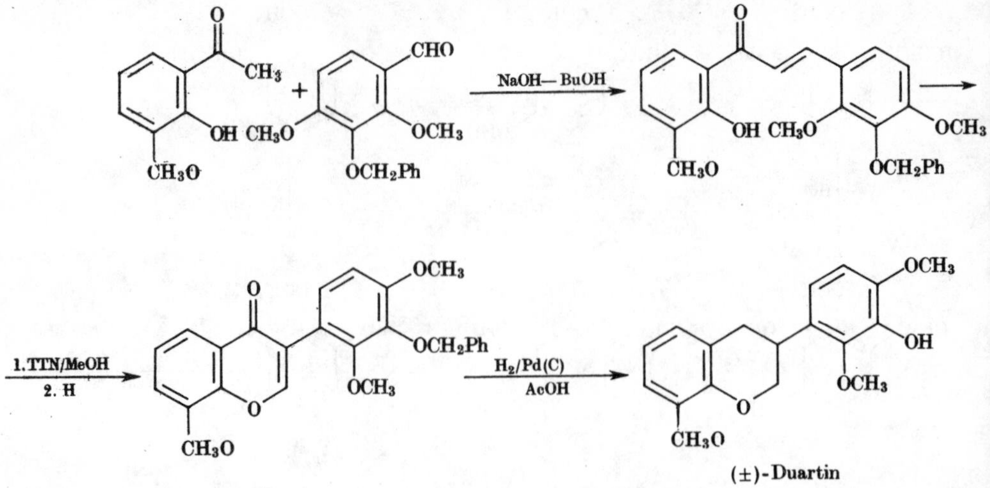

Fig. 22

$R_1 = PhCH_2$; $R_2 = CH_3$). The synthesis of the pure material presented some problems. The Gatterman reaction of 2,3-dimethoxyphenol gave only the *o*-hydroxyaldehyde. Benzylation of pyrogallolaldehyde was non-selective and afforded the required 4-benzyl ether only in 1 % yield, and thus this reaction could not be utilized as the source of the starting material.

Finally, Vilsmeyer formylation of 1-benzyloxy-2,-3-dimethoxybenzene gave an inseparable mixture of isomeric aldehydes (P, $R_1 = CH_3$, $R_2 = PhCH_2$ and $R_1 = PhCH_2$, $R_2 = CH_3$). This mixture was condensed directly with the appropriate acetophenone to give the mixture of isomeric chalcones. This was directly oxidized and rearranged in methanol with thallic nitrate to a pair of isomeric isoflavones (R and S), which could at last be separated. In order to decide about their structure, we prepared by an unambiguous synthesis, starting from a known aldehyde, the 2′-benzyloxy isomer (S), thus identifying one of the products. In consequence, the other one (R) had to be the 4′-benzyloxyiso-flavone derivative. Hydrogenation of the latter in acetic acid afforded the corresponding 4′-hydroxyisoflavan, which ought to have been identical with racemic laxifloran. Their identity could not be established, as the mass spectra of all the three isomers differed only in minute details, and comparison of these spectra with that of natural laxifloran was inconclusive.

The synthesis of racemic duartin (Fig. 22) was rather conventional, except for the chalcone condensation which had to be carried out in hot butanol because of the poor solubility of the corresponding acetophenone.

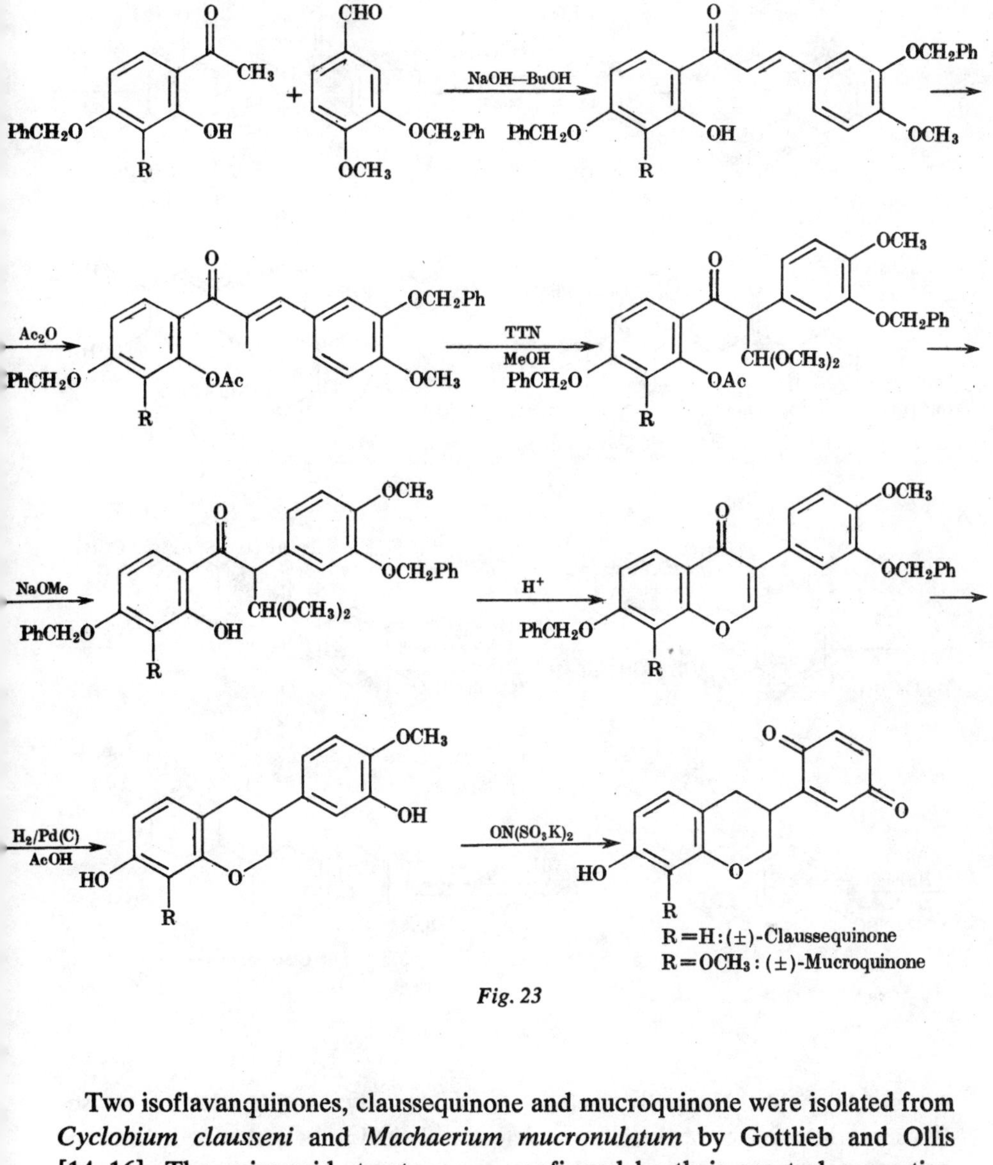

R=H: (±)-Claussequinone
R=OCH_3: (±)-Mucroquinone

Fig. 23

Two isoflavanquinones, claussequinone and mucroquinone were isolated from *Cyclobium clausseni* and *Machaerium mucronulatum* by Gottlieb and Ollis [14, 16]. The quinonoid structure was confirmed by their spectral properties. Their synthesis (Fig. 23) was achieved by condensing the appropriately substituted acetophenones (R = H or CH_3O) with benzylisovanillin to the appropriate chalcones. As the solubility of the chalcones in methanol was insufficient for carrying out the oxidative rearrangement satisfactorily, they were acetylated in the 2′-position, which resulted in oxidative rearrangements occurring with

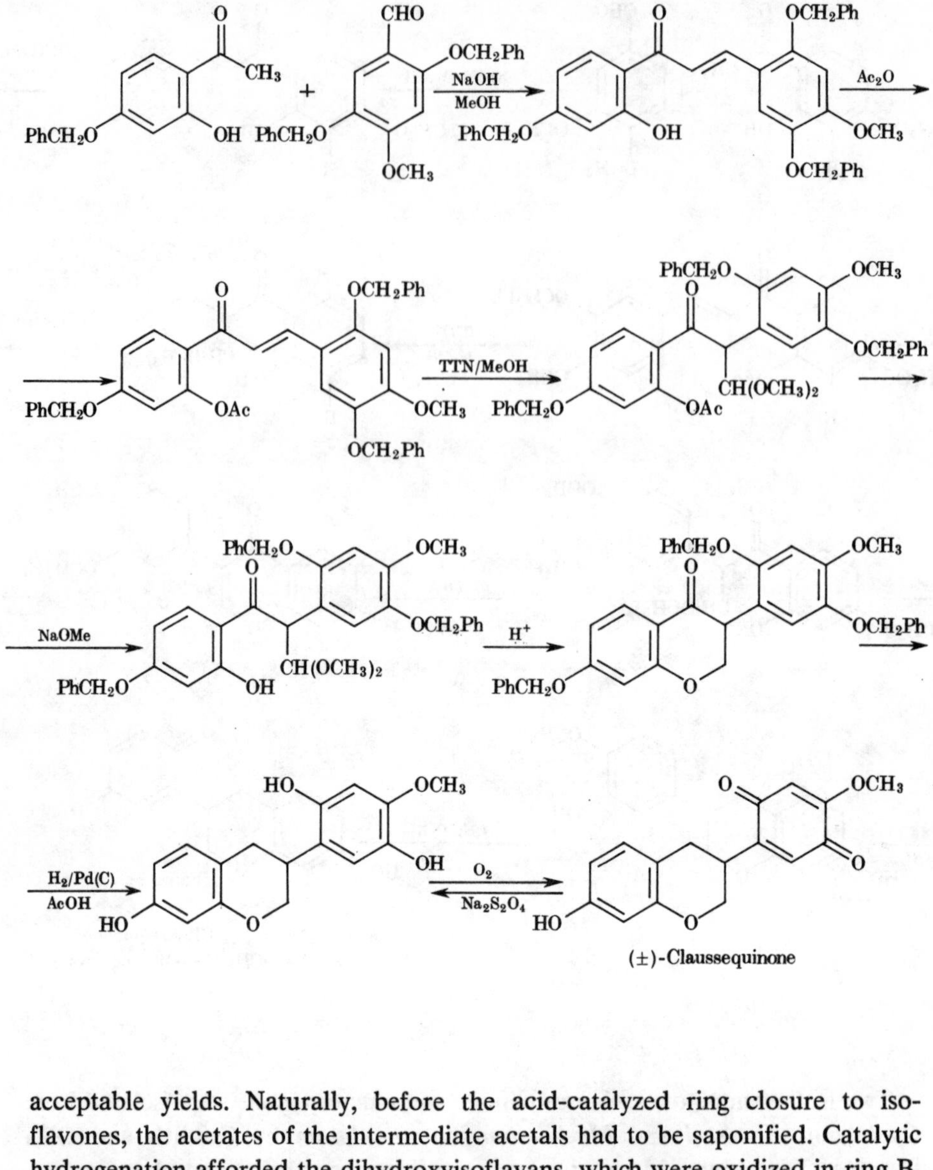

NaOH
MeOH
Ac_2O
TTN/MeOH
NaOMe
H^+
$H_2/Pd(C)$
AcOH
O_2
$Na_2S_2O_4$

(±)-Claussequinone

acceptable yields. Naturally, before the acid-catalyzed ring closure to isoflavones, the acetates of the intermediate acetals had to be saponified. Catalytic hydrogenation afforded the dihydroxyisoflavans, which were oxidized in ring B, in *para* position to the hydroxyl already present, by Fremy's salt.

An alternative way to claussequinone is shown in Fig. 24. This starts with a derivative containing two benzyl-groups in *para* position. This would give ultimately a quinol, the oxidation of which to quinone was expected to be simple. The sequence, leading to the tribenzyloxyisoflavone was similar to the one shown in Fig. 23. Catalytic hydrogenation of the tribenzyloxyisoflavone in acetic acid should have resulted after the uptake of 6 equivalents of hydrogen in the corresponding trihydroxyisoflavan.

Catalytic hydrogenation was duly carried out until the calculated hydrogen absorption, but thin-layer chromatography showed the presence of a by-product. Moreover, on chromatography of the crude product on a silica gel column, the main product disappeared to give place to the former by-product. The vivid yellow colour of this compound, the presence of a carbonyl signal in the IR spectrum and, finally, mixed melting point determination with claussequinone confirmed, that during chromatography the trihydroxyisoflavan suffered oxidation by air. The unstable quinol could be prepared from claussequinone by reduction with sodium dithionite; its UV spectrum proved the quinol structure, but another spectrum, taken from the same solution a few hours later was already characteristic of the quinone.

Thus it cannot be excluded that claussequinone and mucroquinone are artefacts formed from the original quinols during their isolation. In the paper describing their isolation [14] reference was made to unpublished syntheses of (±)-vestitol, (±)-duartin, (±)-mucronulatol and (±)-mucroquinone.

REFERENCES

1. Farkas, L., Gottsegen, Á., Nógrádi, M. and Antus, S., *J. Chem. Soc. (C), 1971,* 1994.
2. Nógrádi, M., Farkas, L. and Ollis, W. D., *Chem. Ber., 103,* 999 (1970).
3. Ollis, W. D., Ormand, K. L. and Sutherland, I. O., *Chem. Comm., 1968,* 1237.
4. McKillop, A., Swann, B. P. and Taylor, E. C., *Tetrahedron Letters, 1970,* 5281; McKillop, A., Swann, B. P., Ford, M. E. and Taylor, E. C., *J. Am. Chem. Soc., 95,* 3641 (1973).
5. Falshaw, C. P., Ollis, W. D., Moore, J. A. and Magnus, K., *Tetrahedron, 22,* 333 (1966).
6. Sibata, S. and Nishikawa, Y., *Chem. Pharm. Bull., 11,* 167 (1963).
7. Adityachaudhury, N. and Gupta, P. K., *Chem. and Ind., 1970,* 745 and 1113.
8. Nakayama, M., Harano, T. and Fukui, K., *Experientia, 27,* 361 (1971).
9. Gilbert, A. H., McGookin, A. and Robertson, A., *J. Chem. Soc., 1957,* 3740.
10. Chatterjee, J. N. and Prasad, N., *Chem. Ber., 97,* 1252 (1964).
11. Donnelly, D. M. X. and Fitzgerald, M. A., *Phytochemistry, 10,* 3147 (1971).
12. Rall, G. J. H., Engelbrecht, J. P. and Brink, A. J., *Tetrahedron, 26,* 5007 (1970).
13. Marrian, G. F. and Beall, D., *Biochem. J., 29,* 1586 (1935).
14. Kurosawa, K., Ollis, W. D., Redman, B. T., Sutherland, I. O., Braga de Oliveira, A., Gottlieb, O. R. and Magalhaes Alves, H., *Chem. Comm., 1968,* 1263.
15. Pelter, A. and Amenechi, P. I., *J. Chem. Soc. (C), 1969,* 887.
16. Braga de Oliveira, A., Gottlieb, O. R., Gonçlaves, T. M. and Ollis, W. D., *Anais. Acad. brasil. Cienc., 43,* 129 (1971) (*C.A. 76,* 110253 (1972)).

SYNTHESIS OF 1-THIAFLAVANONE SULFONE DERIVATIVES

by

J. BÁLINT,* M. RÁKOSI** and R. BOGNÁR**

* BIOGAL Pharmaceutical Factory, Debrecen

** Faculty Research Group of Antibiotic Chemistry of the Hungarian Academy of Sciences Debrecen, Hungary

1-Thiaflavanone is one of the simplest representatives of sulfur-containing flavonoids. This compound was used as the starting material in our recent experiments, the results of which will be briefly reported in this paper.

1-Thiaflavanone has been prepared by slight modification of Arndt's method (Fig. 1).

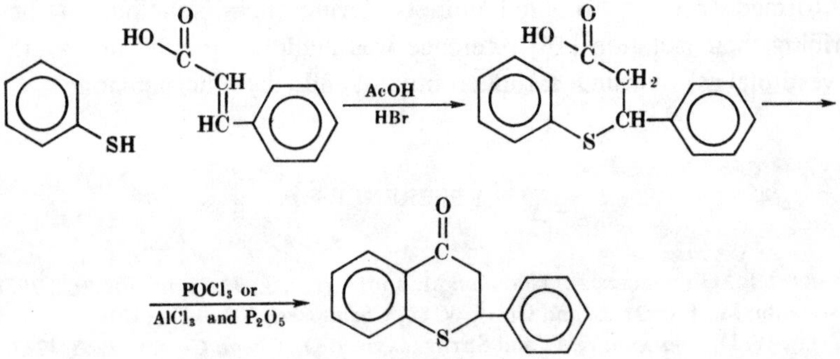

AcOH, HBr; $POCl_3$ or $AlCl_3$ and P_2O_5

Fig. 1

In the condensation reaction of thiophenol with cinnamic acid, the sulfur atom becomes attached to the β-carbon atom of cinnamic acid; the intermediate compound, β-phenylmercaptodihydrocinnamic acid, upon ring closure with phosphoryl chloride affords 1-thiaflavanone with the elimination of water.

Reactions similar to those made on flavanone [1, 2] have been performed with this compound formerly [3]. This work has now been extended to include 1-thiaflavanone sulfone, the oxidized derivative of 1-thiaflavanone.

1-Thiaflavanone sulfone is obtainable in a fair yield from 1-thiaflavanone by oxidation with hydrogen peroxide in acetic anhydride (Fig. 2).

This compound can be converted to 1-thiaflavanone sulfone oxime by a method similar to that described by Bognár *et al.* [4] in the case of flavanone, and later for the preparation of 1-thiaflavanone oxime [5], using hydroxyl-

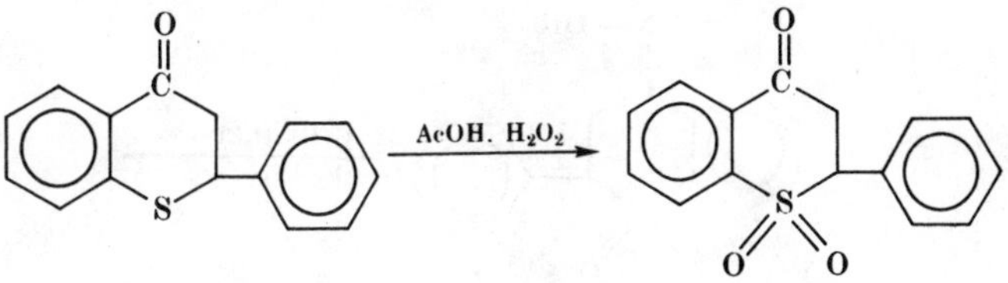

AcOH. H_2O_2

Fig. 2

$NH_2 \cdot OH \cdot HCl$
a. NaOH
b. NaOAc
c. Pyridine

CH_3^+—Cl^-

red.
a. Pd/C
b. $LiAlH_4$
c. Diborane
d. Zn + NH_4OH

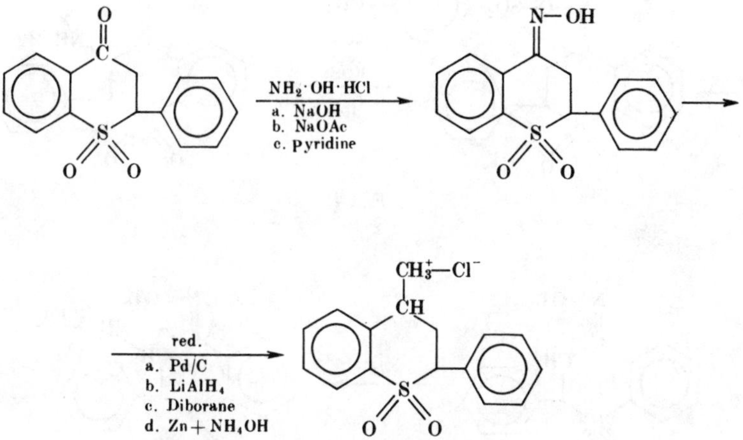

Fig. 3

amine hydrochloride (a) in basic ethyl alcohol; (b) in aqueous ethyl alcohol containing sodium acetate, and (c) in dry pyridine, according to Gulati and Ray [6] (Fig. 3).

All the three methods afford the same 1-thiaflavanon sulfone oxime as light drab crystals.

This oxime is reduced to 4-amino-1-thiaflavan sulfone in glacial acetic acid with hydrogen, in the presence of palladized charcoal, or in ethereal solution with lithium aluminium hydride. The reduction product can be converted into its hydrochloride salt with hydrogen chloride.

The reaction is accomplished equally well by the use of diborane or zinc dust in ammonium hydroxide.

The preparation of 3-amino-1-thiaflavanone sulfone has also been attempted in our experiments from 1-thiaflavanone sulfone oxime using the Neber rearrangement [7, 8] introduced by Bognár *et al.* [9] to the field of flavonoids. This method of synthesizing α-aminoketones proved suitable for the preparation

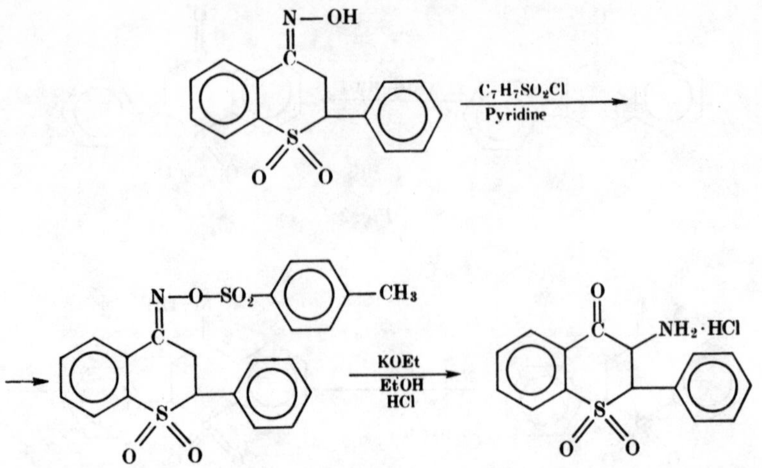
N—OH
C
S
O O
C₇H₇SO₂Cl
Pyridine
N—O—SO₂
CH₃
KOEt
EtOH
HCl
O
C
NH₂·HCl

Fig. 4

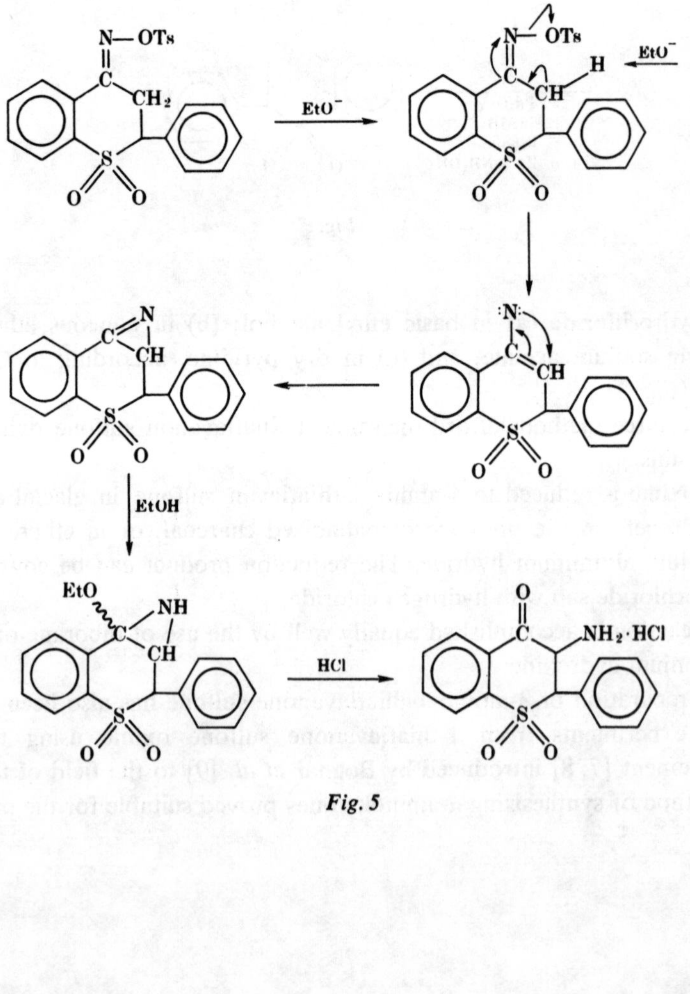
N—OTs
CH₂
EtO⁻
H
CH
N
:N:
EtOH
EtO
NH
HCl
NH₂·HCl

Fig. 5

of 3-amino-1-thiaflavanone sulfone, just like formerly in the case of 3-amino-1-thiaflavanone [10].

The steps of this synthesis are as follows. When 1-thiaflavanone sulfone oxime is treated with *p*-toluenesulfonyl chloride in dry pyridine, 1-thiaflavanone sulfone oxime tosylate is obtained as a red crystalline compound (Fig. 4).

Potassium ethoxide in dry ethanol converts this compound to the 3-amino-1-thiaflavanone sulfone *via* Neber rearrangement. The product can be isolated in the form of its hydrochloride, as pale needles.

The probable mechanism of the reaction is shown in Fig. 5.

According to Parcell, the alkoxyethylenimine is formed in a base-catalyzed direct 1,2-addition of alcohol to the azirine and, contrary to the theory of House and Berkowitz, the splitting of the tosyloxy group precedes the attack by the alkoxide ion.

On the basis of this mechanism it is obvious that the amino group is formed on an electrophilic carbon atom, and in the case of cyclic α-aminoketones the amino group must be in *equatorial* position.

REFERENCES

1. Kállay, F., Janzsó, G. and Koczor, I., *Tetrahedron, 21,* 19 (1965).
2. Kállay, F., Janzsó, G. and Koczor, I., *Tetrahedron, 23,* 4317 (1967).
3. Bálint, J., Reactions of the Basic Compounds of Sulfur-containing Flavonoids. Dissertation, Debrecen 1971.
4. Bognár, R., Rákosi, M., Fletcher, H., Kehoe, D., Philbin, E. M. and Wheeler, T. S., *Tetrahedron, 18,* 135 (1962).
5. Bognár, R., Rákosi, M. and Bálint, J., *Tetrahedron Letters, 1964,* 137.
6. Gulati, K. G. and Ray, J. N., *Current Sci., 5,* 75 (1936); *C. A., 30,* 82144 (1936).
7. Neber, P. W. and Burgard, A., *Ann., 493,* 281 (1932).
8. Neber, P. W. and Huh, G., *Ann., 515,* 283 (1935).
9. Bognár, R., O'Brien, C., Philbin, E. M., Ushioda, S. and Wheeler, T. S., *Chem. and Ind., 1960,* 1186.
10. Bognár, R. and Rákosi, M., *Ann., 693,* 225 (1966).

REACTIONS OF ISOFLAVONES WITH CARBONYL REAGENTS

by
V. SZABÓ and J. BORDA
Institute of Applied Chemistry, Kossuth Lajos University
Debrecen, Hungary

Conversions with carbonyl reagents of many flavonoids are reported in the literature [1], but such reactions of isoflavones have not been studied so far.

Judged on the basis of its structure, the bond order of the carbonyl group in isoflavone is near to two, so it must give carbonyl reactions more readily than flavone. Simultaneously, the nucleophilic sensitivity of the C-2 atom provides opportunity for competing reactions leading to the formation of pyrazole, isoxazole or dioxime derivatives. As it is to be expected, it is possible to prepare different types of products under different reaction conditions and using reagents of different nucleophilicity.

In the present paper a survey is given on the reactions of isoflavones with hydrazine and hydroxylamine.

Isoflavone reacts with hydrazine in both neutral and basic media to give pyrazole derivatives in high yields (Fig. 1).

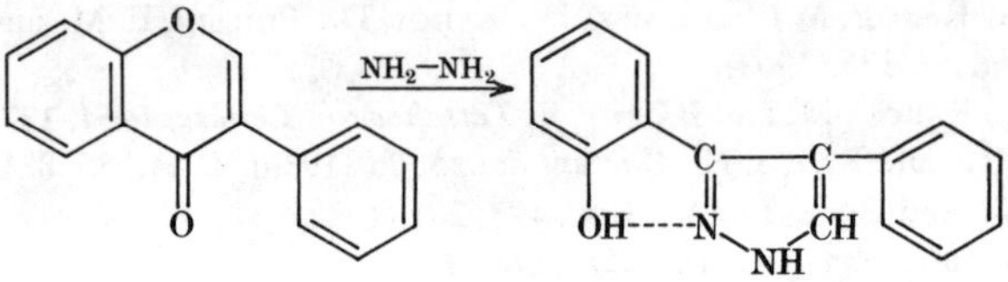

Fig. 1

No hydrazone was formed under the reaction conditions studied. From non-substituted isoflavone the same 3-(2-hydroxyphenyl)-4-phenylpyrazole is obtained as prepared by Baker, Harborne and Clarke from isoflav-4-thione [2].

Substituents in the C-2 position (methyl, trifluoromethyl) — depending on their inductive effects — considerably reduce the rate of the formation of pyrazole. This observation supports the assumption [2] that the point of attack of hydrazine — just as in the case of the 4-thione — is the C-2 atom.

The reaction with hydroxylamine of isoflavones is more complex. Non-substituted isoflavone gives a homogeneous, well-crystallized product in 80–90 % yield in different solvents and at different pH values (Table I).

Table I

Conditions of the Reaction of Isoflavone with Hydroxylamine

	Solvent	pH	Temperature, °C	Yield, %	M.p., °C
1.	EtOH : H_2O = 60 : 40	5-6	80	75	161-164
2.	Dioxan : H_2O = 60 : 40	9.0	30	85	161-164
3.	Dioxan : H_2O = 60 : 40	9.5	25	80	161-164
4.	Dioxan : H_2O = 60 : 40	9.5	40	87	161-164
5.	Dioxan : H_2O = 60 : 40	10.5	30	76	161-164
6.	EtOH : H_2O = 90 : 10	10-11	20	87	161-164
7.	EtOH	7	20	90	155-160*
8.	EtOH	9-10	20	60	159-161*
9.	Pyridine	–	115	80	150-160**

Reagent: $NH_2OH.HCl$ + buffer
In the case of 7 : NH_2OH + NaOEt
In the case of 8 : NH_2OH (free base)

* four components
** three components

This compound contains a nitrogen atom and an acetylable hydroxyl group. Taking into account the carbonyl reactivity of the isoflavone molecule, the conditions of isoxazole formation and the higher yields and higher rate of formation at alkaline pH values, one can expect five products with different structures, shown in Fig. 2.

Structures II and III were ruled out, since the product gave negative $FeCl_3$ reaction; IV was also impossible owing to its elemental analysis and the lack of an expectable further carbonyl reaction; in fact, the same compound — containing one nitrogen atom — was obtained in the reaction with three or eight moles of hydroxylamine. IV has a structure analogous with 2-hydroxyisoflavanone, so it can be stable neither in alkaline nor in acidic media, whereas the compound in question begins to show a slow change above pH ~ 11 only (Fig. 3), and it is very stable in acidic media.

These facts leave structures I or V. The NMR spectra (Table II) (identity of the H-2 signal of isoflavone and the proton signals of our compounds, as well as the appearance of a peak assigned to the oxime proton [3]) confirm the structure I.

Table II

NMR Spectra of Isoflavone and Isoflavone Oxime

Compound	H–2 signal, ppm	–O–H signal, ppm
Isoflavone	8.6	–
Oxime	9.0	10.1
Oxime acetate	8.5	–

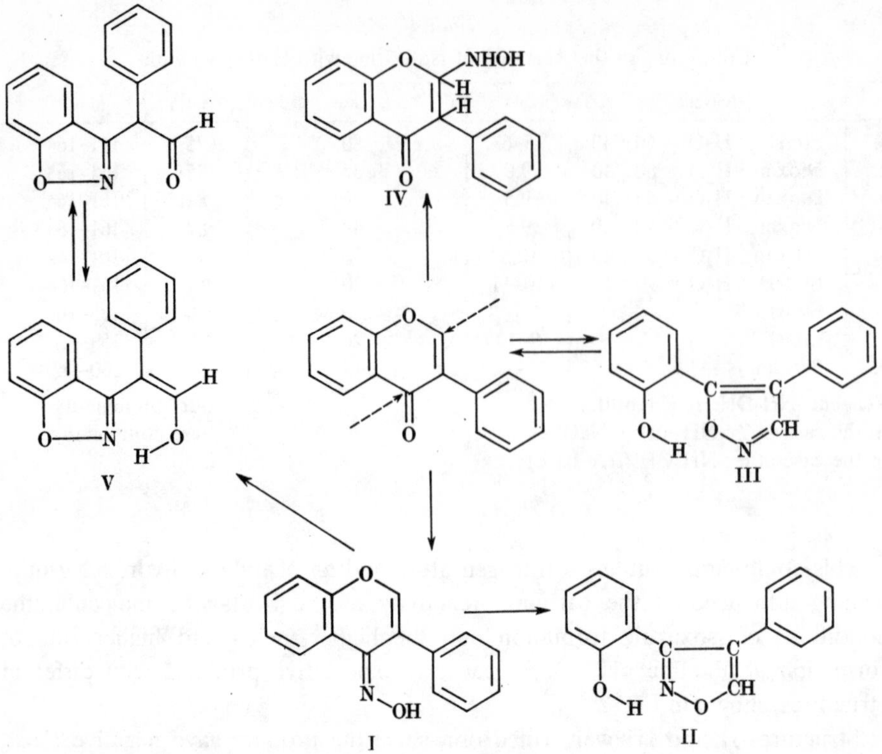

Fig. 2

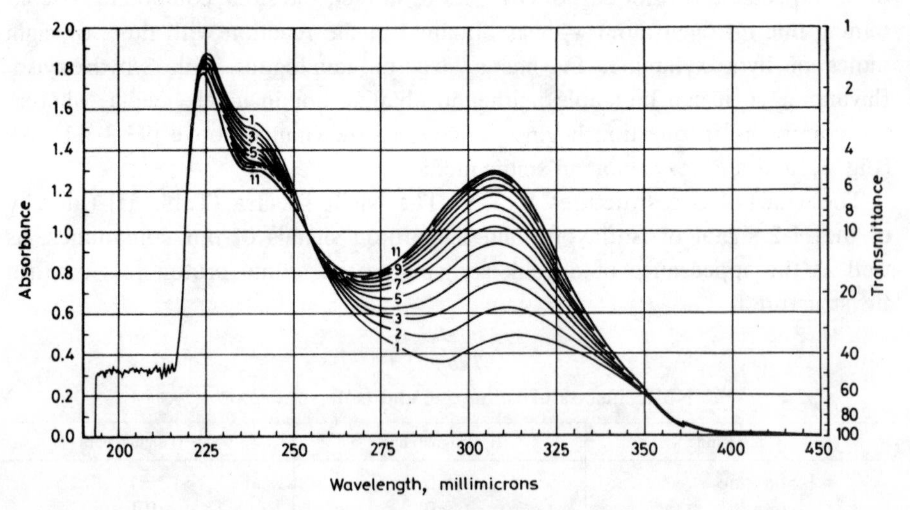

Fig. 3. Isoflavone oxime in N NaOH.
Concentration: 5×10^{-4} M, reference: N NaOH, path length: 2 mm, 7: 1 hr, 11: 3 hrs

For lack of comparative standards, structure V cannot be ruled out with certainty, but it has little probability as shown by the absence of a $FeCl_3$ reaction of the isolated product; such a reaction could be expected if formula V were correct.

Hydrogenation of the compound in ethanol containing a small amount of water, in the presence of Pd-C, gave isoflavone or its hydrogenated derivatives in good yields. In view of analogies in the literature, this reaction may take place with both V and I, as shown in Fig. 4 [3, 4].

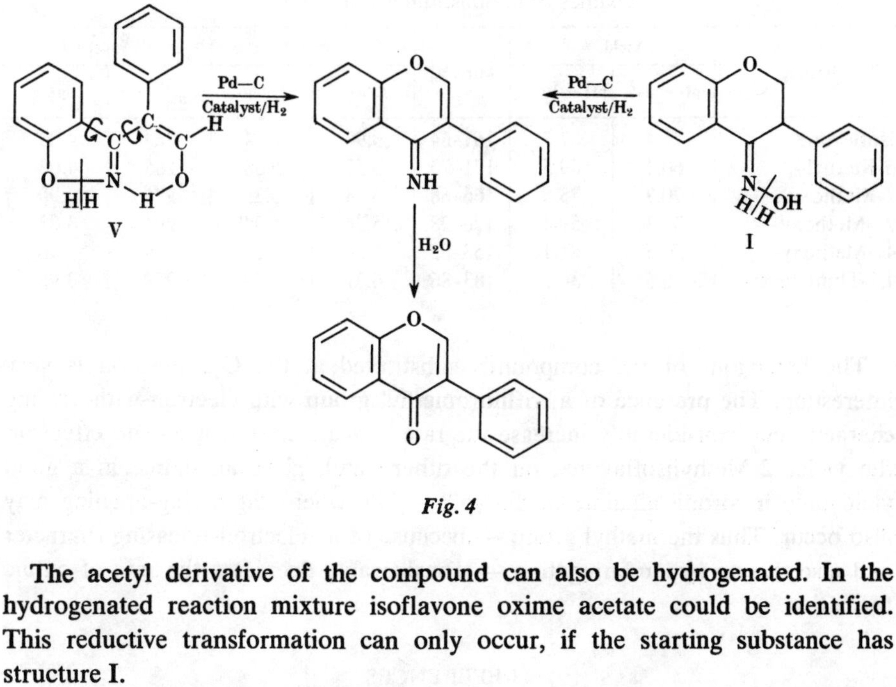

Pd—C Catalyst/H_2 ; V ; NH ; H_2O ; Pd—C Catalyst/H_2 ; I

Fig. 4

The acetyl derivative of the compound can also be hydrogenated. In the hydrogenated reaction mixture isoflavone oxime acetate could be identified. This reductive transformation can only occur, if the starting substance has structure I.

On the basis of these observations it can be established that the product of the reaction of hydroxylamine with isoflavone is the oxime.

As shown in Fig. 3, the heterocyclic ring of isoflavone oxime is cleaved in alkaline medium, similarly to that of isoflavone [5], and can be closed by acidification. However, in strong basic medium (0.5 N base in aqueous ethanol) a product of an irreversible reaction (m.p. 70–72°C and positive $FeCl_3$ colour reaction) is obtained which, on the basis of preliminary investigations, is a diphenylisoxazole derivative.

If oxime formation is effected after the previous cleavage of the pyrone ring, the majority (about 60 %) of the product is 2-hydroxydeoxybenzoin oxime. Thus, the dioxime could not be prepared by this method, probably owing to the formation of a β-dicarbonyl enolate. At the same time, with isoflav-4-thione — on the basis of our preliminary studies — there is a possibility for the preparation of a dioxime (and not isoxazole) under the conditions of oxime formation mentioned above.

Oximes can also be prepared in high yields from substituted isoflavones. Some of these compounds are shown in Table III.

Table III

Oximes from Substituted Isoflavones

Oxime	Yield, %		M.p., °C	N, %		UV Spectrum	
	pH ~ 5-6	pH ~ 10-11		Calcd.	Found	λ_{max}, nm	lgε
Isoflavone	75.4	87.7	161-64	5.90	5.64	263	4.03
6-Methyl-	60.1	89.5	171-73	5.57	5.58	265	4.08
7-Methoxy-	70.7	75.5	166-68	5.24	5.35	276	3.90
2′-Methoxy-	47.1	59.4	126-28	5.24	5.24	291	3.83
4′-Methoxy-	53.5	81.1	153-55	5.24	5.20	279	3.96
4,7-Dimethoxy-	70.5	36.2	183-86	4.71	4.89	279	3.96

The behaviour of the compounds substituted at the C-2 position is very interesting. The presence of a trifluoromethyl group with electron-withdrawing character can considerably increase the rate of reaction, but it has no effect on the yield. 2-Methylisoflavone, on the other hand, gives an oxime in a good yield only in strong alkaline media (pH ~ 11), where slight ring-opening may also occur. Thus the methyl group — because of its electron-releasing character and also due to hyperconjugation — considerably decreases the rate of oxime formation.

REFERENCES

1. KÁLLAY, F., "The Reactions of Flavonoid Compounds with Hydrazines" in "Recent Flavonoid Research" (Vol. 5 of "Recent Developments in the Chemistry of Natural Carbon Compounds", eds R. BOGNÁR, V. BRUCKNER, CS. SZÁNTAY), Akadémiai Kiadó, Publishing House of the Hungarian Academy of Sciences, Budapest, 1973, p. 153.
2. BAKER, W., CLARKE, G. G. and HARBORNE, J. B., *J. Chem. Soc., 1954,* 998.
3. EIDEN, F. and LÖWE, W., *Tetrahedron Letters, 1970,* 1439.
4. BREITNER, E., ROGINSKI, E. and RYLANDER, P. N., *J. Chem. Soc., 1959,* 2918.
5. SZABÓ, V. and ZSUGA, M., *Acta Chim. (Budapest),* In the press.

SOME NOVEL REACTIONS ON THE FLAVONOID C-4 ATOM

by
F. KÁLLAY
Research Institute for Organic Chemical Industry
Budapest, Hungary

This paper has the purpose to outline briefly a few recent features in our research work on reactions in which the flavonoid carbonyl group is playing a part.

Synthesis of Pyrazoline O-Glycosides

In the course of our earlier work [1] many chalcones and flavanones were converted into pyrazolines.

NH_2NH_2

This reaction has now been extended for the case when the flavanone, or the corresponding chalcone, contains a sugar moiety attached to oxygen.

Treatment of the chalcone glycosides with hydrazine gives the pyrazoline O-glycosides. The reaction was first effected using a model compound prepared by condensing vanillin-β-D-glucoside tetraacetate (I) with 2′-hydroxyacetophenone in the presence of potassium hydroxide. When the resulting chalcone glucoside (II) is suspended in water and treated with hydrazine hydrate, 3-(2-

hydroxyphenyl)-5-(3-methoxy-4-glucosyloxyphenyl)pyrazoline (III) is obtained in excellent yield.

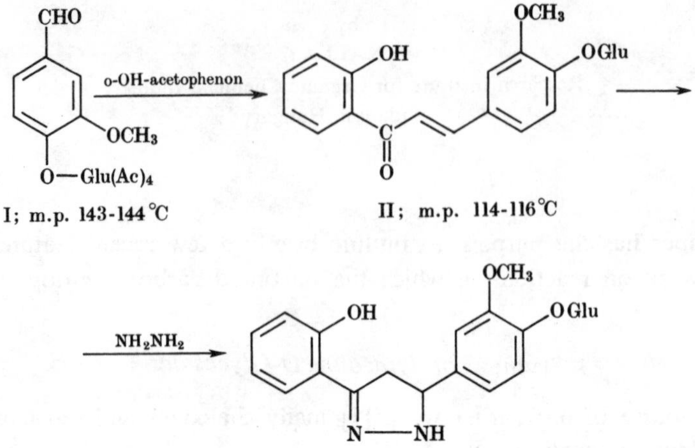

CHO

o-OH-acetophenon

OCH_3

O—Glu(Ac)$_4$

I; m.p. 143-144 °C

OCH_3 OH OGlu

C O

II; m.p. 114-116 °C

NH_2NH_2

OCH_3 OH OGlu

N—NH

III; m.p. 251-253 °C

Another example is the synthesis of 3-(2-hydroxy-4-glucosyloxyphenyl)-5-phenylpyrazoline (V) which contains the sugar moiety in the other benzene ring; this compound is readily prepared by the reaction of 2′-hydroxy-4′-β-D-glucosyloxychalcone (IV)* and hydrazine hydrate in ethanol.

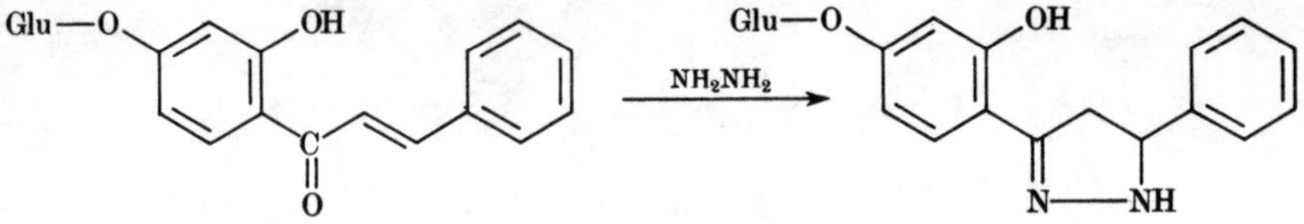

Glu—O OH

C O

IV

NH_2NH_2

Glu—O OH

N—NH

V; m.p. 113-115°C

In order to examine the scope of the reaction for application to natural chalcone glycosides, in a further experiment hesperidin methylchalcone (VI), a physiologically active compound, was used as the starting material. After reaction with hydrazine hydrate in ethanol, isolation of the product is greatly facilitated by acetylation to 1-acetyl-3-(2-methoxy-4-hexaacetylrutinosyloxy-6-acetoxyphenyl)-5-(3-acetoxy-4-methoxyphenyl)pyrazoline (VII).

* For a sample of the starting material our thanks are due to Dr. M. Rákosi

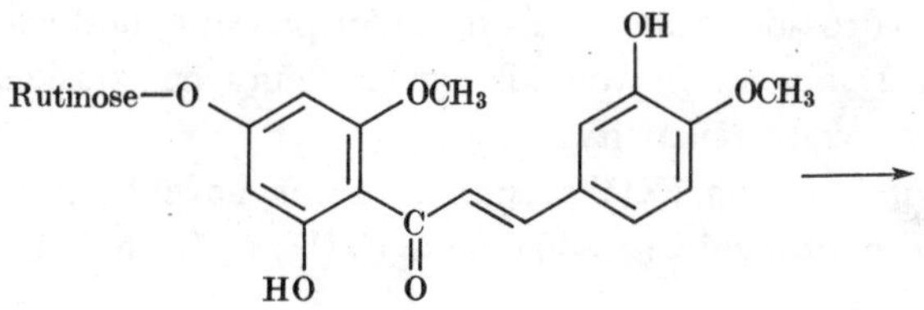

Rutinose—O OCH$_3$ OH OCH$_3$ HO O

VI

(Ac)$_6$-rutinose—O OCH$_3$ OAc OCH$_3$

1. NH_2NH_2
2. Ac_2O/Py

AcO N—N—Ac

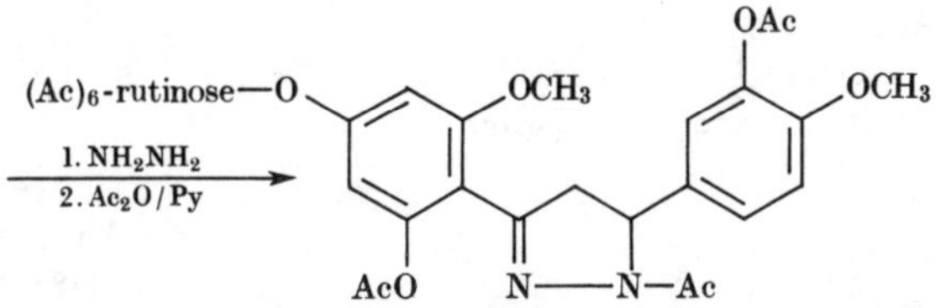

VII; m.p. 126-128°C

This product can be deacetylated with barium hydroxide to the chromatographically pure parent compound, but the latter is hygroscopic and very difficult to crystallize.

The structures of the new glycosides were proved in each case by analysis, IR spectroscopy and hydrolysis to the aglycone pyrazolines.

These simple experiments might be of interest, as the products are — to our knowledge — the first oxygen-glycosyl-pyrazoline derivatives; direct glycosylation of the parent nitrogen-heterocyclic compounds gives N-glycosides, as it has been shown on the example of some pyrazoles and imidazoles [2].

As expected, the glycosides have much better solubility in water than the corresponding aglycones; this can be of practical advantage if their pharmacological testing or use is considered.

Reaction of the Flavonoid Carbonyl Group with Amines

On the basis of their carbonyl reactions [1, 3] and the reactions with Grignard reagents [4], flavanones may be regarded as ketones, and reactions revealing ketonic features also of the flavone carbonyl group were accomplished in our experiments in 1968 [5].

The reaction of a ketone and amine should give the Schiff base, but this direct reaction of flavonoids has not been achieved so far. Aliphatic ketones react more slowly than aldehydes, and aromatic ketones are even less reactive; Weingarten *et al.* [6] reported that ketimines could not be prepared at all with amines from hindered ketones by any of the conventional methods, such as dehydration with potassium hydroxide, heat treatment under pressure, and water removal as an azeotrope in a solvent. Flavonoids can certainly be considered similar to these ketones of very low reactivity.

The flavone imines (e.g. XI) prepared so far have been synthesized [7] indirectly from *o*-hydroxydibenzoylmethane (VIII) or from 4-thionflavone (X).

VIII — n-BuNH$_2$, 100°C; 3hr in MeOH → IX; m.p. 148°C (d.)

IX — AcOH, 100°C; 10 min. → XI

X — n-BuNH$_2$, 100°C; 1.5 hr. in MeOH → XI; m.p. 44-45°C

The first reaction goes through the interesting intermediate 2-hydroxyflavanone n-butylimine (IX), which is an isolable stable compound; the conversion of 4-thionflavanone is reversible, i.e. flavone n-butylimine gives 4-thionflavone on treatment with hydrogen sulfide.

Our purpose has been to find a direct synthesis applicable to flavonoids from which the 4-thiono derivative cannot be prepared and the corresponding 1,3-diketone is not readily available.

In 1967 Weingarten and White [8] reported that titanium tetrachloride was satisfactory in effecting the conversion of secondary amines and ketones into enamines. This method was later extended for the preparation of ketimines from dimethylcyclohexanone, isophorone and *dl*-camphor. Three years later Moretti and Torre [9] showed that N-alkyl ketimines were obtainable by this method from benzophenone and *p*-halo-substituted benzophenones.

According to our experiments, the use of titanium tetrachloride can effect the desired reaction between the flavonoid carbonyl group and amines, the first example being the preparation of flavanone n-butylimine (XII).

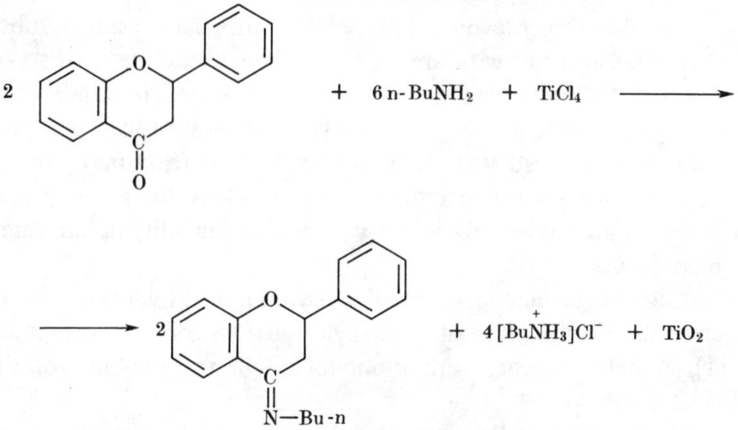

2 [flavanone] + 6 n-BuNH_2 + $TiCl_4$ ⟶

⟶ 2 [XII] + 4 $[BuNH_3]^+Cl^-$ + TiO_2

XII; m.p. 94-96°C

The general conditions are refluxing the reagents in dry benzene solution, when progress of the conversion is indicated by the separation of the alkylammonium chloride, and can be detected by thin-layer chromatography, or IR spectroscopy of the isolated product, showing a C=N band (in the case of XII, at 1630 cm^{-1}) instead of the original C=O band. The compound, however, is not very stable to air and must be rapidly dried in vacuum.

It was interesting to see whether flavone also would show a similar behaviour, as the carbonyl reactions with substituted hydrazines of this compound [5] are known to be more difficult than those of flavanone [10].

Under conditions similar to those mentioned above, flavone readily reacts with n-butylamine to yield the same ketimine (XI) as obtainable by the indirect synthesis from 4-thionflavone [7].

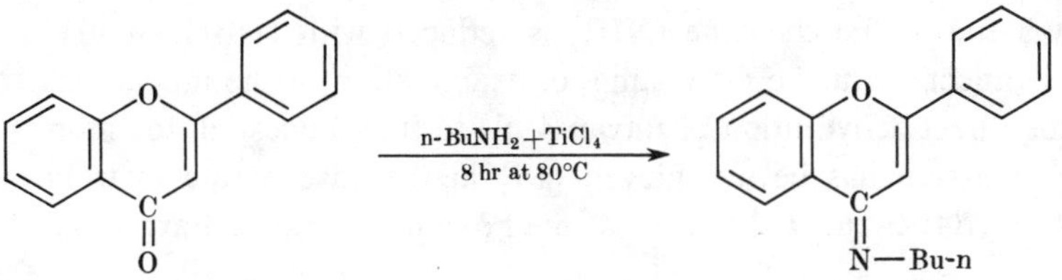

[flavone] —(n-BuNH_2 + $TiCl_4$; 8 hr at 80°C)⟶ [flavone n-butylimine, C=N—Bu-n]

m.p. 44 - 45°C

The product is less sensitive to air and also to hydrolysis than the corresponding flavanone derivative.

When the scope of the reaction was investigated using other flavonoids, definitely no reaction was observed, under the above experimental conditions, in the case of 3-hydroxyflavone. This is not surprising, as this substituted flavone also failed to react with any of the carbonyl reagents tried so far; the presence of the C-3 hydroxyl group in flavone seems to hinder carbonyl reactions so effectively that actually no carbonyl derivative has been prepared.

No success was achieved with 5,7-dihydroxyflavone (chrysine) either, but it has not been decided yet whether the negative result is due to the presence of the C-5 hydroxyl group, or only to the very scarce solubility of chrysine in the reaction medium used.

Returning to the flavanone series, the reaction of 3-hydroxyflavanone has been examined. Though several carbonyl reactions of this compound are known [1], it did not react with n-butylimine in the presence of titanium tetrachloride.

In spite of the apparent limitations in the scope of the reaction revealed so far and mentioned above, the direct synthesis of flavonoid ketimines can be of interest as a possible route to 4-alkylamino- and 4-arylaminoflavonoids. Further interest may be attached to a stereochemical study of the Schiff bases, e.g., by means of cyclization reactions.

Flav-3-enes from Flavan-4-ols

Not long ago flav-3-enes (2H-flavenes) (XV) were considered comparatively inaccessible materials, but several syntheses are known today. They include the dehydrobromination of 4-bromoflavanes [11–13], the $NaBH_4$ reduction of 2′-hydroxychalcones [13–17], the dehydrogenation of cinnamylphenols with 2,3-dichloro-5,6-dicyanobenzoquinone [18], and the reduction of 3-alkoxyflavylium salts with complex metal hydrides [19].

The 4-bromoflavan (XVI) for the above synthesis is prepared [20] from 2′-hydrochalcone (XIII) *via* flavanone (XIIIa) and flavan-4-ol (XIV); and the latter compound is obtained in very good yields as the sole product instead of flav-3-ene, if the chalcone (XIII) is reduced with $NaBH_4$ at pH = 8 [21]. These connections make interesting an examination of the missing interconversion, the direct dehydration of flavan-4-ols to flav-3-enes; all the more so, since the reverse step has been achieved [19] in the case of the hydroboration of 4′-methoxyflav-3-ene to 2,4-*trans*-4′-methoxyflavan-4-ol (a flavan-4-α-ol).

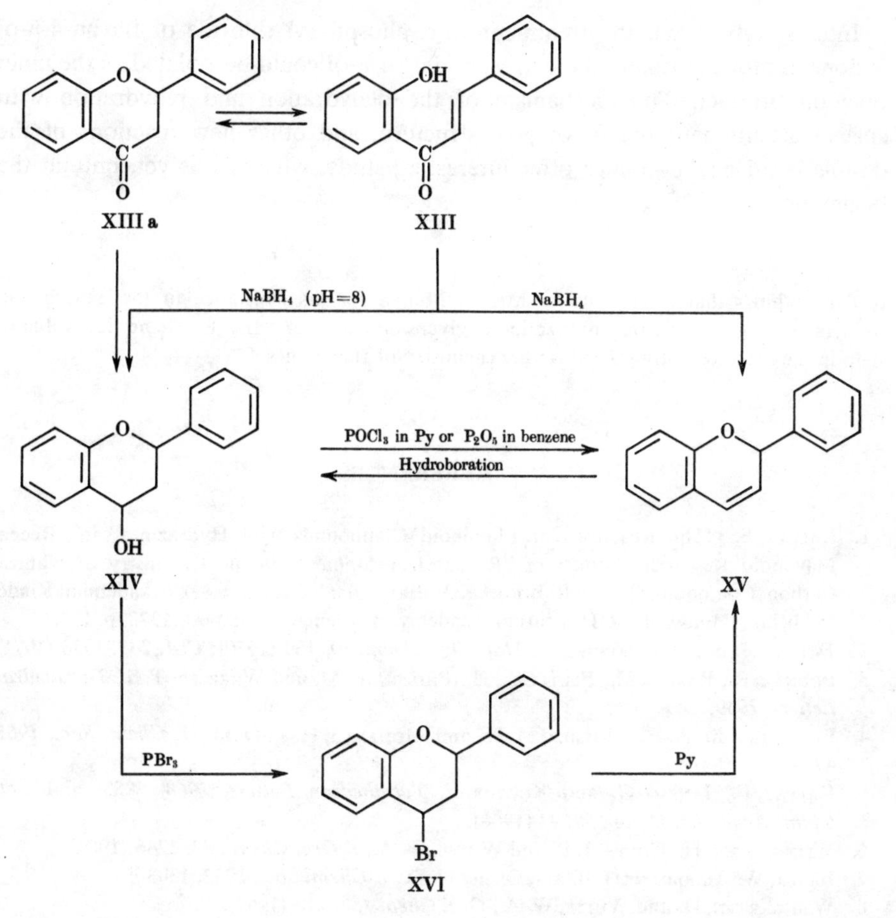

XIII a ⇌ XIII

$NaBH_4$ (pH=8) $NaBH_4$

XIV

$POCl_3$ in Py or P_2O_5 in benzene →

← Hydroboration

XV

PBr_3 → XVI → Py → XV

The dehydration of flavan-4-β-ol has been achieved in our experiments either with phosphorous pentoxide in benzene or with phosphoryl chloride in pyridine.

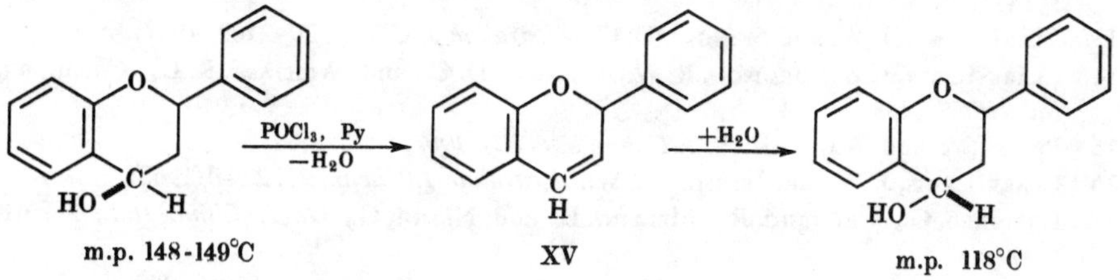

m.p. 148-149°C —$POCl_3$, Py / $-H_2O$→ XV —$+H_2O$→ m.p. 118°C

The product is flav-3-ene, isolable as a pure compound when freshly prepared, but rather unstable on standing.

Interestingly, when the treatment with phosphoryl chloride of flavan-4-β-ol is done in moist pyridine, the corresponding 4-α-ol could be isolated as the main reaction product. The mechanism of the dehydration and rehydration with special attention to the α- or β-ol structure, and other new reactions of the double bond in flav-3-enes offer interesting study, which is, as yet, only at the beginning.

*

The author's thanks are due to Mr. L. Mánya for his assistance in the experiments of synthesizing the various pyrazoline-O-glycosides, and to Mrs. É. Deme for valuable help in the work recently started on the chemistry of flav-3-enes.

REFERENCES

1. Kállay, F., "The Reactions of Flavonoid Compounds with Hydrazines" in "Recent Flavonoid Research" (Vol. 5 of "Recent Developments in the Chemistry of Natural Carbon Compounds", eds R. Bognár, V. Bruckner, Cs. Szántay), Akadémiai Kiadó, Publishing House of the Hungarian Academy of Sciences, Budapest, 1973, p. 153.
2. Jasinka, J. and Sokolowski, J., *Mat. Fiz. Chem., 10,* 169 (1970); *C.A.,74,* 31938 (1971).
3. Bognár, R., Rákosi, M., Fletcher, H., Philbin, E. M. and Wheeler, T. S., *Tetrahedron Letters, 1959,* 4.
4. Elkaschef, M. A.-F., Mossier, M. H. and Mohamed, H.-E.-D. M., *J. Chem. Soc., 1965,* 494.
5. Kállay, F., Janzsó, G. and Koczor, I., *Tetrahedron Letters, 1968,* 3853; and *Acta Chim. Acad. Sci. Hung., 58,* 97 (1968).
6. Weingarten, H., Chupp, J. P. and White, W. A., *J. Org. Chem., 32,* 3246 (1967).
7. Baker, W., Harborne, J. B. and Ollis, W. D., *J. Chem. Soc.,* 1952, 1303.
8. Weingarten, H. and White, W. A., *Org. Chem., 32,* 213 (1967).
9. Moretti, I. and Torre, G., *Synthesis, 1970,* 141.
10. Kállay, F., Janzsó, G. and Koczor, I., *Tetrahedron, 23,* 4317 (1967).
11. Marathe, K. G., Philbin, E. M. and Wheeler, T. S., *Chem. and Ind., 1962,* 1793.
12. Bolger, B. J., Marathe, K. G., Philbin, E. M. and Wheeler, T. S., *Tetrahedron, 23,* 341 (1967).
13. Clark-Lewis, J. W. and Skingle, D. C., *Australian J. Chem., 20,* 2169 (1967).
14. Clark-Lewis, J. W., Jemison, R. W., Skingle, D. C. and Williams, R. L., *Chem. and Ind., 1967,* 1455.
15. Pelter, A. and Stainton, P., *J. Chem. Soc. (C), 1967,* 1933.
16. Clark-Lewis, J. W. and Jemison, R. W., *Australian J. Chem., 21,* 2247 (1968).
17. Cardillo, G., Cricchio, R., Merlini, L. and Nasini, G., *Gazz. Chim. Ital., 99,* 612 (1969).
18. Cardillo, G., Cricchio, R. and Merlini, L., *Tetrahedron Letters, 1969,* 907.
19. Clark-Lewis, J. W. and Baig, M. I., *Australian J. Chem., 24,* 2581 (1971).
20. Shreiner, R. L. and Schaeffer, R. E., *Proc. Iowa Acad. Sci., 61,* 269 (1954).
21. Bognár, R. and Rákosi, M., *Acta Chim. Acad. Sci. Hung., 13,* 217 (1957).

^{13}C-NMR SPECTROSCOPY AS AN EXPEDIENT IN THE ELUCIDATION OF THE CONSTITUTION OF FLAVONOIDS

by

K. WEINGES

Organisch-Chemisches Institut der Universität
Heidelberg, FRG

Pulsed Fourier transform ^{13}C-NMR spectroscopy [1] has become a powerful and indispensable method to evaluate the constitution of natural products, because it allows a direct observation of the carbon skeleton. Compared with proton resonance spectroscopy, there are substantial advantages: 1. The experimental chemical shifts are much larger; 2. no disturbing coupling occurs which would cause certain absorptions to overlap. In the following the importance of this technique will be demonstrated on the example of the structure elucidation of $C_{30}H_{24}O_{12}$-procyanidins. This problem could not be solved by conventional spectroscopic methods.

At the 3rd Hungarian Bioflavonoid Symposium (Debrecen, 1970) I reported [2] about the isolation and constitution of natural $C_{30}H_{24}O_{12}$ (group A) and $C_{30}H_{26}O_{12}$ (group B) procyanidins. At that time I was able to deduct constitution 1 for $C_{30}H_{26}O_{12}$-procyanidins (group B) from ^{1}H-NMR and mass spectroscopical data (Fig. 1).

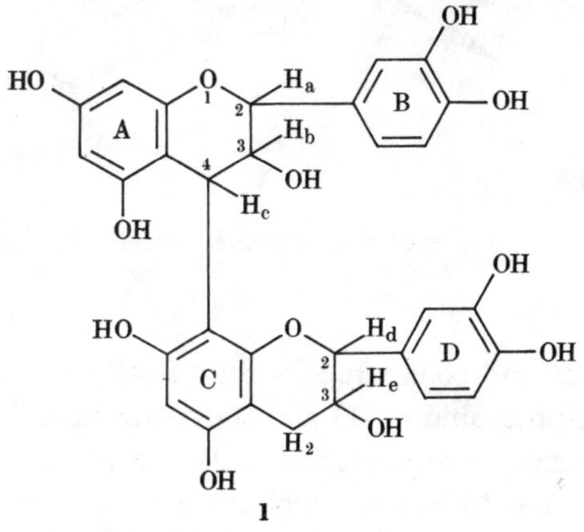

Fig. 1. Constitution of $C_{30}H_{26}O_{12}$-procyanidins (group B)

We could isolate four of the possible 32 optically active forms [3] and evaluated their configurations from the coupling constants of the H_a–H_e protons attached to the aliphatic carbon atoms [4]. The configurations of procyanidins B1–B4 are shown in Fig. 2.

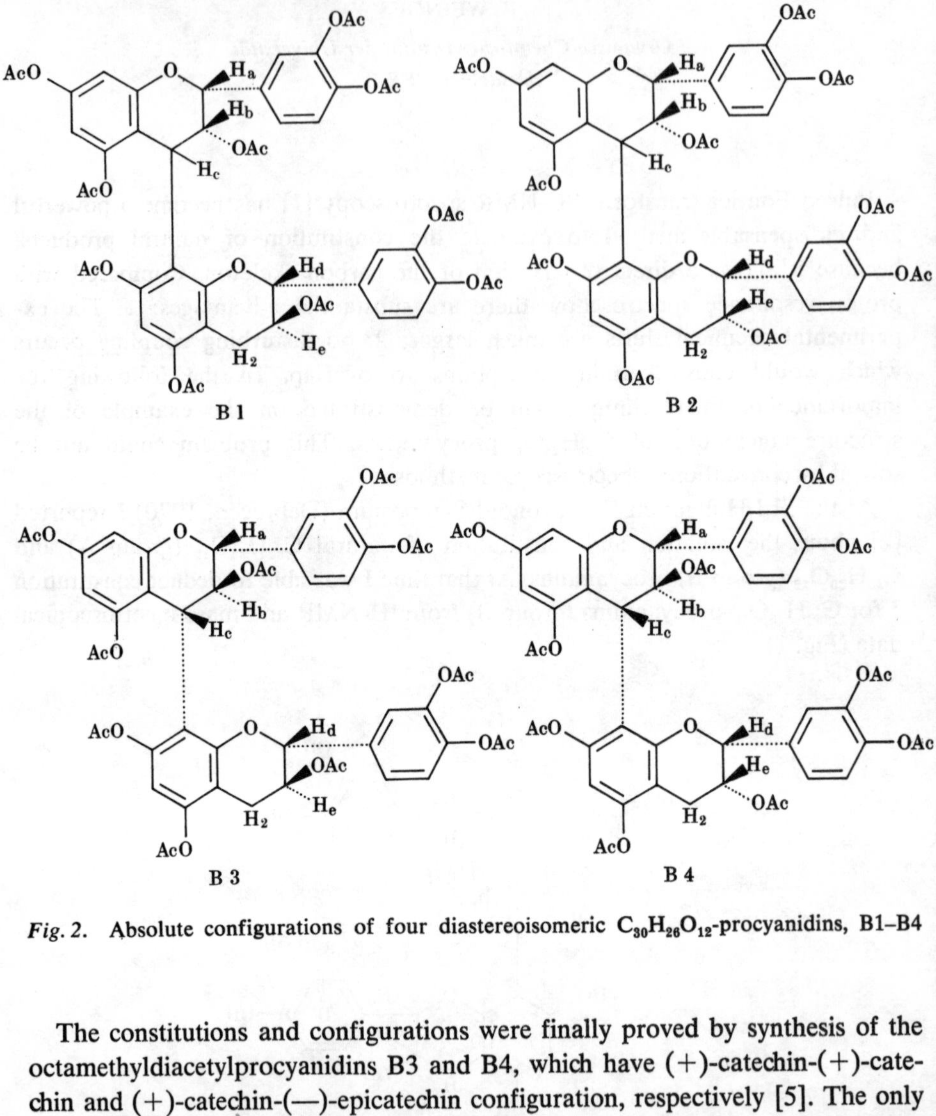

Fig. 2. Absolute configurations of four diastereoisomeric $C_{30}H_{26}O_{12}$-procyanidins, B1–B4

The constitutions and configurations were finally proved by synthesis of the octamethyldiacetylprocyanidins B3 and B4, which have (+)-catechin-(+)-catechin and (+)-catechin-(—)-epicatechin configuration, respectively [5]. The only point of interest, the biogenesis, which we were working on, has recently been published by Thompson, Jaques, Haslam and Tanner [6]. Therefore we have concluded our investigations on $C_{30}H_{26}O_{12}$-procyanidins.

The elucidation of the constitution of $C_{30}H_{24}O_{12}$-procyanidins (group A) turned out to be far more difficult. Here only two isomers, which we have named A1 and A2, are known up to the present. We shall discuss later if they are stereo- or structural isomers. Two constitutions have been reported in the literature. Mayer *et al.* [7] preferred structure 2 (Fig. 3) which embodies two

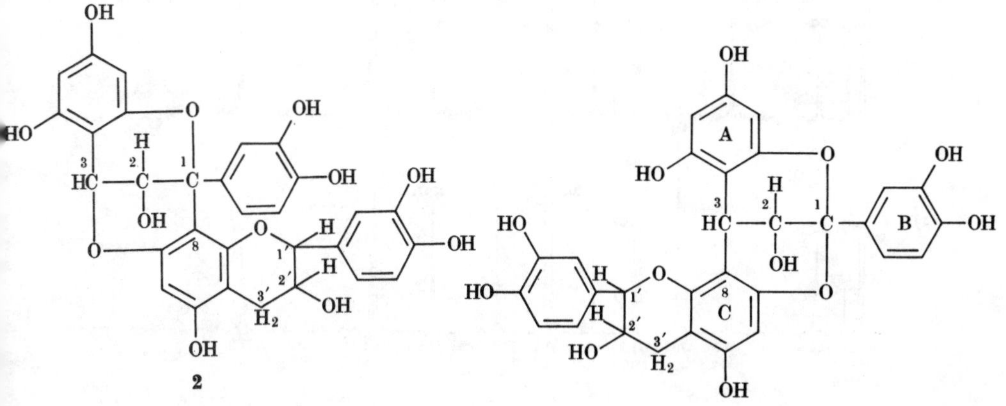

Fig. 3. Propositions of the constitutions of $C_{30}H_{24}O_{12}$-procyanidins, by Mayer *et al.* (2) [7] and Weinges *et al.* (3) [3]
(The numbering of the carbon atoms of the flavonoids used in this paper differs from the ordinary scheme)

benzyl ether groups, while we thought constitution 3 containing a ketal group to be more likely [3]. At that time neither by chemical nor by spectroscopic investigations was it possible to decide conclusively between the two constitutions. Together with Dr. Schilling we have now been able to solve this problem by ^{13}C-NMR spectroscopy [8].

Since no ^{13}C-NMR spectra of constitutionally similar compounds were known to-date, we first had to study those to establish the chemical shifts of the carbon atoms of the flavonoid skeleton. Figure 4 shows the proton-decoupled ^{13}C-NMR spectrum (Fig. 4a) and the off-resonance spectrum (Fig. 4b) of tetramethyl-(+)-catechin. Tetramethylsilane served as internal standard; its peak appears at the computer address 1932. To obtain the δ values, one has to subtract the addresses of the experimental absorption signals from that of the internal standard; e.g., the signal at 1673 has an address of 1932–1673 = 259. One address of the computer is equivalent to 2.44 Hz, so that this signal corresponds to 259 × 2.44 = 631.94 Hz. To obtain the δ values, one has to divide by the frequency of the measurement. In this way a δ value of 631.94 : 22.64 =

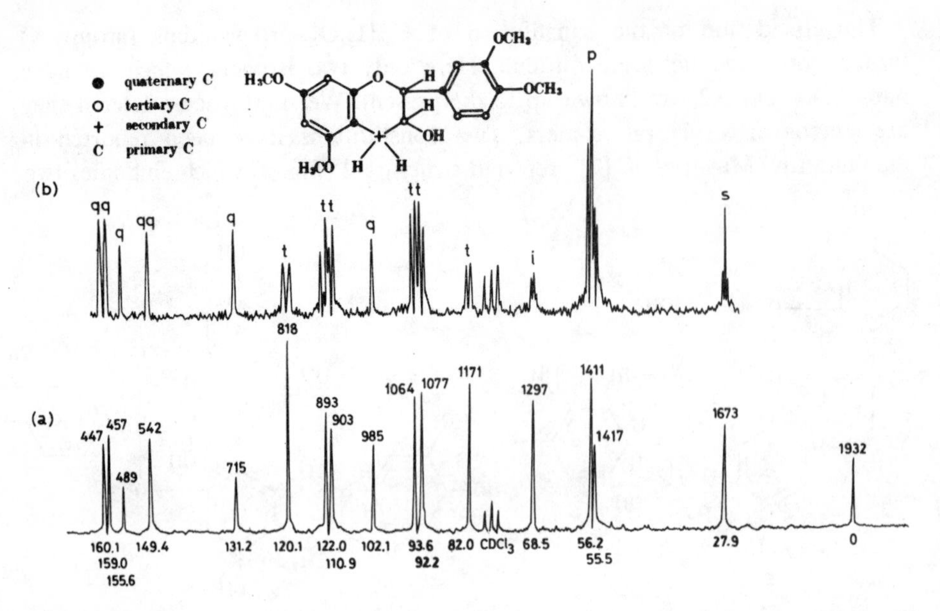

Fig. 4. ^{13}C-NMR spectrum of tetramethyl-(+)-catechin; (a) proton-decoupled, (b) off-resonance

27.9 ppm results for the signal which appears at address 1673. All other δ values may be calculated accordingly.

The off-resonance spectrum of tetramethyl-(+)-catechin (Fig. 4b) yields the coupling of the single carbon atoms with their respective protons. This allows to decide which of the signals belong to a quaternary, tertiary, secondary, or primary carbon atom. Quaternary carbon atoms cannot couple and therefore appear as singlets. Analogously, tertiary C atoms show up as doublets, secondary as triplets, and primary as quartets. As it follows from the constitution of tetramethyl-(+)-catechin, its spectrum (Fig. 4b) has 7 quaternary, 7 tertiary, 1 secondary and 4 primary carbon atoms. This facilitates the task to correlate absorptions with the individual carbon atoms. Since we recorded the spectrum in $CDCl_3$, three additional lines belonging to the chloroform carbon appear. By coupling with deuterium one obtains a triplet $M = 2n \times I + 1 = 2 \times 1 \times 1 + 1 = 3$.

The calculated δ values are transferred on a δ scale in which way we obtain a spectrum, seen in Fig. 5, which is easy to survey. The signals of the single carbon atoms are correlated by the off-resonance spectrum and by comparison with the spectra of methylated catechol and phloroglucinol derivatives. Figure 5b shows the spectrum of tetramethylmonoacetyl-(+)-catechin. The

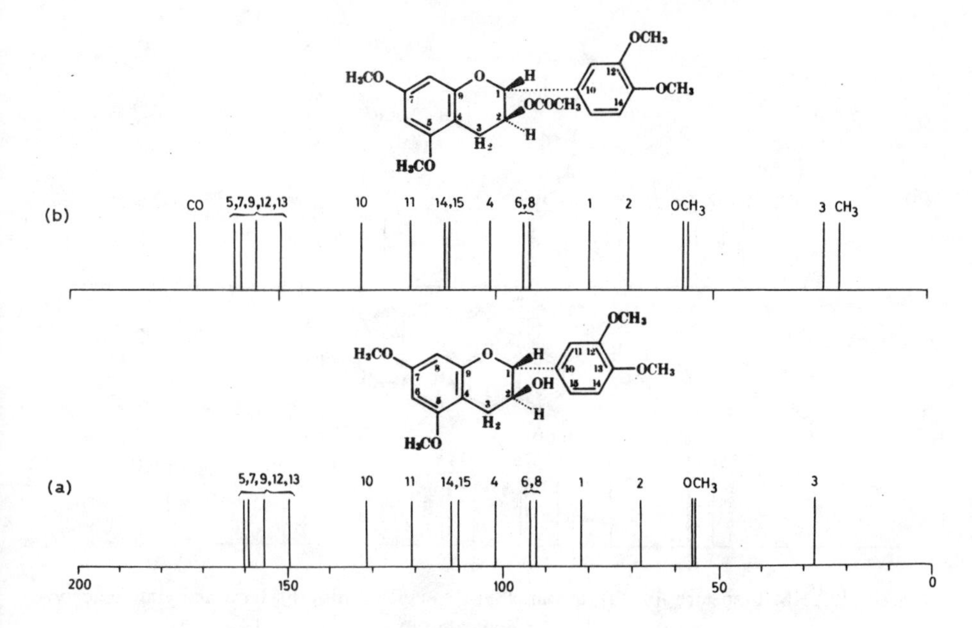

Fig. 5. ¹³C-NMR spectra of (a) tetramethyl-(+)-catechin, (b) tetramethylmonoacetyl-(+)-catechin

presence of the acetyl group causes a shift of the signals of C-1 and C-3 to higher frequency, while that of C-2 appears at somewhat lower frequency, a commonly observed effect. This allows the assignment of the aliphatic carbon atoms which are of special interest for the structure in question.

The spectra of the corresponding (—)-epicatechin derivatives are given in Fig. 6. The correlation of their carbon atoms proceeds analogously. The positions obtained for the carbon atoms compared with those found in the (+)-catechin derivatives exhibit only insignificant shifts. These small differences can be neglected in the structure elucidation, thus carbon atoms 1′, 2′, and 3′ of the "lower" part of the procyanidin molecules **2** and **3** should show practically the same chemical shifts.

To decide upon the constitution of the "upper" halves of **2** and **3**, the spectra of two model substances have been recorded (Fig. 7). Trimethyl-monoacetyl-cyanomaclurin, which in analogy to the proposed structure 2 embodies two benzyl ether groups, is shown in Fig. 7a, whereas Fig. 7b is the spectrum of a synthetic model compound that, corresponding to proposition **3**, contains a ketal group. The chemical shifts of the aliphatic carbon atoms are of interest in both compounds. The carbon atoms 1, 2, 3 of trimethylmonoacetylcyanomaclurin

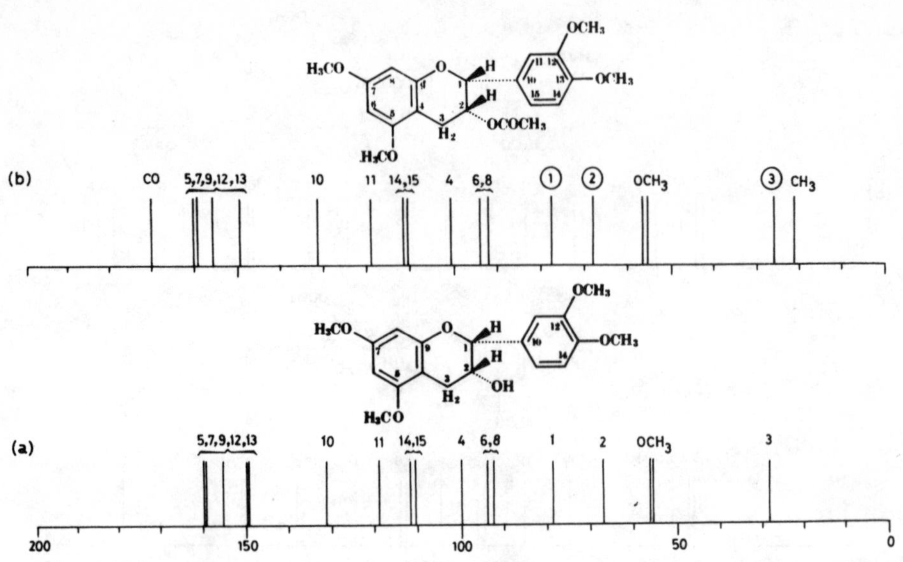

Fig. 6. ^{13}C-NMR spectra of (a) tetramethyl-(−)-epicatechin, (b) tetramethylmonoacetyl-(−)-epicatechin (signals for C atoms 1′, 2′, 3′ (marked by circles) in 2 and 3)

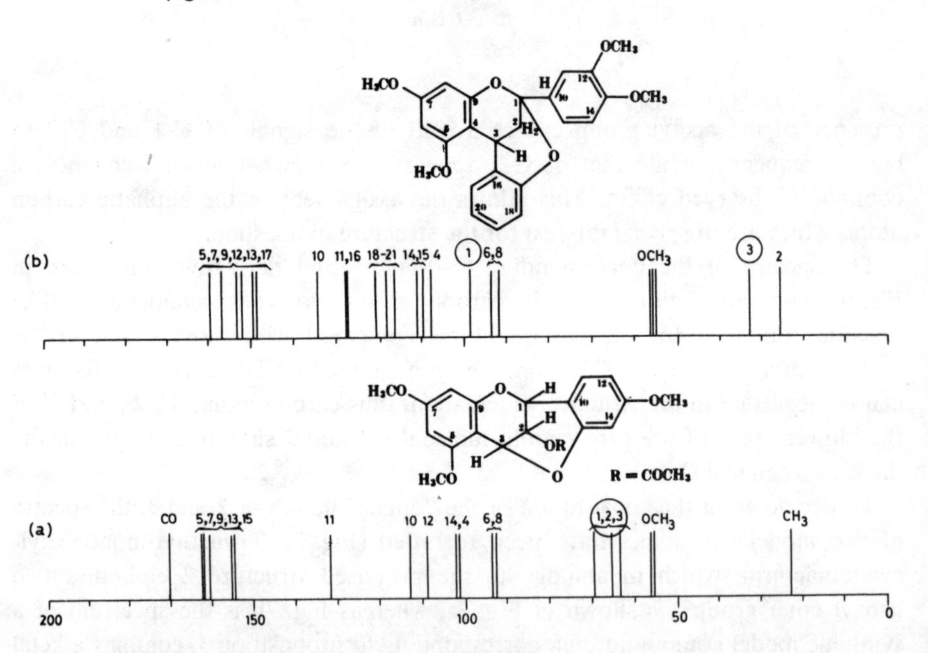

Fig. 7. ^{13}C-NMR spectra of (a) a model compound analogous to proposition 2 and (b) a model compound analogous to 3 (correlation of C atoms 1 and 2 in 3 and 2 and 3 in 2)

have nearly identical chemical shifts and appear at δ values between 60–70 ppm. In contrast, the model substance, allowing conclusion only for C-1 and C-3, yields drastic chemical shift differences. Here C-1 appears at δ = 98 ppm and C-3 at 33 ppm. The latter absorption, however, awaited additional proof. For this purpose the spectra of two more model substances both containing the benzyl ether carbon atom in question were obtained; they are shown in Fig. 8. From

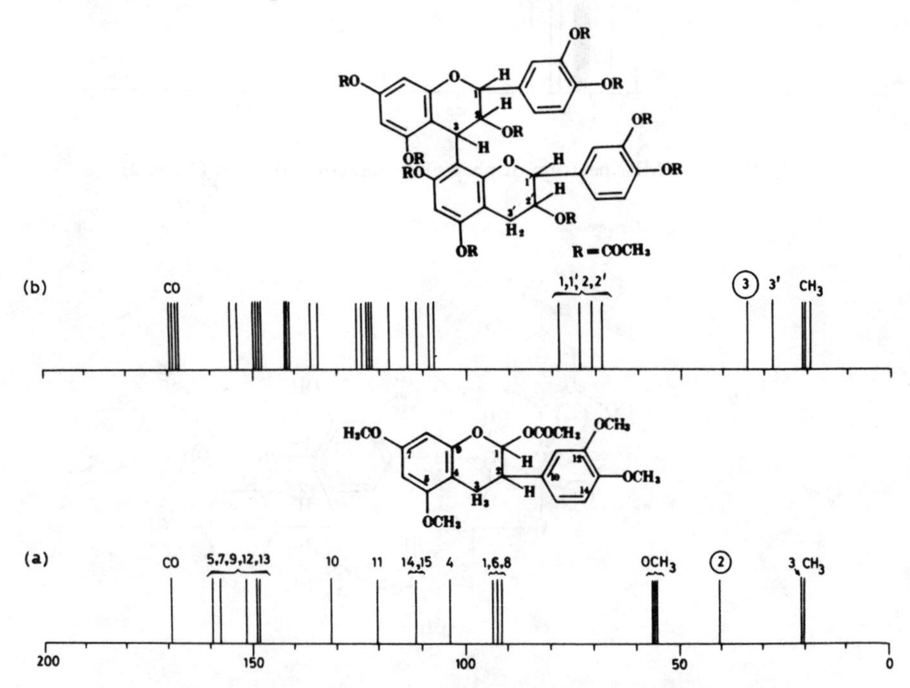

Fig. 8. ^{13}C-NMR spectra of (a) tetramethylmonoacetylisocatechin; (b) decaacetylprocyanidin B1

these one may conclude that tertiary benzyl ether carbon atoms absorb at δ values between 25 and 45 ppm, depending on their chemical neighbourhood.

In order to complete the assignment of chemical shifts to the carbon atoms in the proposed constitutions **2** and **3**, the chemical shift of C-1 in **2** had still to be determined. This again was achieved by means of a suitable model compound, the spectrum of which is shown in Fig. 9.

These data enable us to propose the chemical shifts of the aliphatic carbon atoms for both constitutions in question; in Fig. 10 these values are given with the formulas **2** and **3**. The distinct differences are clearly seen by comparison of the δ values drawn on a δ scale (Fig. 11). By knowing the chemical shifts of

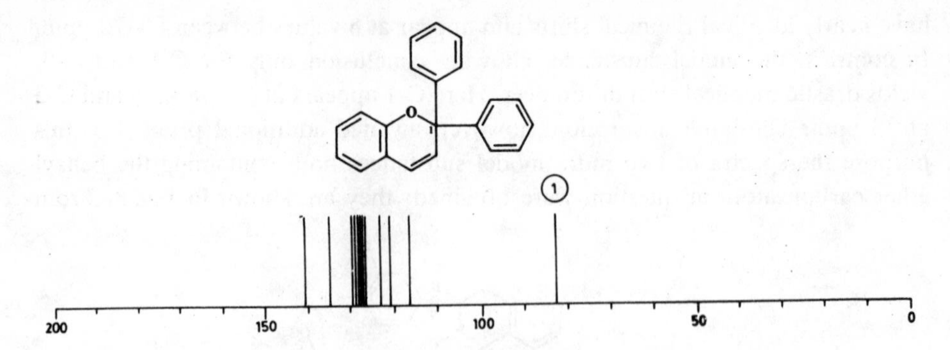

Fig. 9. ^{13}C-NMR spectrum of phenylflavene (correlation of C-1 in 2)

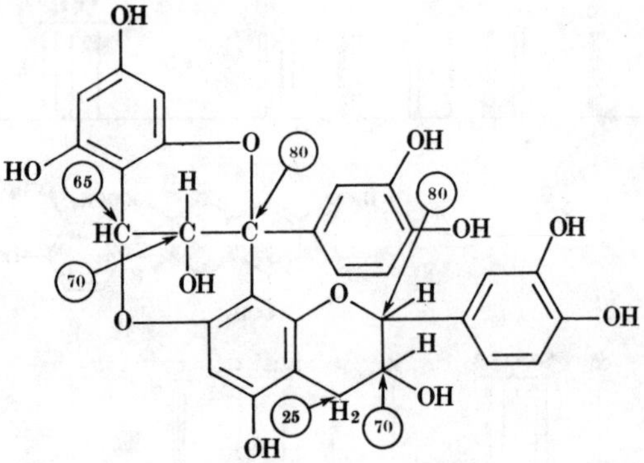

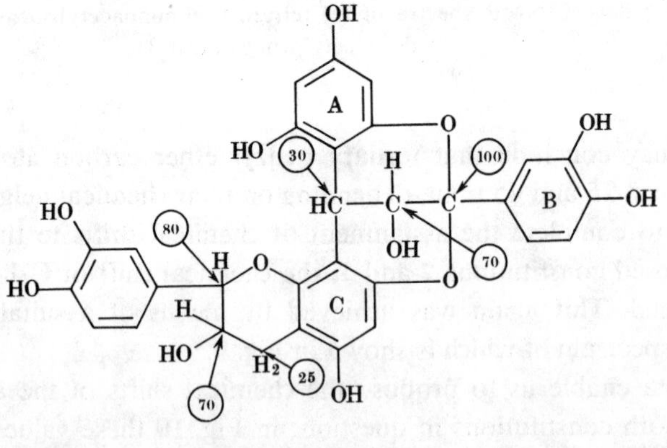

Fig. 10. Chemical shifts of the aliphatic carbon atoms 1–3 and 1′–3′ for the proposed formulas 2 and 3 (*cf.* Figs 6–9)

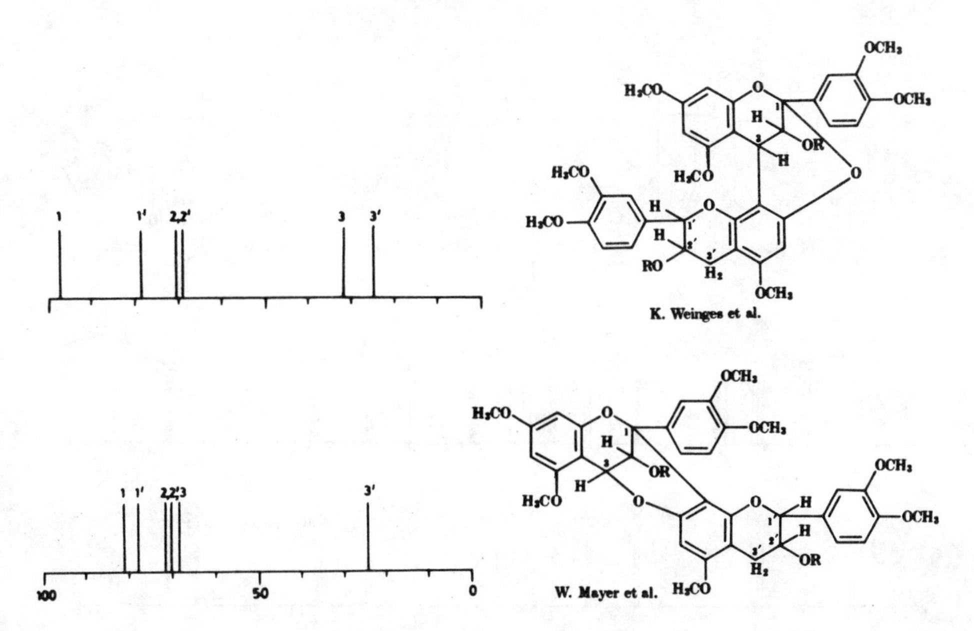

Fig. 11. Chemical shifts as proved by models of the aliphatic carbon atoms 1–3 and 1′–3′ in **2** and **3**, respectively

carbon atoms 1 and 3, it has become possible to determine the constitution of the $C_{30}H_{24}O_{12}$-procyanidins A1 and A2.

The actual ¹³C-NMR spectra of both heptamethyldiacetylprocyanidins A1 and A2 are shown in Fig. 12. They unequivocally prove the constitution **3**, proposed by us, conceding that two additional forms having 3/6′ links are also possible.

The results obtained by ¹³C-NMR spectroscopy are thus in accordance with the presence of a ketal group as required by constitution **3**; this, however, does not decide upon the kind of isomerism, steric or structural, existing between A1 and A2. Based on our experimental results collected so far, we tend to believe that we are dealing with structural isomers. Procyanidin A1 should have a 3/6′ and A2 a 3/8′ link between the two halves of the molecule. Indications thereof are as follows:

1. Both procyanidins (A1 and A2) have the configuration of (—)-epicatechin in the "lower" half of the molecule. This has been proved by careful acid-catalyzed cleavage [3, 7].
2. The "lower" half of the molecule can only be linked in *cis* configuration to the C-1 and C-3 atoms of the "upper" half. This is readily seen on the Stuart models.

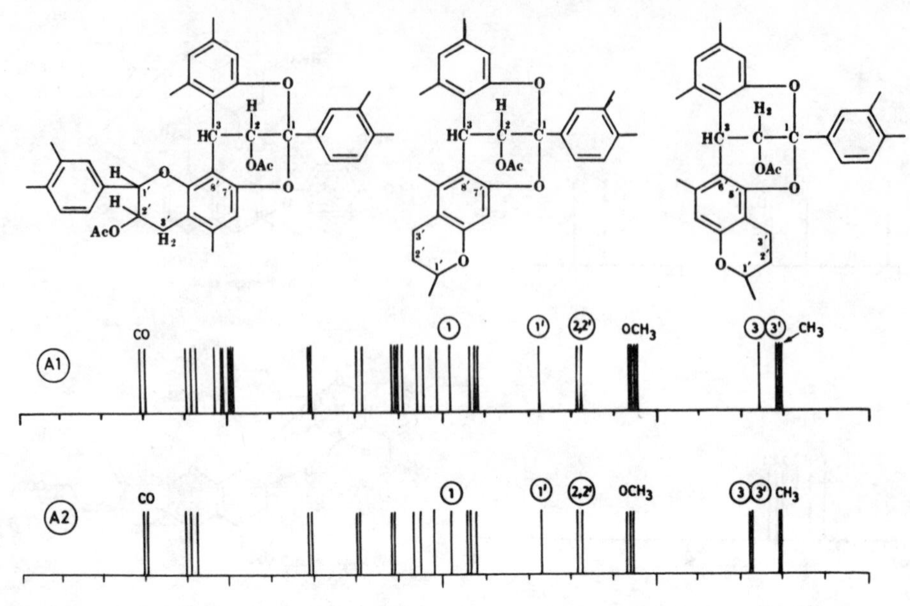

Fig. 12. ^{13}C-NMR spectra of the heptamethyldiacetylprocyanidins A1 and A2

3. Methylation of pure procyanidin A1 gives, besides the heptamethyl ether of A1, also that of A2. Thus methylation must involve either rearrangement or epimerization. The latter process can be excluded. Both procyanidins have (—)-epicatechin configuration in the "lower" part of the molecule, therefore no epimerization at carbon atoms 1′ and 2′ is possible. Furthermore, the *cis*-link (argument 2) between the "lower" and "upper" halves precludes epimerization at C-1 or C-3, respectively. Since epimerization at C-2 during the methylation of polyhydroxyflavanols has never been observed, the configuration at C-2 of 3 should be retained in analogy. Therefore, only a rearrangement could take place, which must have involved the cleavage of the benzyl ether linkage in the "lower" part of the molecule and its closing again with a different hydroxyl group, as this is known to occur in other flavonoids.
4. The proton resonance spectrum of the heptamethyldiacetylprocyanidin A1 (Fig. 13b) shows two methoxyl groups shifted to higher frequency. This is understood only in the case of a 3/6′ link between the two halves of the molecule, since only such a constitution will allow two methoxyl groups to get within the effective sphere of the aromatic rings.

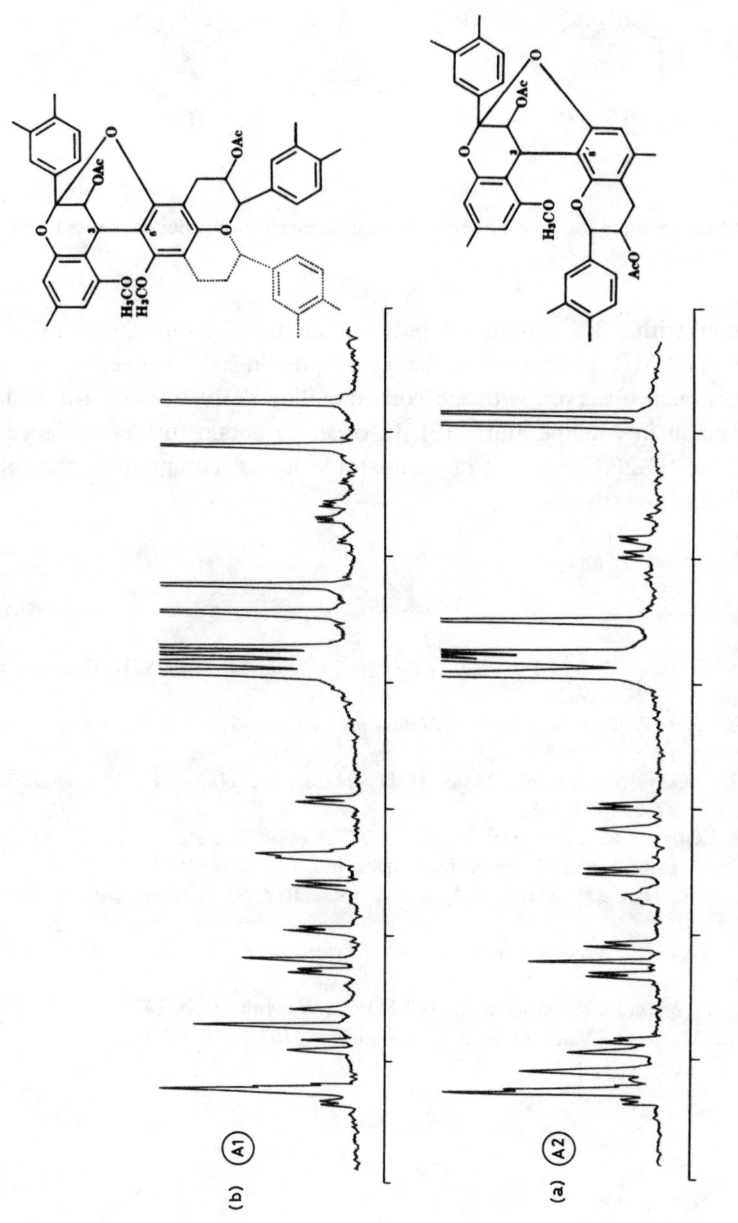

Fig. 13. ^{1}H-NMR spectra of the heptamethyldiacetylprocyanidins (a) A2 and (b) A1

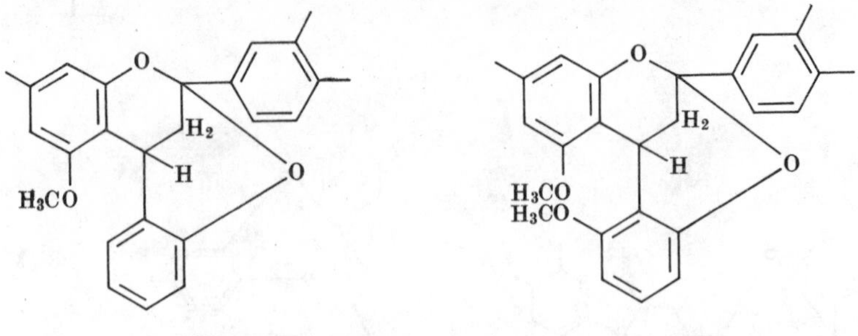

Fig. 14. Model compounds for proving the structural isomerism of procyanidins A1 and A2

In agreement with a 3/8′ link in A2, only one methoxyl group shows a similar shift in the ^{1}H-NMR spectrum of the corresponding derivative. A similar behaviour has been observed with the corresponding derivatives of 4-aryl-substituted flavans at low temperatures [9]. In order to obtain further evidence in this respect, the ^{1}H-NMR spectra of comparable model compounds, shown in Fig. 14, are being investigated.

REFERENCES

1. Breitmaier, E., Jung, G. and Voelter, W., *Angew. Chem., 83,* 659 (1971); Bremser, W., *Chemiker-Ztg., 97,* 248 (1973).
2. Weinges, K., *Acta Universitatis Debreceniensis de Ludovico Kossuth Nominatae, Series Physica et Chimica, 1971,* 249.
3. Weinges, K., Kaltenhäuser, W., Marx, H.-D., Nader, E., Nader, F., Perner, J. and Seiler, D., *Ann., 711,* 184 (1968).
4. Weinges, K., Göritz, K. and Nader, F., *Ann., 715,* 164 (1968).
5. Weinges, K., Perner, J. and Marx, H.-D., *Chem. Ber., 103,* 2344 (1970).
6. Thompson, R. S., Jaques, D., Haslam, E. and Tanner, R. J. N., *J. Chem. Soc. (London), 1972,* 1387.
7. Mayer, W., Goll, L., von Arndt, E. M. and Mannschreck, A., *Tetrahedron Letters, 1966,* 429.
8. Schilling, G., Weinges, K., Müller, O. and Mayer, W., *Ann., 1973,* 1471.
9. Weinges, K., Marx, H.-D. and Göritz, K., *Chem. Ber., 103,* 2336 (1970).

THE CHEMISTRY AND BIOCHEMISTRY OF PLANT PROANTHOCYANIDINS

by

E. HASLAM

Department of Chemistry, University of Sheffield
Sheffield, UK

Introduction

Around the turn of this century several observations were reported of the reddening of plant tissues when treated with mineral acid. Although the first observations of this type have been variously attributed to Willstatter or Tswett, records show that the scientist Robert Boyle first made this observation in 1844 [1]. Rosenheim [2] attempted the first isolation from *Vitis vinifera* of the colourless substances responsible for this reaction and he named them *"leuco-anthocyanins"* since he was able to show that the red colour which they yielded with acid was due to the pigment cyanidin (I). Rosenheim also made the first structural proposals for these compounds and he suggested that they were glucosylated derivatives of the type shown (II).

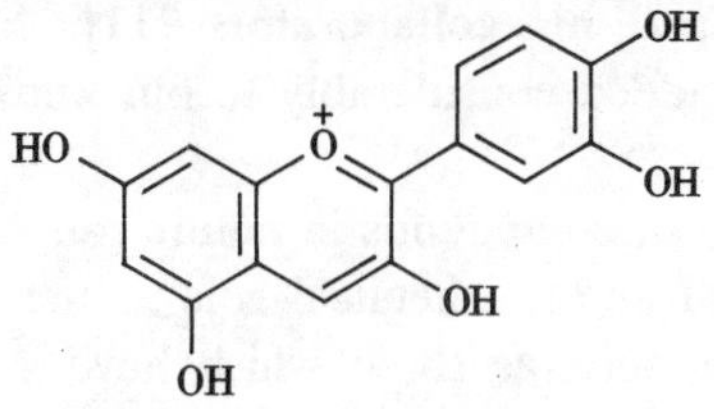

I 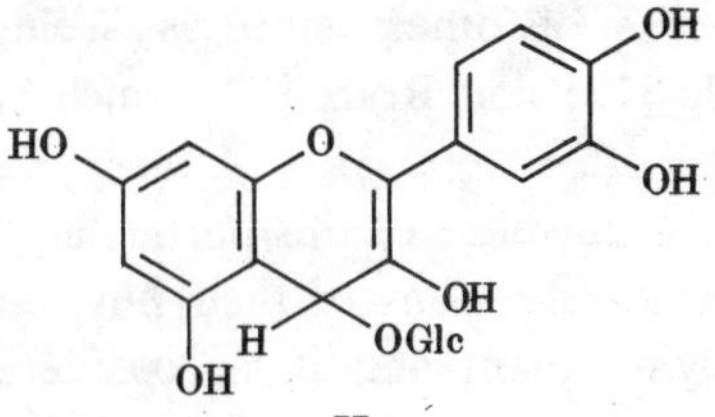 II

In the 1930's in Oxford the Robinsons [3–6] conducted surveys of the distribution of these compounds in higher plants and these were greatly extended some twenty years later in Cambridge by Bate-Smith [7, 8]. Besides confirming and extending the earlier observations, Bate-Smith also drew attention to the fact that *"leuco-anthocyanins"* were confined mainly to plants with a woody habit of growth. Bate-Smith and Swain [9] also noted the close similarity in systematic distribution between *"leuco-anthocyanins"* and the class of substance rather indefinitely defined as tannins in the botanical literature. They indeed suggested that *"leuco-anthocyanins"* were most commonly responsible for the broad range of reactions (precipitation of gelatin and alkaloids, astringent taste and the formation of amorphous polymeric phlobaphens with acid) generally attributed in plants to the presence of tannins. Also in the 1950's came, with the recognition that their structures did not contain sugar residues, a revision of the nomenclature to *"leucoanthocyanidin"*. In the same period the belief that they possessed a flavan-3,4-diol type of structure (III) gained uncritical acceptance.

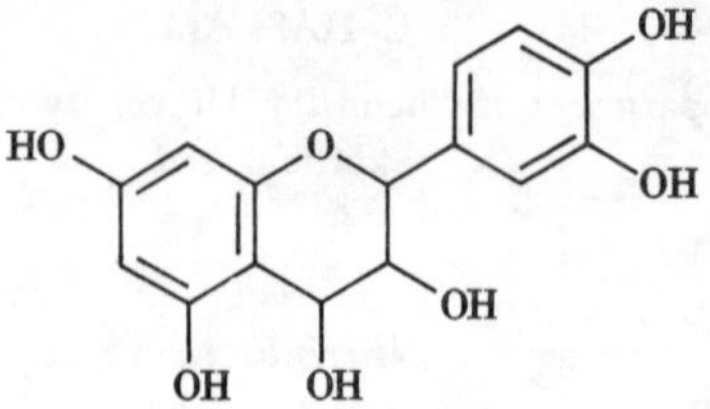

III

The major chemical advance in this field did not, however, occur until the late 1950's and the years thereafter. It was at this time that the general terminology of *proanthocyanidin* (which is employed here) was introduced by Freudenberg and Weinges and came into use for this class of compound.

Our own interests in these substances, which form the basis of this paper, derived from our earlier work on the chemistry of vegetable tannins [10] of the hydrolyzable class and on the problem of the nature of the condensed tannins. In this review some of our own contributions to this area of study are outlined. In doing so it is therefore only possible to mention in passing the very important researches of others such as Weinges and his collaborators [11], Marini-Bettòlo [12] and Roux [13] which have added considerably to our knowledge in this field.

The occurrence of this particular group of compounds in nature raises many interesting questions of their physiological and biochemical role in the life of the higher plant and it is considerations such as these which have strongly influenced our particular approach to this subject. Our initial investigations have concentrated on the chemistry of *procyanidins* (those members of this group which give cyanidin on acid treatment) and related derivatives which are found in the vegetative tissues of higher plants. We have excluded from examination those procyanidins which occur in the bark and heartwood of trees, since in later work we have extended our studies to an investigation of the biosynthesis of the procyanidins and some of their biochemical properties, in particular their role as vegetable tannins.

Isolation and Structure Determination

The isolation of plant procyanidins in their free phenolic forms has been greatly assisted by the use of the dextran gel Sephadex LH-20 as a chromatographic support. In this way some fifteen plant procyanidins have been obtained from plant sources and characterized by the usual methods (analysis, ^{1}H-NMR, ORD and CD, mass spectrometry of their methyl ethers and chemical degradation). Only one of these substances was isolated in the free phenolic form in the crystalline state (proanthocyanidin A-2) and the remainder were obtained as colourless amorphous powders which slowly darken in light. Paper chromatography played a very important part in the early studies (Fig. 1), both as a means of characterization and as a means of carrying out detailed surveys of the distribution of the individual procyanidins in plants [14].

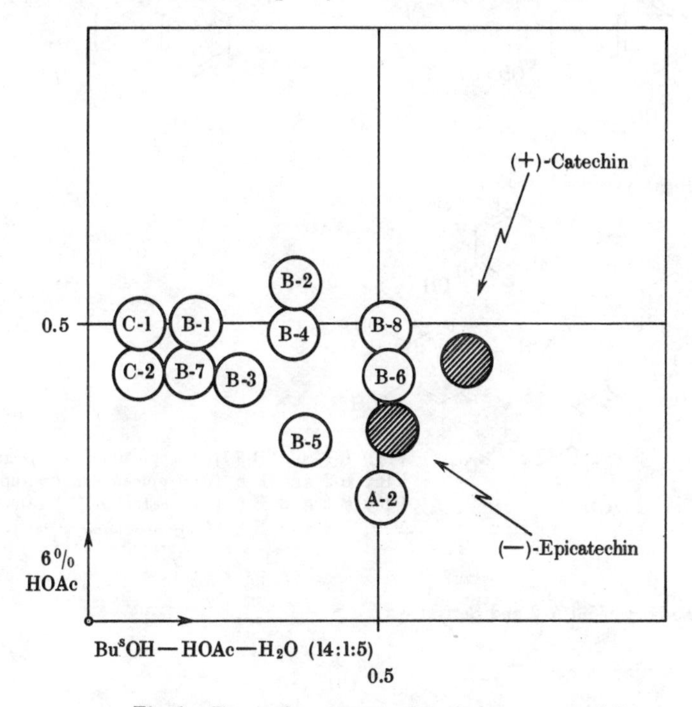

Fig. 1. Paper chromatography of plant procyanidins

These surveys drew attention in particular to the distinctive patterns of co-occurrence of the procyanidins, and to the fact that they always were found in plant tissues alongside (+)-catechin (IV) and/or (—)-epicatechin (V). In retrospect this observation is readily rationalized when the structures of the major types of procyanidins which occur in plants are examined. One group of procyanidins — group B — possess structures VI, which are essentially singly linked flavan-3-ol (IV or V) dimers. These occur in pairs, e.g. B-2 and B-5, which probably only differ in the position (4–6′ or 4–8′) of the single inter-flavan bond or, alternatively, in the absolute stereochemistry at C-4. Eight dimeric procyanidins of this type have been found in plant tissues and, in addition to these, two trimeric substances based on this same structural pattern. The other major group are those based on the doubly linked proanthocyanidin A-2 structure and include the parent dimer A-2 (VII) and trimeric structures derived by the addition of further (+)-catechin or (—)-epicatechin units.

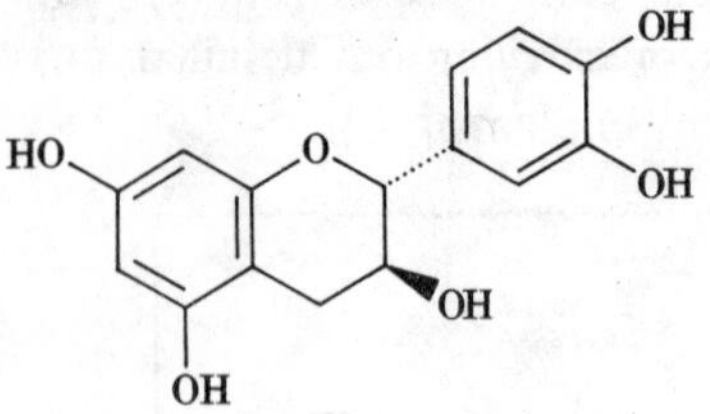

IV

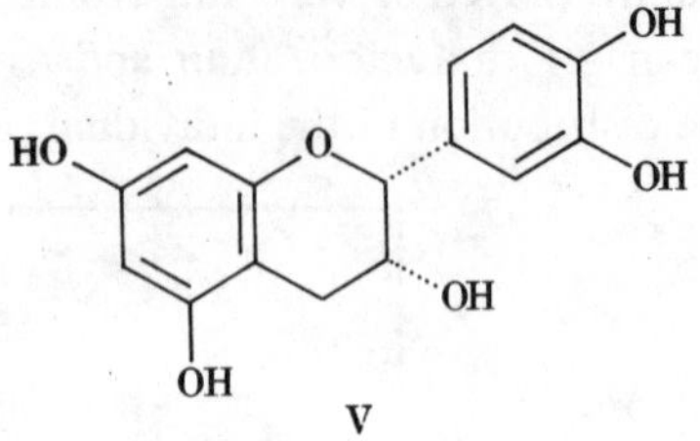

V

Group B-procyanidins

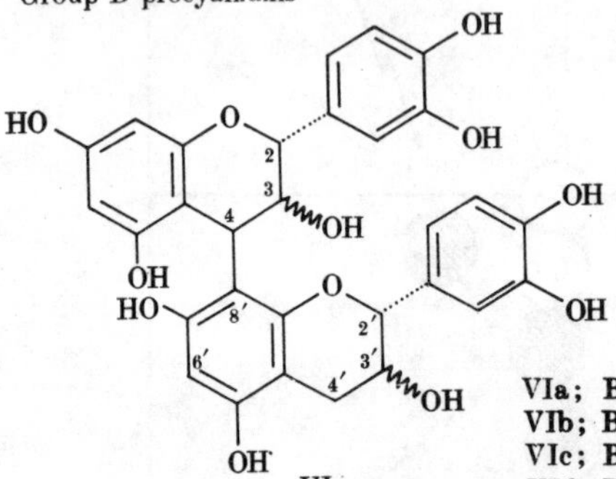

VI

VIa; **B-1** and B-7 [(—)-epicatechin-(+)-catechin]
VIb; **B-2** and B-5 [(—)-epicatechin-(—)-epicatechin]
VIc; **B-3 and B-6** [(+)-catechin-(+)-catechin]
VId; B-4 and **B-8** [(+)-catechin-(—)-epicatechin]

Proanthocyanidin A-2 and derivatives

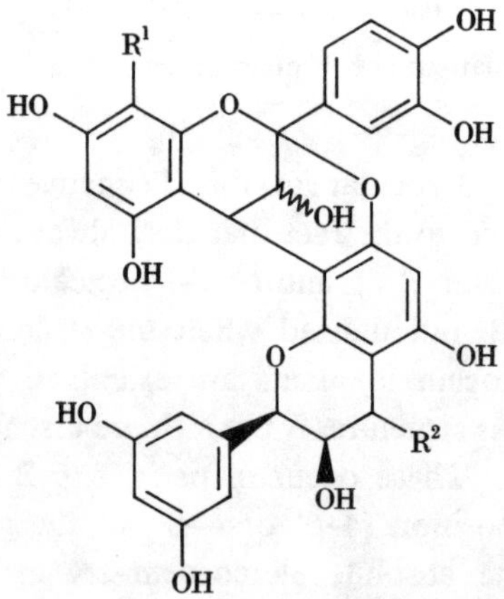

(VII; $R^1 = R^2 = H$)
(VIII; $R^2 = H$, $R^1 =$ (—)-epicatechin)
(IX; $R^1 = H$, $R^2 =$ (—)-epicatechin)
(X; $R^1 = H$, $R^2 =$ (+)-catechin)

The acid-catalyzed degradation of group B procyanidins and proanthocyanidin A-2 and its derivatives to give cyanidin may be readily rationalized in terms of the structural features which they possess. Fission of the interflavan linkage may thus be envisaged as shown in Fig. 2. The "upper" flavan unit is liberated as a resonance-stabilized carbonium ion (XI) which then gives cyanidin by proton loss and oxidation. The "lower" flavan unit is simultaneously released as (+)-catechin or (—)-epicatechin.

VI - Procyanidin (B-type)

XI

(+)-Catechin
or (—)-epicatechin

I
Cyanidin

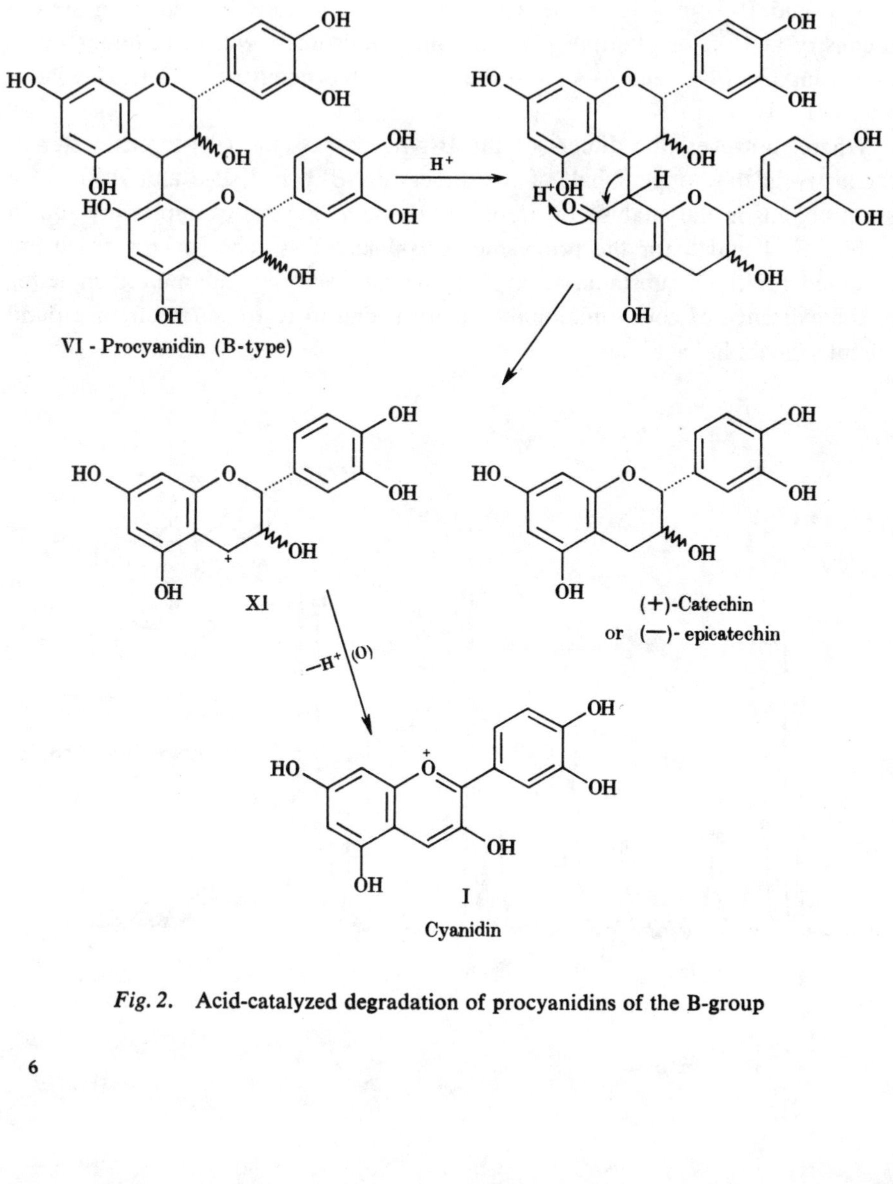

Fig. 2. Acid-catalyzed degradation of procyanidins of the B-group

The usual physical methods of structure determination (^{1}H-NMR and mass spectroscopy) are quite adequate for the procyanidin dimers of the B-group in which the "upper" half of the dimer has the (—)-epicatechin stereochemistry (i.e. B-1, B-2, B-5 and B-7). These measurements are sufficient to permit definition of the structural features of these dimers and, assuming the *2(R),2′(R)* configuration (as in (+)-catechin and (—)-epicatechin), the stereochemistry at the 3 and 3′ positions can also be defined. Certain structural features cannot, however, be readily determined for these dimers. In particular these include the absolute stereochemistry at C-4 and the position of the linkage at C-6′ or C-8′ (as depicted) in ring A of the "lower" flavan unit. Thus the dimeric pairs, such as B-2 and B-5, may have the same interflavan linkage but differing stereochemistry at C-4, or alternatively, they may be isomers with a different interflavan linkage (4–6′ and 4–8′) and the same stereochemistry at C-4. This latter alternative is favoured.

Where, however, the dimers of the B-type possess the (+)-catechin stereochemistry in the "upper" half of the dimer (i.e. B-3, B-4, B-6 and B-8), there are problems in the analysis of the ^{1}H-NMR data. Typical examples are shown in Figs 3, 4 and 5 for the procyanidins B-3 and B-4. The spectra which are observed for these substances may only be satisfactorily rationalized in terms of the existence of conformational isomerism due to restricted rotation around the interflavan linkage.

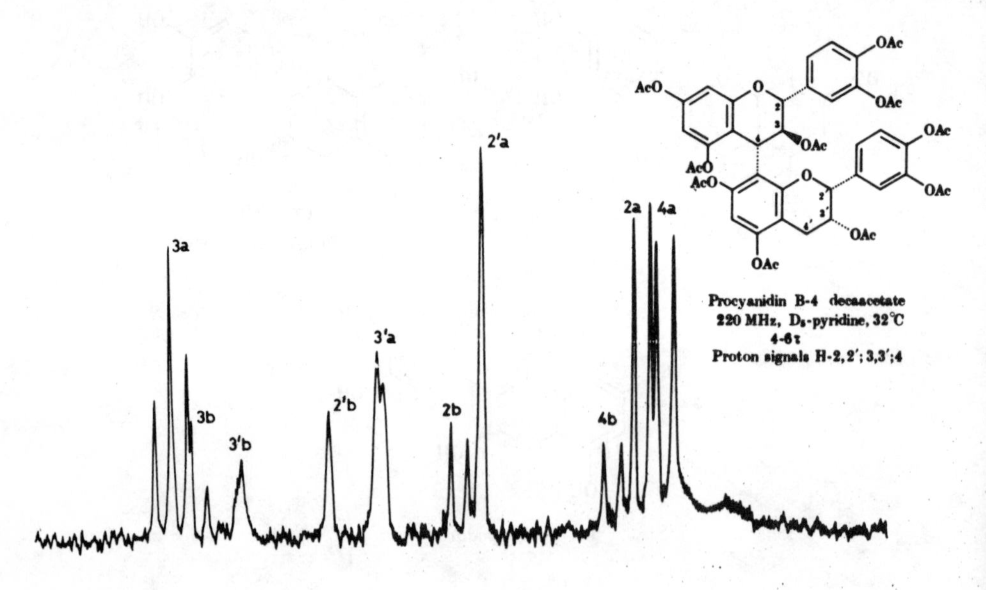

Fig. 3

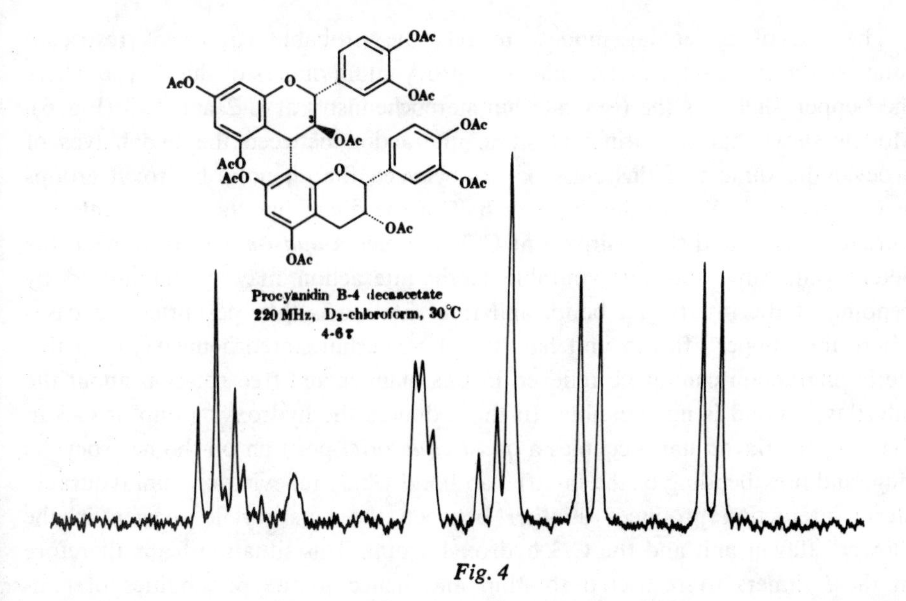

Fig. 4

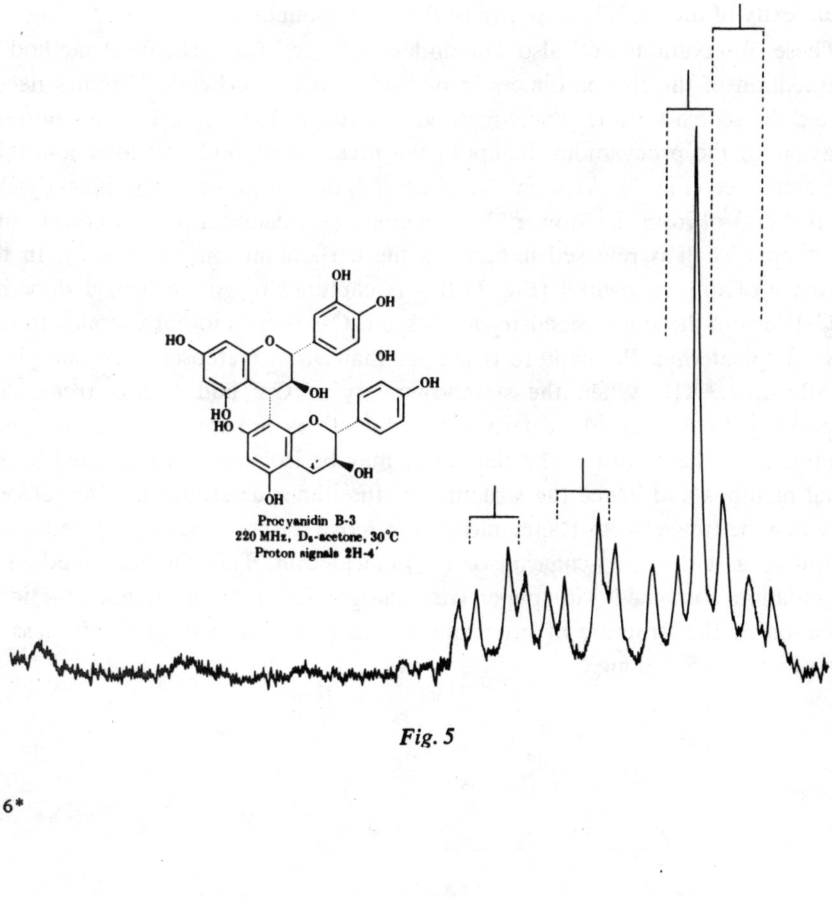

OH
OH
HO
OH
OH
OH
HO
HO
OH
OH
Procyanidin B-3
220 MHz, D₆-acetone, 30°C
Proton signals 2H-4'

Fig. 5

The use of molecular models reveals the probable causes of restricted rotation about the interflavan linkage in procyanidin dimers of the B-type where the "upper" half has the (+)-catechin stereochemistry at C-2 and C-3 (Fig. 6). Models show that the principal steric interaction between the two halves of procyanidin dimers of this class occurs between the phenolic hydroxyl groups at C-5 and C-7′. Where the "upper" half of the dimer has the (—)-epicatechin stereochemistry and the hydroxyl at C-3 occupies a *quasi-axial* position on the heterocyclic ring, this unfavourable steric interaction may be minimized by bending of the interflavan bond, and free rotation is thus permitted. In cases where the "upper" flavan unit has the (+)-catechin stereochemistry, then this steric interaction cannot be relieved in this manner and free rotation about the interflavan bond is not possible. In these dimers the hydroxyl group at C-3 in the "upper" flavan unit occupies a *quasi-equatorial* position on the heterocyclic ring, and now bending of the interflavan bond whilst relieving one unfavourable steric interaction produces another between the heterocyclic oxygen in the "lower" flavan unit and the C-3 hydroxyl group. This situation leads therefore in these dimers to restricted rotation and hence to the possibilities of conformational isomerism. This is probably the principal factor responsible for the complexity of the ^{1}H-NMR spectra of these compounds.

These observations and also our underlying need for a chemical method of degradation of the B-type dimers in our subsequent biochemical studies necessitated the search for a chemical method of structural analysis. For this purpose cleavage of the procyanidin dimers in the presence of acid and toluene-α-thiol was employed (Fig. 7). Acid-catalyzed degradation of procyanidin dimers yields the flavan-3-ol from the "lower" half as either (+)-catechin or (—)-epicatechin; the "upper" half is released initially as the carbonium ion XI (Fig. 2). In the presence of toluene α-thiol (Fig. 7) this is captured to give a benzyl thioether at C-4; where the stereochemistry at C-2 and C-3 is *cis* and corresponds to that of (—)-epicatechin, the capture is greater than 90 % stereoselective and yields the thioether XII. When the stereochemistry at C-2 and C-3 is *trans,* corresponding to that of (+)-catechin, two thioethers are formed and XIV predominates in the mixture. The thioethers may be isolated and identified by the usual methods and hence the structure of the dimer determined. Alternatively, they may be treated with Raney nickel to remove the thioether group and hence identified as either (+)-catechin or (—)-epicatechin. This chemical method of degradation combined with paper chromatographic analysis permits the identification of the structure of procyanidin dimers and trimers of the B-class on quantities of 0.5–1.0 mg.

(—)-Epicatechin "upper-half"

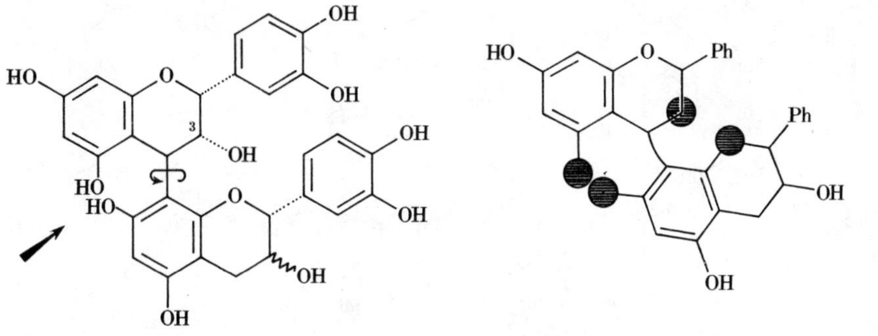

B-2 & 5 EC—EC
B-1 & 7 EC—C
Satisfactory structural analysis by proton-NMR

(+)-Catechin "upper-half"

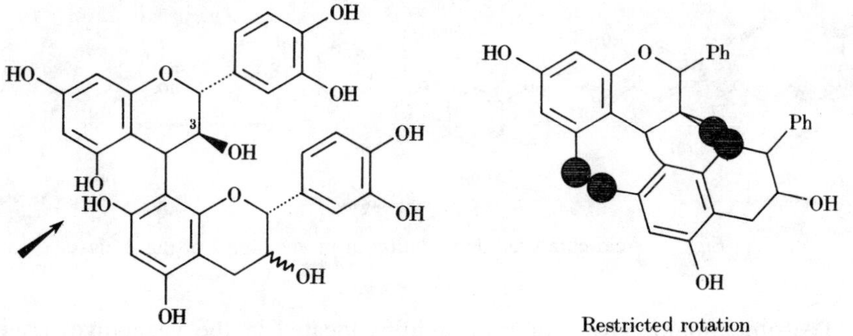

Restricted rotation

B-3 & 6 C—C
B-4 & 8 C—EC

Fig. 6

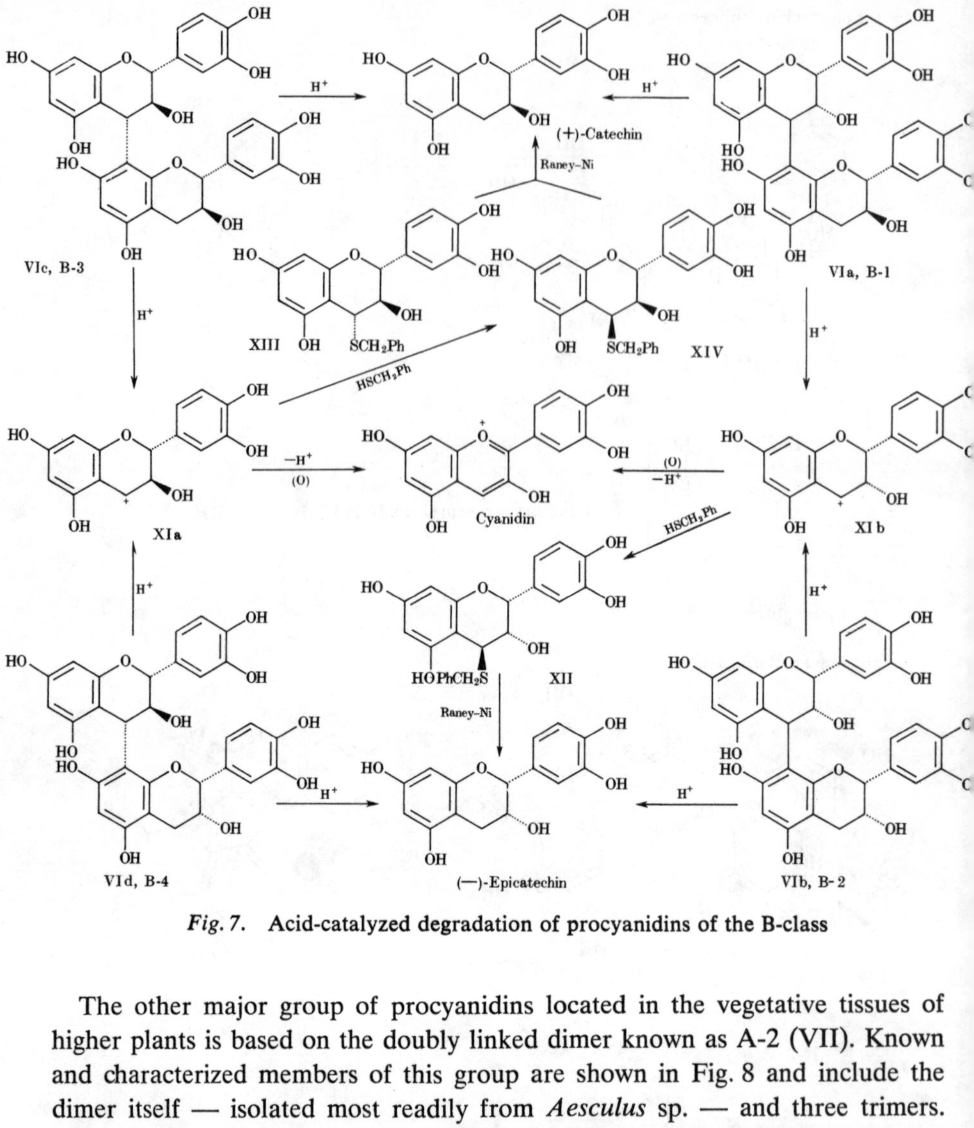

Fig. 7. Acid-catalyzed degradation of procyanidins of the B-class

The other major group of procyanidins located in the vegetative tissues of higher plants is based on the doubly linked dimer known as A-2 (VII). Known and characterized members of this group are shown in Fig. 8 and include the dimer itself — isolated most readily from *Aesculus* sp. — and three trimers. The compound A-2 (VII), whilst it is clearly related both structurally and biogenetically to the procyanidin dimers of the B-group discussed above, is named as a proanthocyanidin, since it gives rise to both cyanidin (I) and the pigment XV on treatment with acid (Fig. 8). The proanthocyanidin A-2 (VII) is also itself surprisingly resistant to degradation by toluene-α-thiol and acid and the structure of the various trimers was deduced principally by the action of toluene-α-thiol and identification of the products.

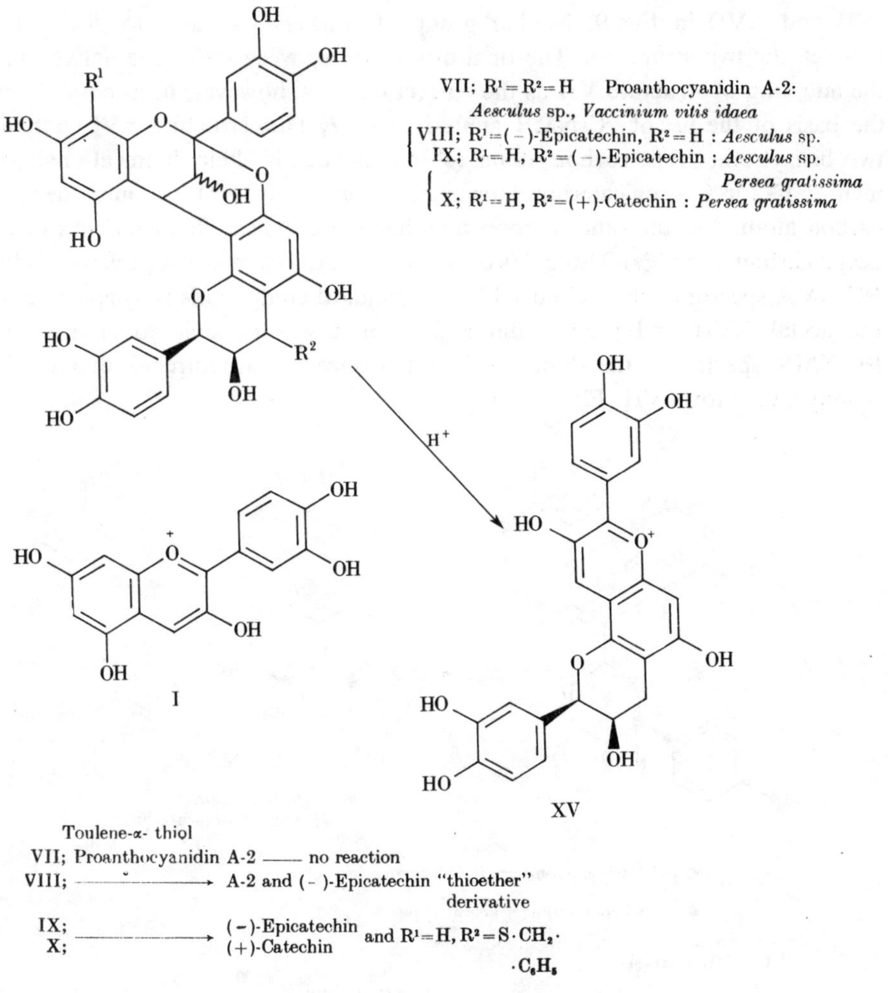

VII; $R^1 = R^2 = H$ Proanthocyanidin A-2: *Aesculus* sp., *Vaccinium vitis idaea*
VIII; R^1 = (−)-Epicatechin, $R^2 = H$: *Aesculus* sp.
IX; $R^1 = H$, R^2 = (−)-Epicatechin : *Aesculus* sp. *Persea gratissima*
X; $R^1 = H$, R^2 = (+)-Catechin : *Persea gratissima*

I

H^+

XV

Toulene-α- thiol
VII; Proanthocyanidin A-2 —— no reaction
VIII; ——→ A-2 and (−)-Epicatechin "thioether" derivative
IX; X; ——→ (−)-Epicatechin (+)-Catechin and $R^1 = H$, $R^2 = S \cdot CH_2 \cdot C_6H_5$

Fig. 8. Proanthocyanidin A-2 and its derivatives

The major structural problem associated with this group of proanthocyanidins has been the identification of the proanthocyanidin A-2 (VII) itself. Mayer, Goll, von Arndt and Mannschreck [15] first isolated this novel proanthocyanidin from the shells of unripe horse chestnuts *(Aesculus hippocastanum)* and Weinges and his collaborators [11] later isolated the same compound from other fruit including the mountain cranberry *(Vaccinium vitis idaea)*. ^{1}H-NMR and mass spectrometry of the proanthocyanidin and its derivatives led to the proposal of two alternative structures whose carbon–oxygen structural skeletons are shown

(VII and XVI) in Fig. 9. Neither group of workers was able to distinguish between the two structures. The final distinction between these alternatives and the adoption of structure VII as the correct one has, however, been possible on the basis of the use of ^{13}C-NMR analysis [16, 17]. Thus structure XVI contains two benzyl ether type carbon atoms (◐), analogous in their chemical environment to C-2 in (—)-epicatechin. However, the alternative VII has only one such carbon atom, and the other carbon now has the chemical characteristics of an acetal carbon atom (●). Using this distinction and the chemical shift values in the ^{13}C-NMR spectra of the ^{13}C nuclei in the standard compounds (—)-epicatechin, the acetal XVII and procyanidin B-2, then it was possible to analyze the ^{13}C-NMR spectra of the dimer A-2 and deduce its structure as that of the doubly linked form VII (Fig. 9).

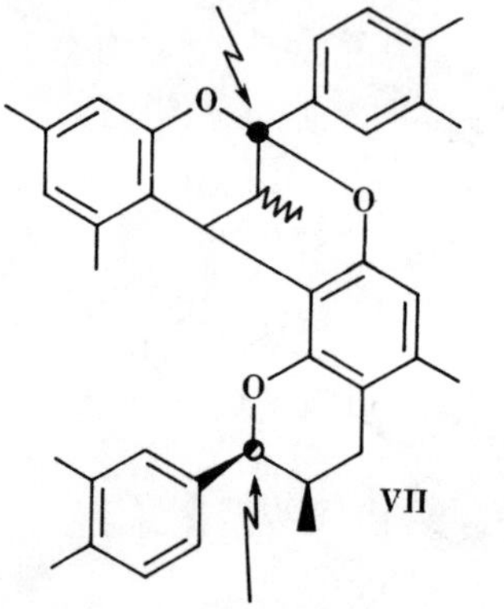

VII

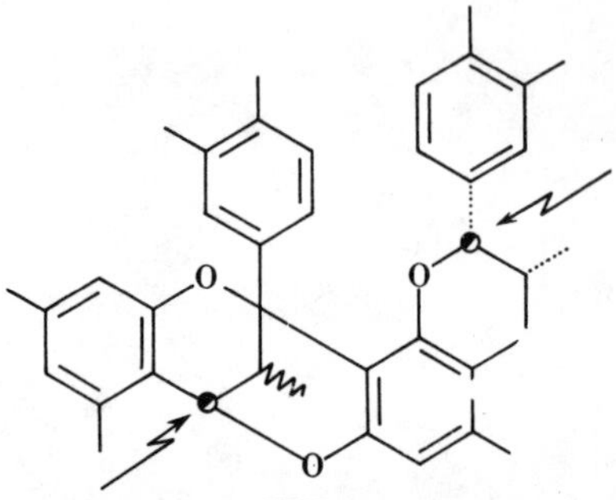

XVI

◐ 2 Carbon atoms *cf.* C-2 (—)-epicatechin

◐ 1 Carbon atom — *cf.* C-2 (—)-epicatechin

● 1 Carbon atom — acetal type

^{13}C-NMR analysis

Carbon atom	◐ 2	3	4
Chemical shift, ppm, TMS	78.1	65.1	25.1

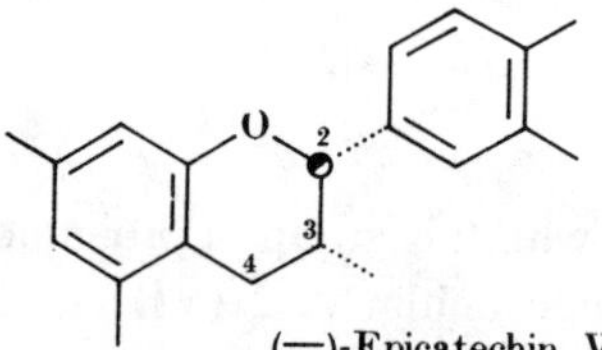

(—)-Epicatechin V

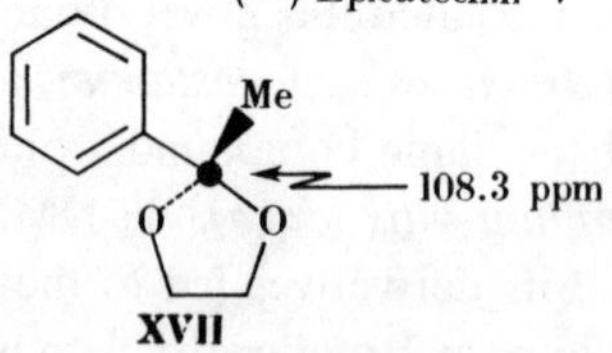

Me

108.3 ppm

XVII

Fig. 9. Structure of proanthocyanidin A-2

More recently our attention has been diverted from the purely structural problems associated with this group of compounds to those of their biosynthesis and their physiological role in the plant. The earlier plant surveys which were carried out drew attention to a number of important observations which any theory of biogenesis must explain. Thus the procyanidins (+)-catechin and (—)-epicatechin always co-occur in the free unglycosylated forms and wherever (+)-catechin or (—)-epicatechin predominates in the plant extract then so also do particular procyanidin dimers. Thus when there is a predominance of (+)-catechin, there is also a preponderance of the dimers B-1, B-3 and the trimer C-2. Alternatively, where (—)-epicatechin predominates, so also do the dimers B-1, B-2 and B-4 and the trimer C-1. Similarly, no flavan-3,4-diols

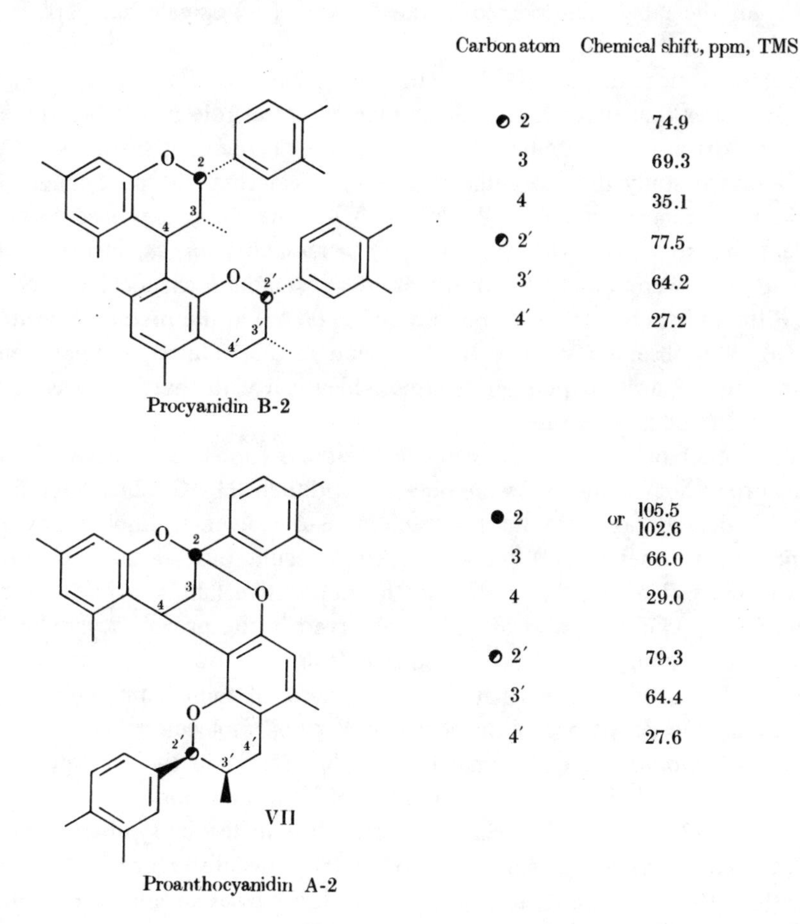

Carbon atom Chemical shift, ppm, TMS
2 74.9
3 69.3
4 35.1
2′ 77.5
3′ 64.2
4′ 27.2
Procyanidin B-2
2 105.5 or 102.6
3 66.0
4 29.0
2′ 79.3
3′ 64.4
4′ 27.6
VII
Proanthocyanidin A-2

Fig. 9. (continued)

(e.g. III) were found in the vegetative tissues examined. These observations prompted the establishment of an early working hypothesis that the procyanidins were derived by some form of oxidation process from (—)-epicatechin and/or (+)-catechin.

Our biosynthetic work has concentrated principally on the biosynthesis of (—)-epicatechin (V), procyanidin B-2 (VIb, (—)-epicatechin–(—)-epicatechin) and proanthocyanidin A-2 (VII) in *Aesculus carnea* and procyanidin B-4 (VId, (+)-catechin–(—)-epicatechin) in *Rubus* sp. One of the earliest observations made was that generally ^{14}C-labelled (—)-epicatechin was readily incorporated into the young shoots of *Rubus idaeus* and into the procyanidin B-4 (VId). The latter was isolated and degraded using the toluene-α-thiol reaction, and the majority of the label was located in the "lower" (—)-epicatechin part of the molecule.

Grisebach's elegant work [18] on the biosynthesis of flavonoids in higher plants indicates that the cinnamic acids play a central role in the biosynthesis, and later work using the embryo fruit of *Aesculus carnea* has made use of this observation to study the biosynthesis of (—)-epicatechin, the procyanidin B-2 and the proanthocyanidin A-2. 3-[^{3}H]-3-[^{14}C]-Cinnamic acid proved to be an excellent precursor of (—)-epicatechin (V) in *Aesculus carnea;* incorporations of up to 5 % were observed. Degradation by standard chemical procedures showed the radioactive labels to be located ($>$ 90 %) at the predicted positions (Fig. 10). The observed ratio of $^{3}H/^{14}C$ found in the isolated (—)-epicatechin was, in three separate experiments, almost identical with that of the cinnamic acid administered as substrate.

In the same biosynthetic experiments in *Aesculus carnea* the procyanidin B-2 and the proanthocyanidin A-2 were obtained and their $^{3}H/^{14}C$ ratios determined. Chemical degradation of the procyanidin B-2 (toluene-α-thiol, separation, treatment with Raney-nickel) to give the two molecules of (—)-epicatechin from the two "halves" of the dimer showed the radioactive labels to be located as indicated (Fig. 11). The ratio of $^{3}H/^{14}C$ observed in the proanthocyanidin A-2 was at first puzzling. It had been predicted from the known structure of A-2 (VII) and from the concept that the biosynthesis of these compounds might involve a simple dehydrogenation of two "catechin" molecules that the $^{3}H/^{14}C$ ratio in A-2 would be approximately half that found in (—)-epicatechin or procyanidin B-2 (VIb). This deduction follows from the theory that one tritium atom from C-2 in (—)-epicatechin should be lost in the biosynthetic process. The explanation which is suggested to explain this difference is that both "halves" of the dimers, although very similar in chemical structure, are derived from different biochemical sources and in the biosynthetic experiments are labelled to differing extents. On this basis calculation shows that the ratio of

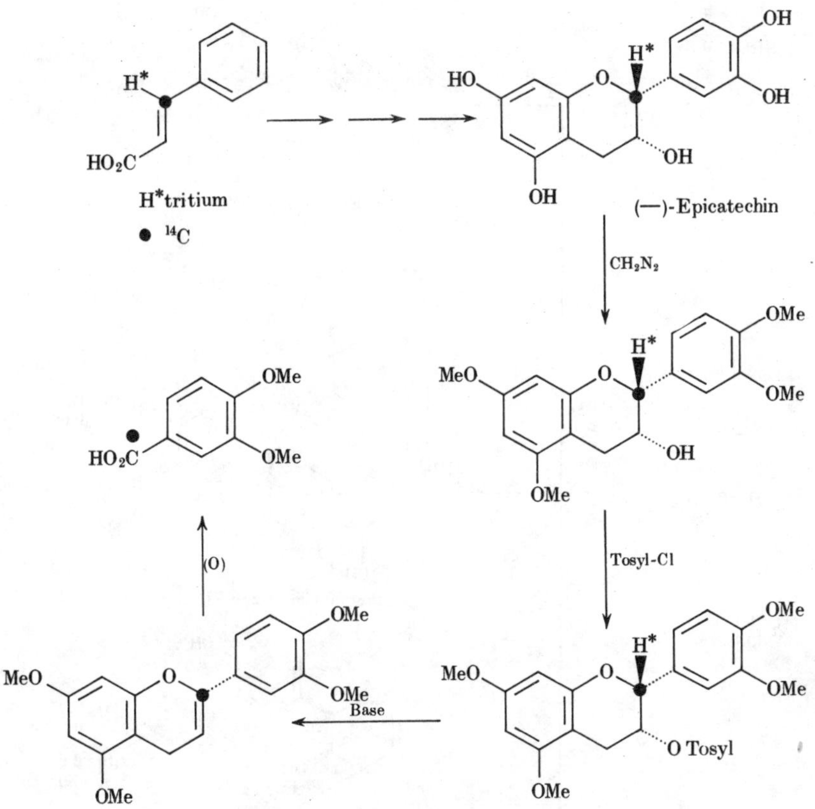

Fig. 10. Biosynthesis of (–)-epicatechin in *Aesculus carnea*

activities in the "upper" as compared to the "lower" flavan unit of the dimer A-2 is in the region of 4:1 or 5:1. Chemical degradation of the dimer B-2 supports this view. The ratio of specific activities of the two halves of this dimer, determined experimentally, is here between 2:1 and 3:1. The conclusion which has been drawn is that the biosynthesis of these substances is not in fact a straightforward dehydrogenation process.

In the light of this biosynthetic work, new proposals for the biogenesis of these substances have been formulated. These proposals (Fig. 12) also suggest a link between proanthocyanidin and anthocyanidin biogenesis and this is based on purely circumstantial evidence. Thus the autumnal changes in the colour of leaves and the ripening of fruits appear to reflect in many cases a switch from proanthocyanidin to anthocyanidin metabolism by the plant.

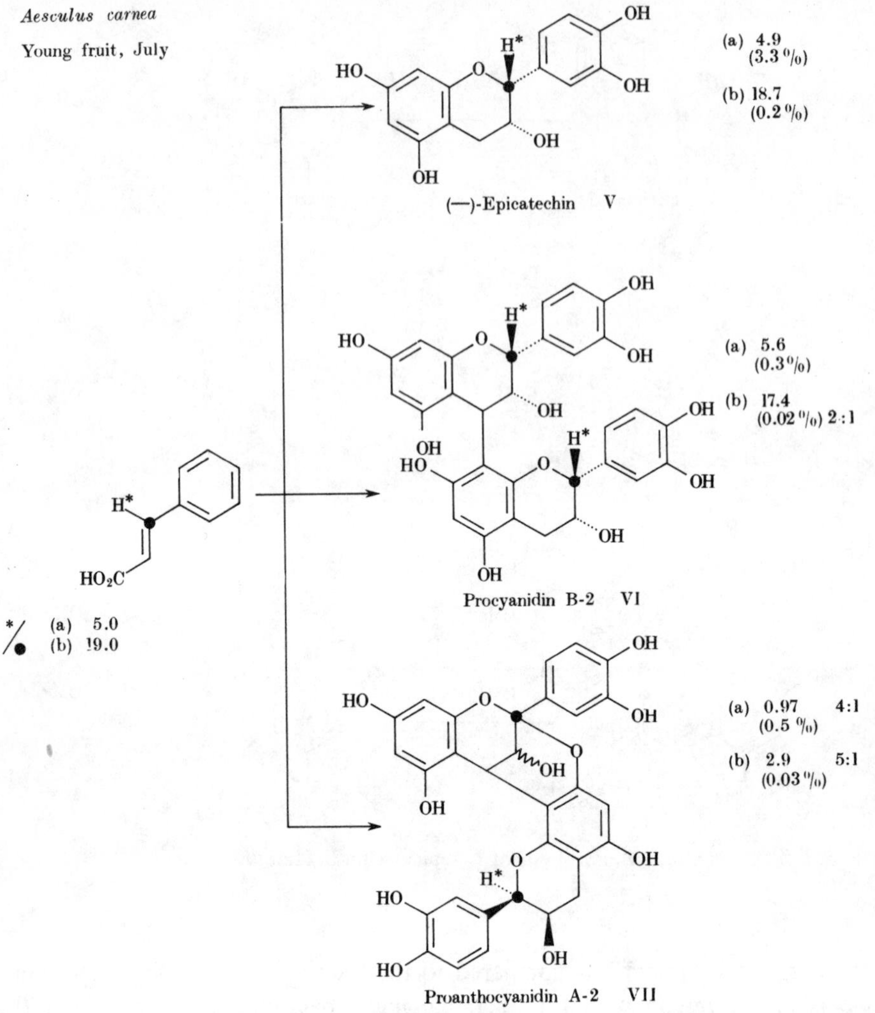

Aesculus carnea

Young fruit, July

(—)-Epicatechin V

(a) 4.9 (3.3 %)

(b) 18.7 (0.2 %)

Procyanidin B-2 VI

(a) 5.6 (0.3 %)

(b) 17.4 (0.02 %) 2 : 1

Proanthocyanidin A-2 VII

(a) 0.97 (0.5 %) 4:1

(b) 2.9 (0.03 %) 5:1

*/● (a) 5.0 (b) 19.0

Fig. 11. **Biosynthesis of procyanidins in *Aesculus carnea***

Grisebach and his collaborators have shown that the $C_6 \cdot C_3$ fragment of the carbon skeleton of both cyanidin (I) and (—)-epicatechin (V) is derived from the related cinnamic acid *via* the chalcone ⇌ flavanone intermediate (XVIII). The exact details of the chemical transformations from the chalcone ⇌ flavanone stage are not yet clear, but one of the possible routes to both cyanidin (I) and the *flavanol ((+)-catechin or (—)-epicatechin) is shown in Fig. 12. Here a central extended quinone type of intermediate (XIX) has been postulated.

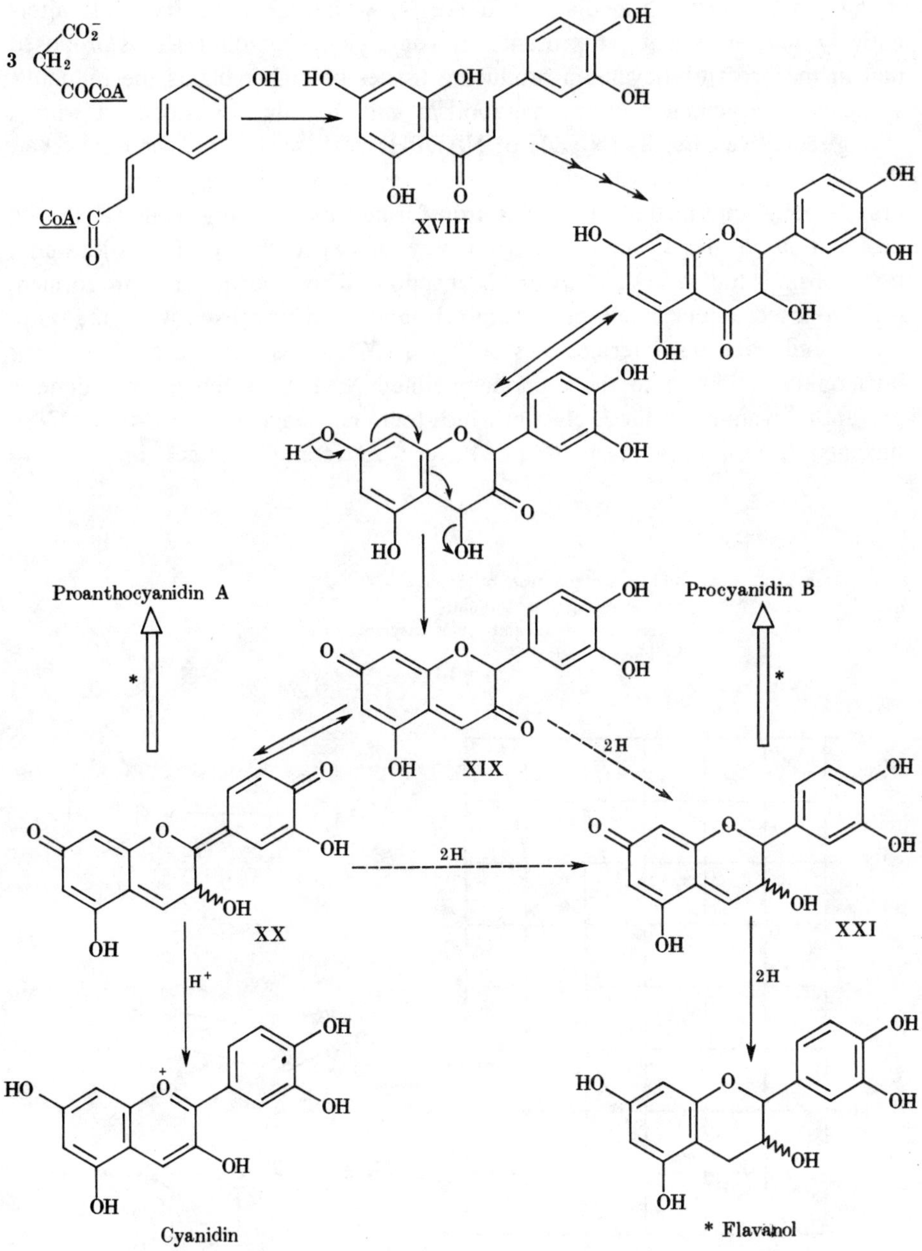

XVIII

Proanthocyanidin A

Procyanidin B

XIX

XX

XXI

Cyanidin

* Flavanol

Fig. 12. Biosynthesis of procyanidins — a hypothesis

Reduction of this intermediate (two stages) would give the flavanol; alternatively, rearrangement and protonation would yield cyanidin (I). It is proposed that in the pro(antho)cyanidin-producing tissues of higher plants the reductive sequence predominates in the metabolism and that this is associated with a strong reductive capacity (NADH or NADPH) in these tissues. When, however, the tissues ripen, or at senescence, this process declines in importance, and the branch point intermediate (XIX) is transformed by rearrangement (XX) and protonation to give cyanidin (I). It is now suggested that pro(antho)cyanidin metabolism is linked to this network of reactions. These polyphenols are formed, it is proposed, under conditions of high flavan-3-ol metabolism, when the latter reacts and captures intermediates XX or XXI to its own formation. Thus nucleophilic addition to the quinone methide XXI gives the normal dimeric pro(antho)cyanidin of the B class in which there is a single C-4 to C-6′ (or C-8′) linkage. Alternatively, carbon and oxygen nucleophilic attack on the two

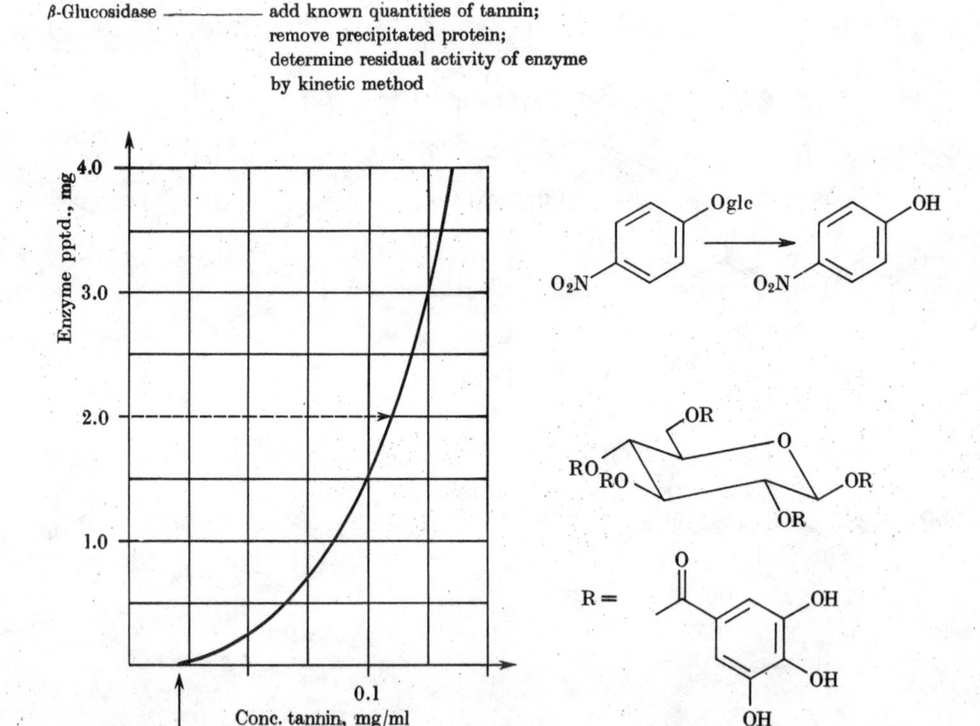

Fig. 13. Phenol–protein interactions

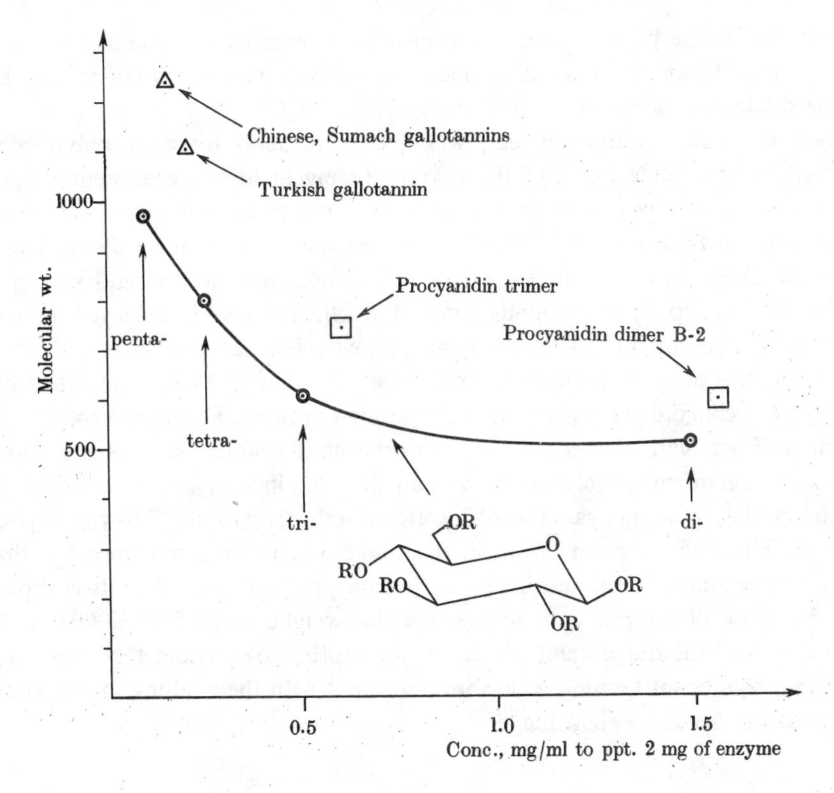

Fig. 14

quinone methide groups in the intermediate XX would give analogously the doubly linked proanthocyanidins such as A-2 (VII).

Proanthocyanidins have been linked with the astringent properties of plant tissues for a long time. Bate-Smith and Swain [19] in 1953 were probably the first to draw unequivocal attention to this probable relationship and they suggested that many of the earlier references in the botanical literature to "tannins" were in fact due to the presence of these particular types of polyphenols. Only very recently, with the advent of methods of isolation of the phenolic forms of the pro(antho)cyanidins has it become possible to attempt to verify this suggestion.

We have now done so by measuring the capacity of various synthetic and naturally occurring polyphenols to precipitate the enzyme β-glucosidase from aqueous solution. Precipitation curves were obtained for each substance and in their overall shape they show many similarities. For each substance a threshold value was obtained which indicated the minimum concentration of polyphenol required to initiate precipitation of the protein. Thereafter the graphs display a rapid change to an approximately linear form, such as that shown in Fig. 13 for the synthetic β-pentagalloyl-D-glucose.

From these observations values of the concentration for each polyphenol required to precipitate 50 % of the enzyme (2 mg in these experiments) were determined. These values plotted against the molecular weight of the polyphenol are shown in Fig. 14. This graph shows very clearly in the galloylglucose series the dependence of tanning action upon molecular weight and size and that in this group of compounds maximum effectiveness is achieved in the β-pentagalloyl-D-glucose structure. Monogalloyl-D-glucose derivatives did not possess tanning actions and measurements similarly showed that (—)-epicatechin and (+)-catechin did not possess the ability to tan proteins. However, pro(antho)-cyanidin dimers and trimers did so. Thus procyanidin dimers such as B-2 had a tanning action quantitatively similar to a digalloyl-D-glucose, and trimers VIII–X based on the proanthocyanidin A-2 were as effective as β-1,3,6-trigalloyl-D-glucose. These observations confirm the suggestion made some time ago that maximum tanning action, and hence astringency, of plant polyphenols is shown only by those of intermediate size (molecular weight range 500–3,000). They also show that the dimers and trimers of the pro(antho)cyanidin type possess a small but significant tanning action in agreement with their falling at the lower end of this molecular weight range.

REFERENCES

1. Swain, T. and Bate-Smith, E. C., "The Chemistry of Vegetable Tannins", Society of Leather Trades' Chemists, Croydon, 1956, p. 109.
2. Rosenheim, O., *Biochem. J., 14,* 278 (1920).
3. Robinson, R. and Robinson, G. M., *Biochem. J., 27,* 206 (1933).
4. Robinson, R., *Nature, 137,* 172 (1936).
5. Robinson, G. M., *J. Chem. Soc., 1937,* 1157.
6. Robinson, R. and Robinson, G. M., *J. Am. Chem. Soc., 61,* 1605 (1939).
7. Bate-Smith, E. C., *Biochem. J., 58,* 122 (1954).
8. Bate-Smith, E. C. and Lerner, N. H., *Biochem. J., 58,* 126 (1954).
9. Bate-Smith, E. C. and Swain, T., *Chem. and Ind., 1953,* 377.
10. Haslam, E., "Chemistry of Vegetable Tannins", Academic Press, London and New York, 1965.
11. Weinges, K., Bahr, W., Ebert, W., Goritz, K. and Marx, H.-D., *Fortschritte Chem. Org. Naturstoffe, 27,* 158 (1969).

12. Marini-Bettòlo, G. B. and Delle Monache, F., *Commentarii, Pontificia Academia Scientiarum, 2,* 1 (1973).
13. Roux, D. G., *Phytochemistry, 11,* 1219 (1972).
14. Thompson, R. S., Jacques, D., Haslam, E. and Tanner, R. J. N., *J. Chem. Soc., 1972,* 1387.
15. Mayer, W., Goll, L., von Arndt, E. M. and Mannschreck, A., *Tetrahedron Letters, 1966,* 429.
16. Jacques, D., Haslam, E., Bedford, G. R. and Greatbanks, D., *Chemical Communications, 1973,* 518.
17. Schilling, G., Weinges, K., Müller, O. and Mayer, W., *Ann., 1973,* 1471.
18. Grisebach, H., "Recent Advances in Phytochemistry" (eds T. J. Mabry, V. C. Runeckles and R. E. Alston), Appleton Century Crofts, New York, 1968, p. 379.
19. Swain, T. and Bate-Smith, E. C., *Chem. and Ind., 1953,* 377.

RECENT EXPERIMENTS ON THE SYNTHESIS OF 4-AMINOFLAVAN

by

M. RÁKOSI, A. L.-TŐKÉS and R. BOGNÁR

Institute of Organic Chemistry, Kossuth Lajos University and Antibiotic Research Group of the Hungarian Academy of Sciences
Debrecen, Hungary

In the present paper a brief survey is given of our experiments dealing with the synthesis of 2,4-*cis*-4-aminoflavan and its stereoisomer 2,4-*trans*-4-aminoflavan.

The synthesis of 4-aminoflavan was first reported in our paper [1] in 1959. This new type of flavan derivatives was prepared from the parent compound, flavanone, in cooperation with the Department of Organic Chemistry in Dublin directed by the late Professor Wheeler [2].

The starting point of our work was the fact that the carbonyl group of the dihydro-γ-pyrone ring in flavanone is capable of undergoing carbonyl reactions. Thus, with hydroxylamine the desired oxime can be prepared in several ways and in excellent yields. The corresponding amine, 4-aminoflavan, is obtained in high yield by hydrogenation of the oxime in 80 % acetic acid in the presence of palladium-charcoal or platinum oxide until two moles of hydrogen has been absorbed, and also by reduction with lithium aluminium hydride in tetrahydrofuran solution (Fig. 1). The amine was isolated as its salt with hydrochloric or acetic acid.

It has been observed that during the catalytic hydrogenation of flavanone oxime, reduction of the oximino group is not the sole reaction. Absorption of a third mole of hydrogen can lead to hydrogenolysis with cleavage of the heterocyclic ring to yield 1-(*o*-hydroxyphenyl)-1-amino-3-phenylpropane. The structure of the latter compound was proved in the following way. *o*-Hydroxy-β-phenylpropiophenone — prepared by the catalytic hydrogenation of *o*-hydroxychalcone in the course of our earlier reduction experiments [3] — was converted into the oxime and this was then catalytically reduced to the corresponding amine. As proved by the physical properties and IR spectra, the amine obtained from the oxime of known structure was identical with 1-(*o*-hydroxyphenyl)-1-amino-3-phenylpropane prepared by the reduction and hydrogenolysis of flavanone oxime.

A few years later Merten and Müller [4] synthesized 4-aminoflavan in another way. Salicylaldehyde and ethyl carbamate were allowed to react in the

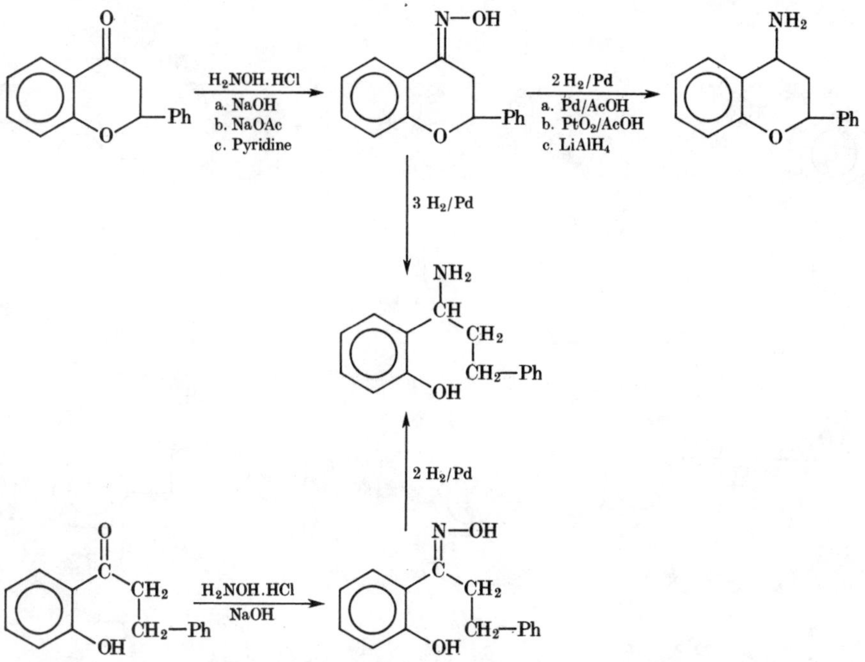

Fig. 1

presence of boron trifluoride etherate and the resulting salicylidene-bis-(ethyl carbamate) was condensed with styrene. The reaction of the produced 4-ethoxy-carbonylaminoflavan with phthalic anhydride resulted in 4-phthalimidoflavan, which gave 4-aminoflavan on hydrazinolysis (Fig. 2).

The physical properties of 4-aminoflavan and the derivatives synthesized by the authors mentioned above were identical with the data of the products prepared by us.

4-Aminoflavan was also made by Dudykina and Zagorewsky [5], and Ito *et al.* [6, 7] by the lithium aluminium hydride reduction of flavanone oxime. Under the conditions of reduction with lithium aluminium hydride a ring expansion reaction was also observed, which gave, besides 4-aminoflavan, 2-phenyl-2,3,4,5-tetrahydro-1,5-benzoxazepine as a by-product (Fig. 3). According to the observations of the Japanese authors, the ratio of formation of 4-aminoflavan and the benzoxazepine derivative depends on the substituents at C-2.

The stereochemistry of 4-aminoflavan and its derivatives obtainable in the way described above was investigated by NMR and CD measurements and the 2,4-*cis* configuration of the compounds was proved [8, 9] (Fig. 4).

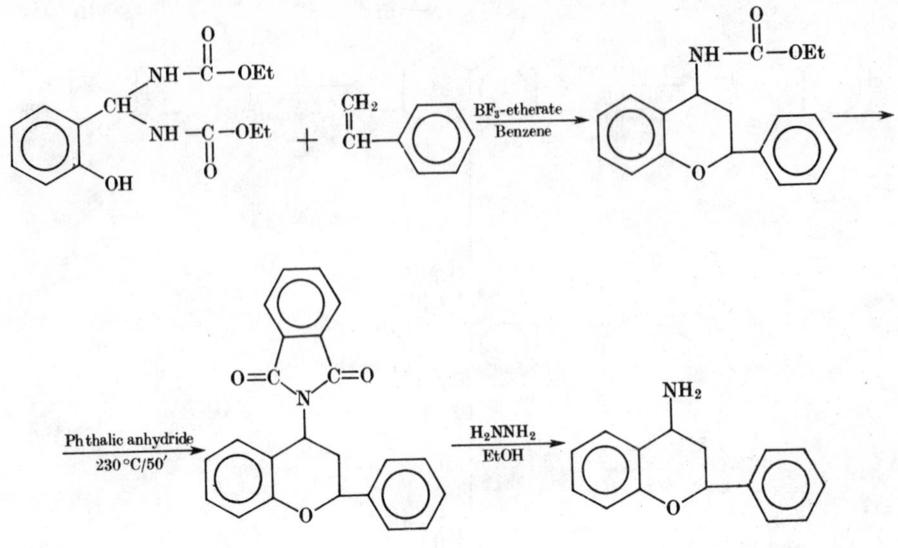

NH—C(=O)—OEt
CH
NH—C(=O)—OEt
OH
+
CH_2
CH—
BF_3-etherate
Benzene
NH—C(=O)—OEt
O

Phthalic anhydride
230 °C/50′
O=C
N
C=O
O
H_2NNH_2
EtOH
NH_2
O

Fig. 2

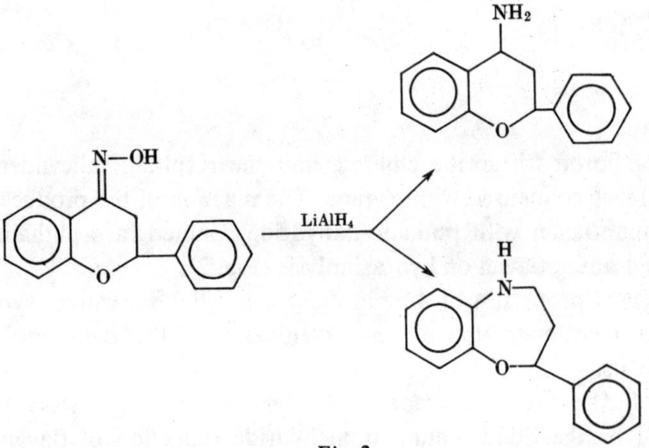

N—OH
O
$LiAlH_4$
NH_2
O
H
N
O

Fig. 3

H
Ph
O
H
NH_2
≡
NH_2
O
Ph

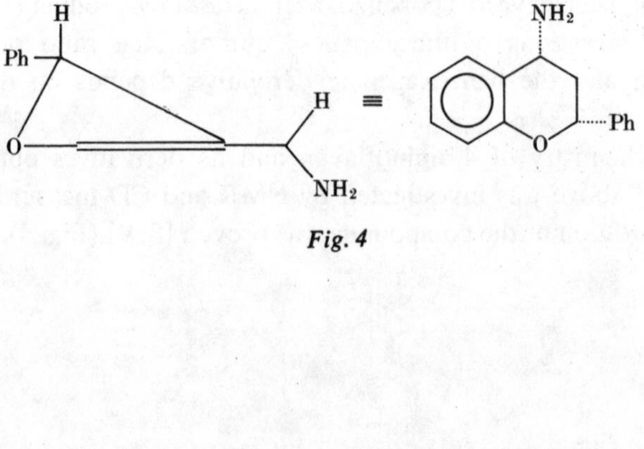

Fig. 4

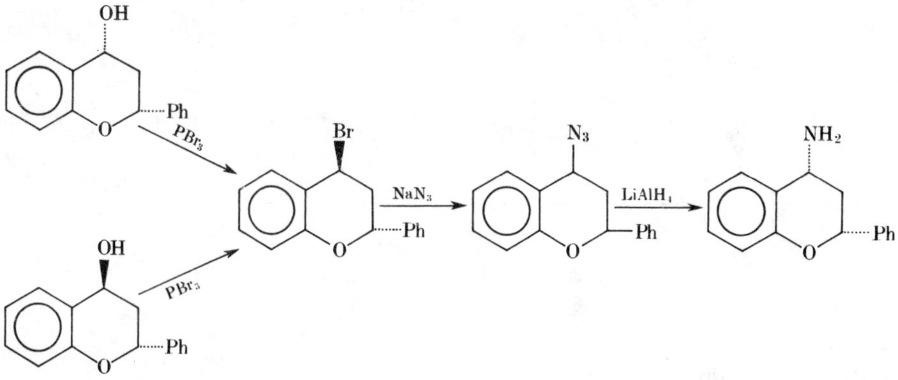

Fig. 5

4-Aminoflavan is an important starting material for the synthesis of flavan derivatives substituted in the C-4 position. The reactions of its reactive primary amino group give rise to several new derivatives of theoretical and practical interest [10–13]. A part of these reactions — primarily the results of the experiments with the optically active derivatives — will be reported in another paper.

As it has been mentioned, each synthetic method described above gives the same product, the racemate of 2,4-*cis*-4-aminoflavan. As an extension of our experiments on the chemistry of 4-aminoflavans, the possibilities of synthesizing the hitherto unknown 2,4-*trans*-4-aminoflavan isomer were explored. The synthesis of the desired compound was attempted in several ways.

One of the possibilities is the reaction route shown in Fig. 5.

In the course of our previous work, reaction methods giving homogeneous products in high yields had been elaborated for the synthesis of the starting materials, the racemates of 2,4-*cis*-4-hydroxyflavan and 2,4-*trans*-4-hydroxyflavan [1–3, 14]. We wished to prepare the stereoisomeric 4-bromoflavans, the 2,4-*cis*- and 2,4-*trans*-bromo derivatives from both 4-hydroxyflavans with phosphoric tribromide. However, this method was not suitable for the synthesis of 2,4-*trans*-aminoflavan, because both 4-hydroxyflavan stereoisomers gave the same 4-bromoflavan derivative. The identity of the products was proved, besides the physico-chemical constants, by IR and NMR spectroscopy. The coupling constants of the NMR spectra allow to draw the conclusion that the bromine atom in the 4-bromoflavan obtained from either 2,4-*cis*- or 2,4-*trans*-hydroxyflavan, has *axial* configuration, thus the compound is analogous to 2,4-*trans*-4-hydroxyflavan. The reason for this is that the energetically and sterically favourable position of the two substituents of great steric demand, the C-2

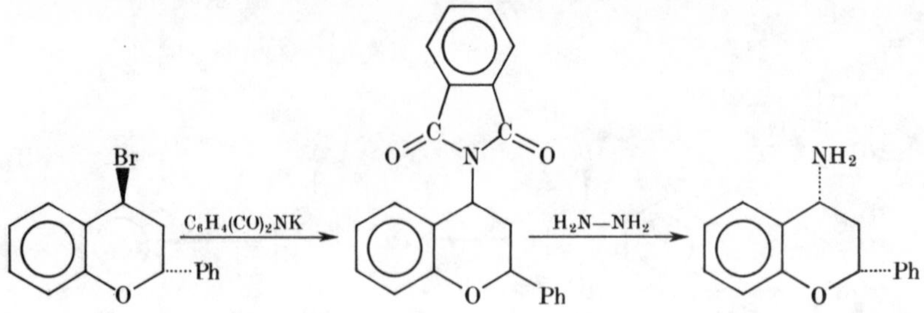

Fig. 6

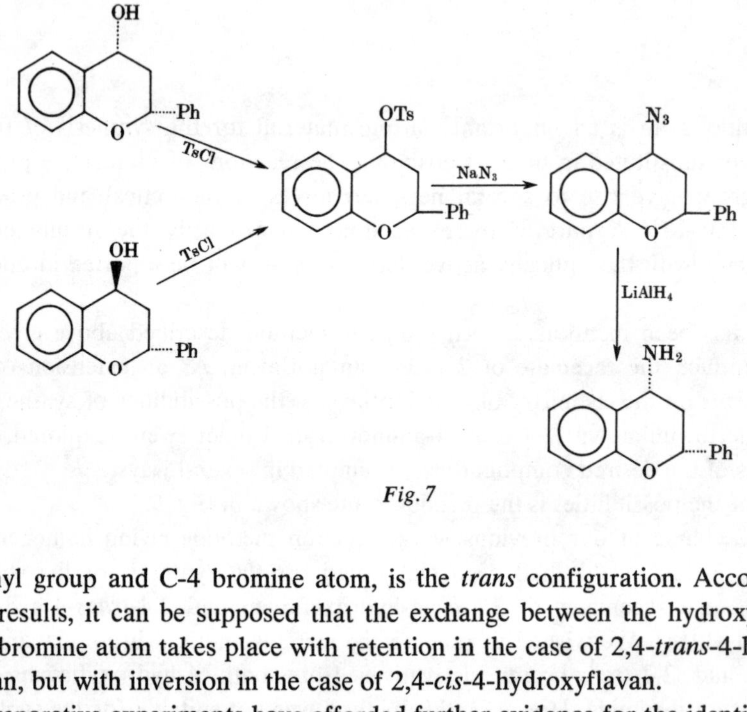

Fig. 7

phenyl group and C-4 bromine atom, is the *trans* configuration. According to our results, it can be supposed that the exchange between the hydroxyl group and bromine atom takes place with retention in the case of 2,4-*trans*-4-hydroxyflavan, but with inversion in the case of 2,4-*cis*-4-hydroxyflavan.

Preparative experiments have afforded further evidence for the identity of the prepared 4-bromoflavans. The 4-bromoflavan obtained from either of the two 4-hydroxyflavan isomers was transformed into the 4-azido derivative, and the azide was reduced to 4-aminoflavan. As expected, the same azido- and amino derivatives were obtained.

The reaction with potassium phthalimide of 4-bromoflavan obtained from either of the two 4-hydroxyflavan isomers and subsequent hydrazinolysis according to the procedure of Merten and Müller [4] gave also 2,4-*cis*-4-aminoflavan (Fig. 6).

Since the synthesis of the desired 2,4-*trans*-4-aminoflavan was not successful through the 4-bromoflavan intermediates, new possibilities were sought to prepare this isomer starting directly with the two 4-hydroxyflavan isomers.

The tosylates of both 4-hydroxyflavan isomers were prepared, then transformed into 4-azidoflavans with sodium azide. The reduction of the products with lithium aluminium hydride resulted in 4-aminoflavan (Fig. 7).

Unfortunately, this route of the synthesis of 2,4-*trans*-4-aminoflavan was also unsuccessful. The physical constants, IR and NMR spectral data of the amine hydrochloride and its N-acyl derivative obtained from both 4-hydroxyflavan isomers were identical with the data of the 2,4-*cis*-4-aminoflavan hydrochloride and the N-acyl derivative prepared by the reduction of flavanone oxime.

Further experiments are in progress with the aim of finding possibilities for the synthesis of the hitherto unknown 2,4-*trans*-4-aminoflavan.

REFERENCES

1. Bognár, R., Rákosi, M., Fletcher, H., Philbin, E. M. and Wheeler, T. S., *Tetrahedron Letters, 1959,* 4.
2. Fletcher, H., Philbin, E. M., Wheeler, T. S., Bognár, R. and Rákosi, M., *Acta Phys. et Chim. Debrecina, VIII,* 5 (1962); *Magyar Kém. Foly., 68,* 465 (1962).
3. Bognár, R. and Rákosi, M., *Acta Chim. Acad. Sci. Hung., 13,* 217 (1957); *Magyar Kém. Foly., 64,* 111 (1958).
4. Merten, R. and Müller, G., *Chem. Ber., 97,* 682 (1964).
5. Dudykina, N. V. and Zagorewsky, V. A., *Sintez Prirodn. Soedin., ikh Analogov i Fragmentov,* Akad. Nauk SSSR, Otd. Obshch. i Tekhn. Khim., Moscov, 1965; *C.A., 65,* 683 (1966), p. 134.
6. Inoue, N., Yamaguchi, S., Ito, S. and Suzuki, I., *Bull. Chem. Soc. Japan, 41,* 2078 (1968).
7. Ito, S., *Bull. Chem. Soc. Japan, 43,* 1824 (1970).
8. Bognár, R., Clark-Lewis, J. W., L.-Tőkés, A. and Rákosi, M., *Australian J. Chem., 23,* 2015 (1970).
9. Snatzke, G., Snatzke, F., L.-Tőkés, A., Rákosi, M. and Bognár, R., *Tetrahedron, 29,* 909 (1973).
10. Bognár, R. and Farkas, I., *Acta Chim. Acad. Sci. Hung., 35,* 223 (1963).
11. Bognár, R., L.-Tőkés, A. and Rákosi, M., *Acta Chim. Acad. Sci. Hung., 58,* 195 (1968); *Magyar Kém. Foly., 74,* 457 (1968).
12. Bognár, R., L.-Tőkés, A. and Rákosi, M., *Magyar Kém. Foly., 76,* 271 (1970).
13. Rákosi, M., L.-Tőkés, A. and Bognár, R., *Tetrahedron Letters, 1970,* 2305.
14. Bognár, R., Rákosi, M., Fletcher, H., Kehoe, D., Philbin, E. M. and Wheeler, T. S., *Tetrahedron, 18,* 135 (1962).

THE SYNTHESIS OF OPTICALLY ACTIVE FLAVAN DERIVATIVES

by

A. L.-TŐKÉS, M. RÁKOSI and R. BOGNÁR
Institute of Organic Chemistry, Kossuth Lajos University
Debrecen, Hungary

The present paper is a summary of our results concerning the synthesis of optically active 4-substituted flavan derivatives.

The starting compound of these syntheses was (+)-*cis*-4-aminoflavan obtained by the reduction of 4-oximinoflavan [1]. It was resolved into enantiomeric forms by fractional crystallization of its amine salts with (+)-camphor-10-sulfonic acid and with (—)-di-O-benzoyltartaric acid [2]. The pairs of diastereoisomeric salts thus separated were converted into (+)- and (—)-*cis*-4-aminoflavan hydrochlorides, from which the free bases and the optically active N-acetyl and N-benzoyl derivatives were prepared. NMR data unequivocally establish the *cis* configuration of all these 4-substituted flavan derivatives.

On deamination with nitrous acid, according to the method reported by Wheeler and Bognár earlier [1], (—)-4-aminoflavan hydrochloride gives (+)-4α-hydroxyflavan, whereas (+)-4-aminoflavan hydrochloride yields (—)-4α-hydroxyflavan. Deamination of the enantiomers with nitrous acid involves inversion at C-4, and the *trans* configurations of the two optically active 4α-hydroxyflavans produced have been proved by means of IR and NMR spectrometry.

The corresponding O-acetyl derivatives were obtained by acetylation of the antipodes in dry pyridine with acetic anhydride [3].

Sodium dichromate oxidation of the C-4 hydroxyl group of the enantiomeric 4α-hydroxyflavans eliminated one of the two asymmetric centres giving (+)- and (—)-flavanone [3]. The absolute configurations of the flavanone enantiomers were determined by Snatzke *et al.* [4]. Dextrorotatory flavanone was assigned 2*R* configuration by comparison with other flavanones of known absolute configuration. As regards the phenyl ketone $n \rightarrow \pi^*$ chromophore, the 2*R* configuration of the dextrorotatory flavanone could be proved by comparing its CD curve with those of other flavanones. The strongly negative CD indicates independently that the absolute configuration of (+)-flavanone is 2*R*. The sign of this $n \rightarrow \pi^*$ band is independent of the substitution pattern of the aromatic

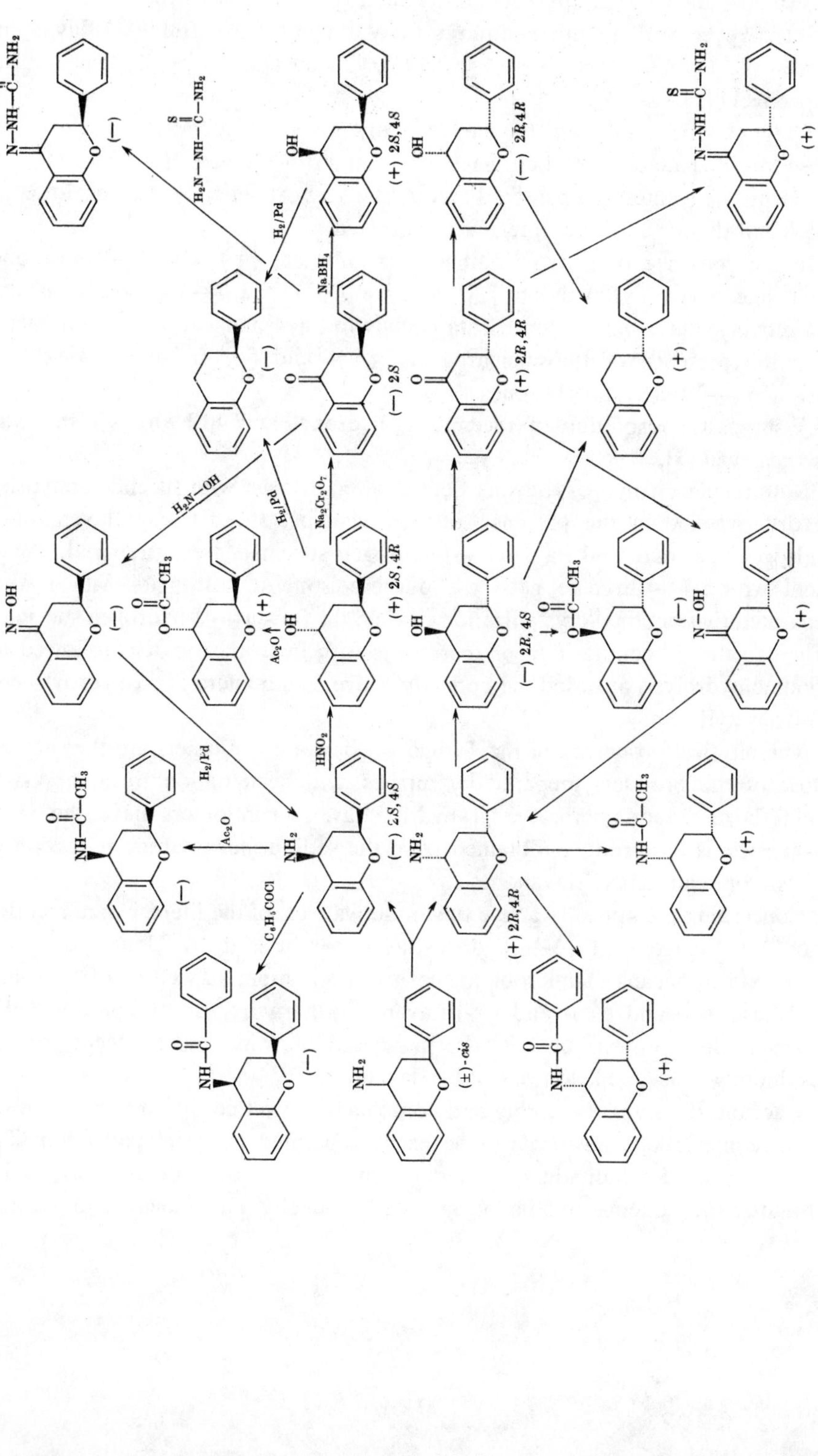
H2N−NH−C(=S)−NH2
H2/Pd
NaBH4
(+) 2S,4S
(−) 2R,4R
(−) 2S
(+) 2R,4R
H2N−OH
Na2Cr2O7
Ac2O
(+) 2S,4R
(−) 2R,4S
HNO2
H2/Pd
(−) 2S,4S
(+) 2R,4R
C6H5COCl
(±)-cis
(+)
(−)

Fig. 1

ring system, and (+)-flavanone can thus serve as the reference compound of known absolute configuration for all the other flavans described.

Selective reduction with sodium borohydride of (+)- and (—)-flavanone yielded the two enantiomers (—)-4β-hydroxyflavan and (+)-4β-hydroxyflavan, respectively [5].

The catalytic reduction of the optically active 4-hydroxyflavans in the presence of palladium-carbon gave (+)- and (—)-flavan. In our first experiments we wanted to open the heterocyclic ring catalytically, but this attempt failed and the flavan enantiomers were obtained.

In the presence of hydroxylamine hydrochloride, (+)- and (—)-flavanone react like racemic flavanone [1], to give (+)- and (—)-4-oximinoflavan, respectively; these were reduced catalytically to the 4-aminoflavan enantiomers.

In the presence of thiosemicarbazide, (+)- and (—)-flavanone yield the (+)- and (—)-thiosemicarbazones.

A successful resolution of racemic 4α-hydroxy- and 4β-hydroxyflavan was also achieved [5].

Both racemic 4-hydroxyflavans were allowed to react with succinic anhydride in dry pyridine in the presence of triethylamine. In this way flavanyl(4α)-hydrogen succinate and flavanyl(4β)-hydrogen succinate were prepared. Treatment with L-(—)-brucine gave the diastereoisomeric salt pairs, which were separated by fractional crystallization to yield the flavanyl(4)-hydrogen succinate brucine salts. Decomposition of these salts with hydrochloric acid followed by alkaline hydrolysis afforded the optically active enantiomers of 4α-hydroxy- and 4β-hydroxyflavan.

The physical properties of the 4α-hydroxyflavan enantiomers are the same as those of the products prepared by nitrous acid deamination from the *cis*-4-aminoflavan enantiomers; the 4β-hydroxyflavan enantiomers have the same properties as the products obtained from the sodium borohydride reduction of the two optically active flavanones.

Concerning the optically active flavan derivatives, of the literary data mention should be made of (—)-4α-hydroxyflavan synthesized by Udupa [6] from flavanone by means of microbiological transformation, as well as of (+)- and (—)-flavanone and (+)- and (—)-flavan. Furthermore, Corey and Mitra [7] reported the synthesis of L(+)-2,3-butanedithiol, which is an agent for the resolution of racemic ketones, such as flavanone.

Racemic flavanone is readily and completely converted into a mixture of the diastereomeric ketals which can be easily separated by recrystallization. The hydrolysis of the individual isomers using mercuric chloride–mercuric oxide promoter in aqueous methanol afforded optically pure levo- and dextro-

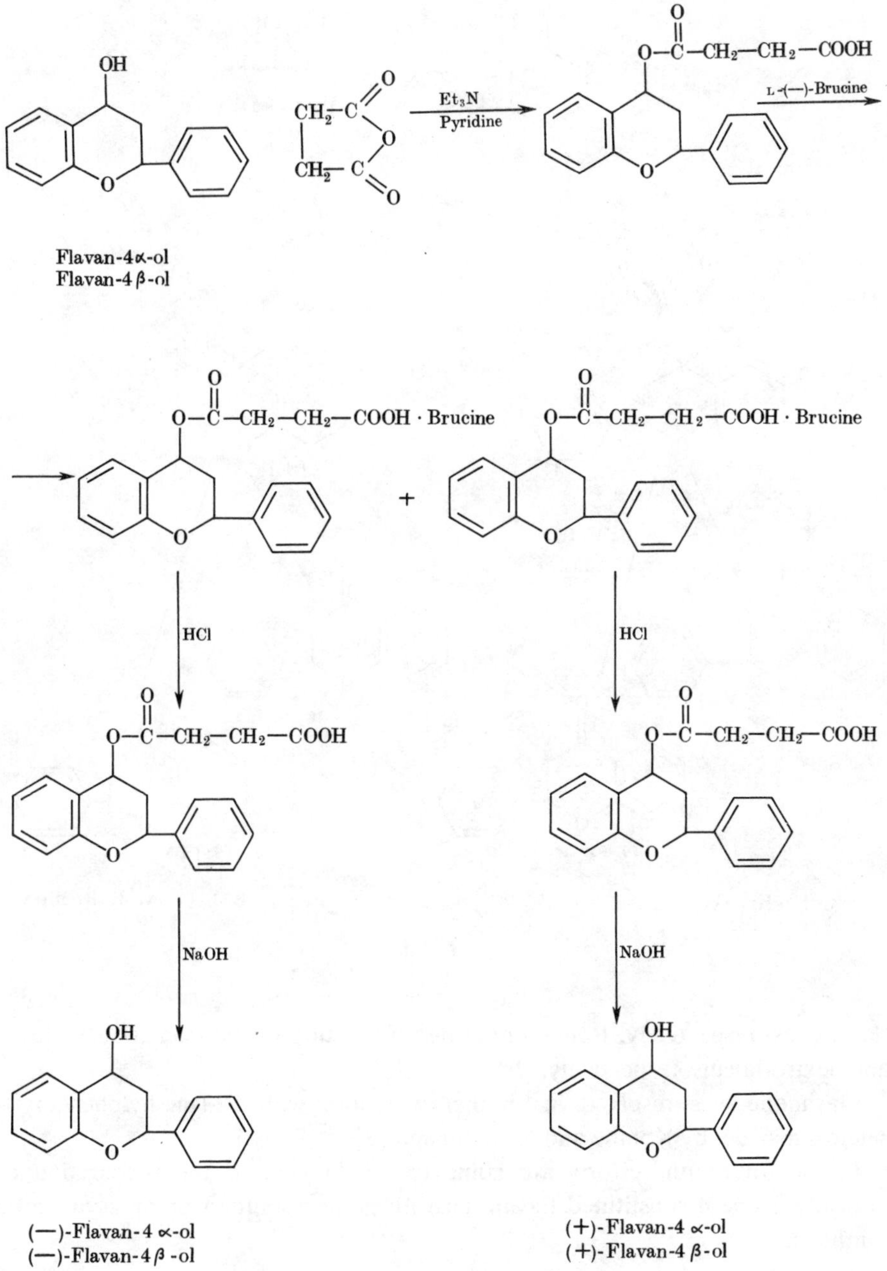

OH

O

CH_2—C

O

CH_2—C

O

Et_3N

Pyridine

O

O—C—CH_2—CH_2—COOH

L-(—)-Brucine

Flavan-4α-ol
Flavan-4β-ol

O—C—CH_2—CH_2—COOH · Brucine

+

O—C—CH_2—CH_2—COOH · Brucine

HCl

HCl

O—C—CH_2—CH_2—COOH

O—C—CH_2—CH_2—COOH

NaOH

NaOH

OH

OH

(—)-Flavan-4 α-ol
(—)-Flavan-4 β-ol

(+)-Flavan-4 α-ol
(+)-Flavan-4 β-ol

Fig. 2

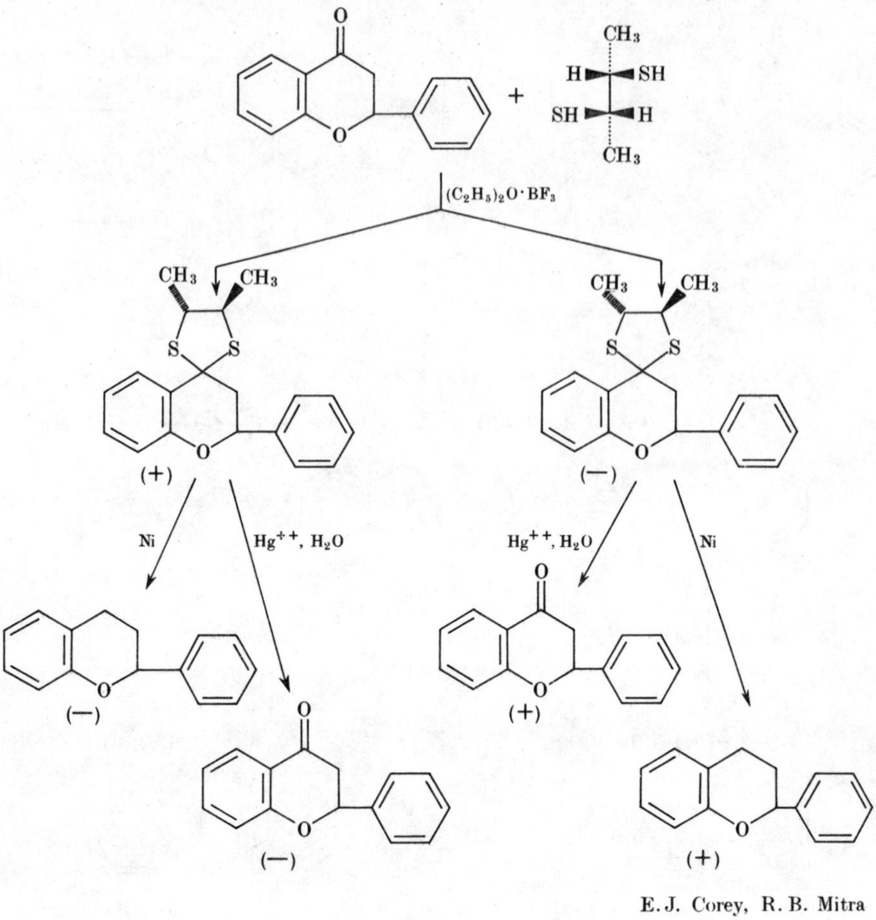

Fig. 3

flavanones, respectively. Raney nickel desulfurization of the isomers gave levo- and dextroflavans, respectively.

Flavanone was resolved in another way too, with 5α-(methylphenethyl)-semioxamazide, by Kotake and Nakaminami [8].

In the future our efforts are going to be directed to the preparation of optically active 4-substituted flavan derivatives by resolution or by asymmetric synthesis.

REFERENCES

1. BOGNÁR, R., RÁKOSI, M., FLETCHER, H., PHILBIN, E. M. and WHEELER, T. S., *Tetrahedron Letters, 1959,* 4; *Magyar Kém. Foly., 68,* 465 (1962).
2. BOGNÁR, R., CLARK-LEWIS, J. W., L.-TŐKÉS, A. and RÁKOSI, M., *Australian J. Chem., 23,* 2015 (1970).
3. RÁKOSI, M., L.-TŐKÉS, A. and BOGNÁR, R., *Tetrahedron Letters, 1970,* 2305.
4. SNATZKE, G., SNATZKE, F., L.-TŐKÉS, A., RÁKOSI, M. and BOGNÁR, R., *Tetrahedron, 29,* 909 (1973).
5. BOGNÁR, R., L.-TŐKÉS, A. and RÁKOSI, M., *Acta Chim. (Budapest), 79,* 357 (1973).
6. UDUPA, S. R., BANERJI, A. and CHADHA, M. S., *Tetrahedron Letters, 1968,* 4003.
7. COREY, E. J. and MITRA, R. B., *J. Am. Chem. Soc., 84,* 2938 (1962).
8. KOTAKE, M. and NAKAMINAMI, G., *Proc. Japan Acad., 29* (1953).

OXIDATION EXPERIMENTS IN THE GROUP OF FLAVONOIDS (ALGAR–FLYNN–OYAMADA OXIDATIONS)

by

GY. LITKEI, R. BOGNÁR, Z. DINYA and É. R.-DÁVID

Institute of Organic Chemistry, Kossuth Lajos University
Debrecen, Hungary

It was forty years ago that Algar and Flynn [1], then Oyamada [2] published the earliest report on the alkaline oxidation of chalcones. Since then numerous researchers investigated this reaction, but its mechanism has not been elucidated so far. First Algar and Flynn [1] and subsequently others [3, 4] supposed that a chalcone epoxide could be the intermediate product of the reaction. It was Bognár and co-workers [5] who first succeeded in preparing and investigating the chalcone epoxides formed in the course of the alkaline oxidation of 2′-O-substituted chalcones. On the basis of detailed examinations [6] these chalcone epoxides proved to be important intermediate products, but they offered only indirect evidence for the mechanism of the Algar–Flynn–Oyamada reaction.

In 1965 Dean *et al.* [7] suggested other possibilities for the explanation of this reaction (Fig. 1). It has been supposed that the chalcone cyclizes before oxidation, or the corresponding flavonoid is formed simultaneously with the oxidation. This hypothesis, however, has not been substantiated so far.

To verify these assumptions, UV spectroscopic investigations have been performed. The reaction conditions were similar to the Algar–Flynn–Oyamada reactions and the results were detected by following the changes in the UV spectra and by TLC. In the present paper a survey of our results is given.

At pH 9.57 and room temperature either the chalcone or flavanone gave an equilibrium mixture of identical composition and no oxidation reaction occurred (Fig. 2).

At different pH values it was observed that the decisive factor in the reaction was the chalcone $\rightleftharpoons$ flavanone equilibrium. The pH-dependence of the rate of isomerization is shown in Fig. 3.

It is apparent that in the first stage of the reaction, starting either with flavanone or the chalcone, the same results are obtained; the hydrogen peroxide does not take part in the reaction. At pH 9.57, however, an abrupt jump can be observed. This is very important, because above this pH value a new band appears in the spectrum at 420 nm. The decrease or increase of this band in

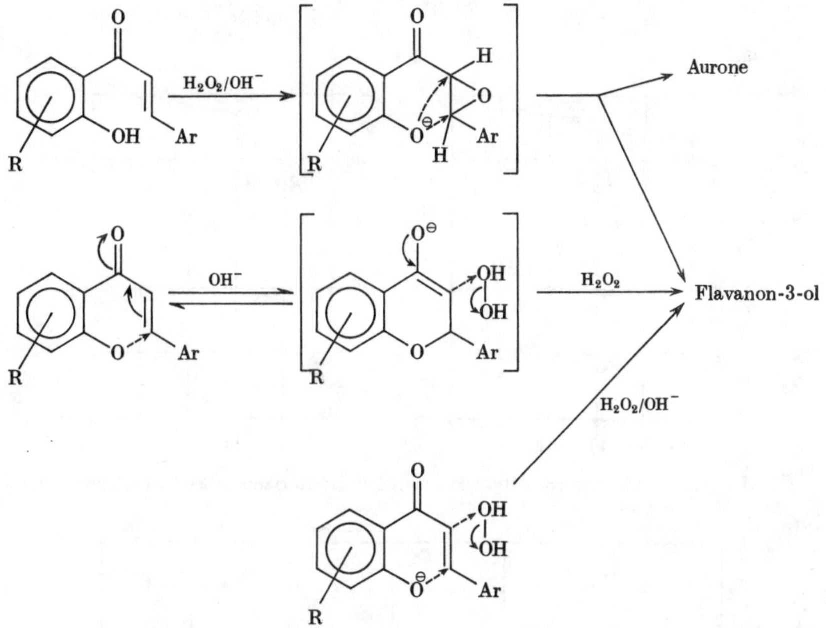

H_2O_2/OH^-

Aurone

OH^-

H_2O_2

Flavanon-3-ol

H_2O_2/OH^-

Fig. 1

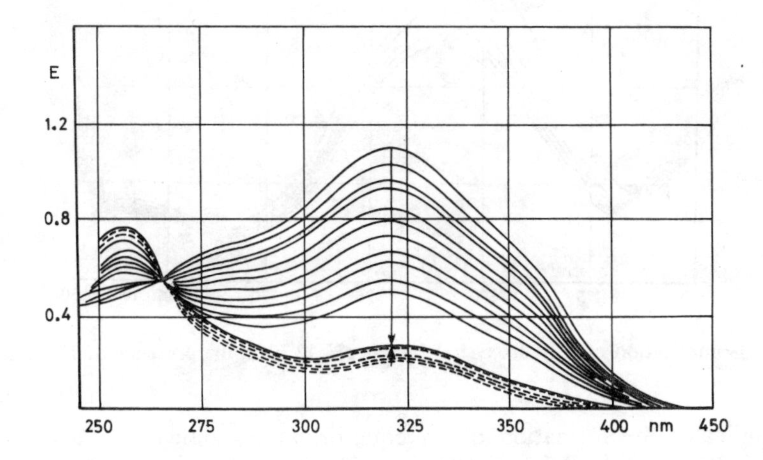

Fig. 2. The chalcone ⇌ flavanone isomerization at pH 9.57.

—— 2′-Hydroxychalcone + Buffer (pH 9.57)

2′-Hydroxychalcone + Buffer (pH 9.57) + H_2O_2

--- Flavanone + Buffer (pH 9.57)

Flavanone + Buffer (pH 9.57) + H_2O_2

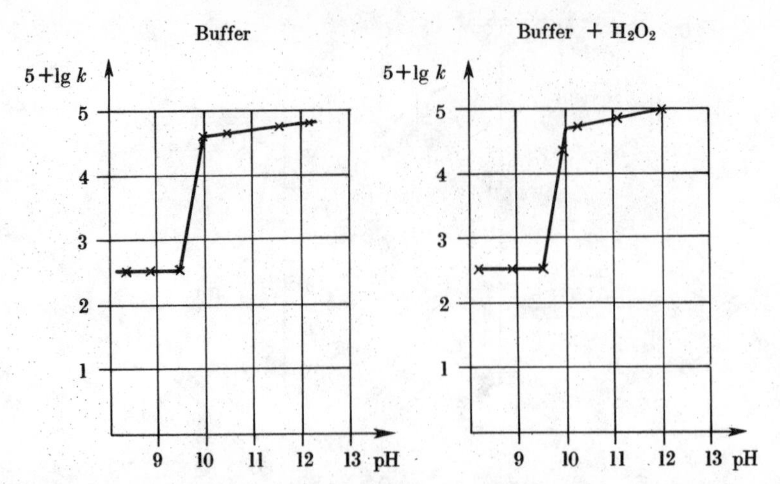

Fig. 3. The flavanone ⇌ 2'hydroxychalcone isomerization as a function of pH

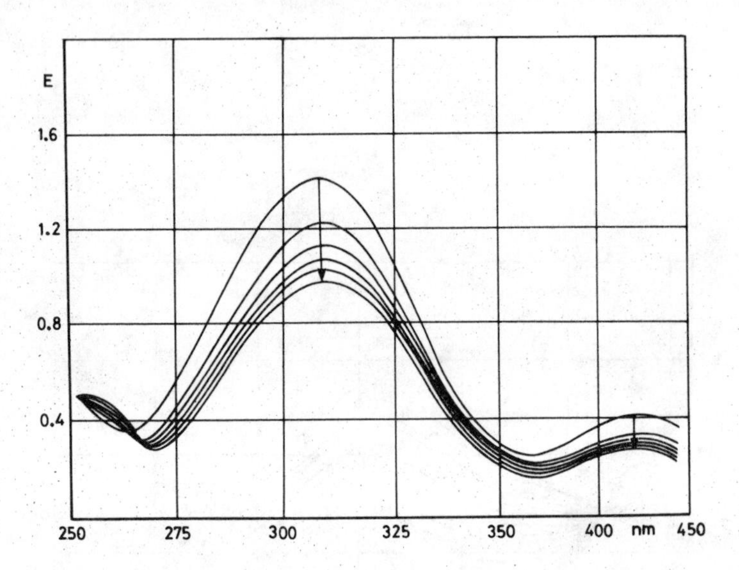

Fig. 4. Isomerization of 2'-hydroxychalcone at pH 12.30 in the presence of H_2O_2 (1st part)

time indicates the formation of an equilibrium, as shown in Figs 4 and 5. Completely identical values were obtained for flavanone and the chalcone.

It is obvious that the equilibrium is rapidly attained, to be followed by a slow, not unequivocally traceable oxidation process.

The appearance of the new band is indicative of the formation of a phenolate anion [8]. This suggestion is supported by our quantum chemical calculations,

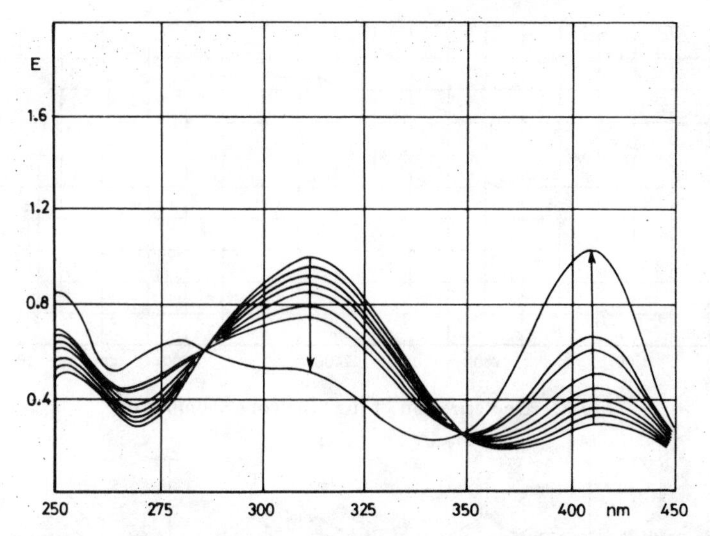

Fig. 5. The change of the equilibrium mixture at pH 12.30 (2nd part)

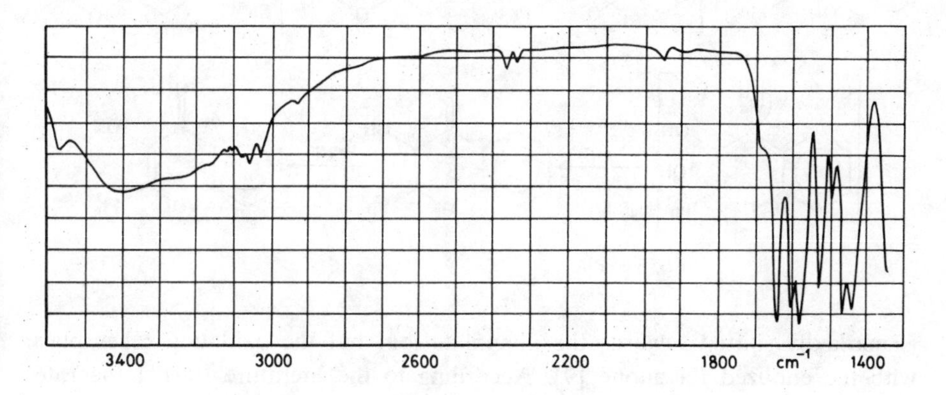

Fig. 6. "Chalcone sodium salt" from 2′-hydroxychalcone

according to which this band should appear at 418 nm – a value being in very good agreement with the band found actually at 420 nm.

In order to determine the components of the equilibrium mixture, the structure of the so-called "chalcone sodium salt" was examined. According to the infrared spectra, this sodium salt, unlike the corresponding chalcone, is not deuterable (Figs 6 and 7) and yields flavanone on treatment with deuterium oxide.

It seems that the equilibrium mixture, the "chalcone sodium salt", is not a homogeneous, salt-like compound, but the sodium complex of 2′-hydroxychalcone and enolized flavanone (Fig. 8). At higher pH values the equilibrium

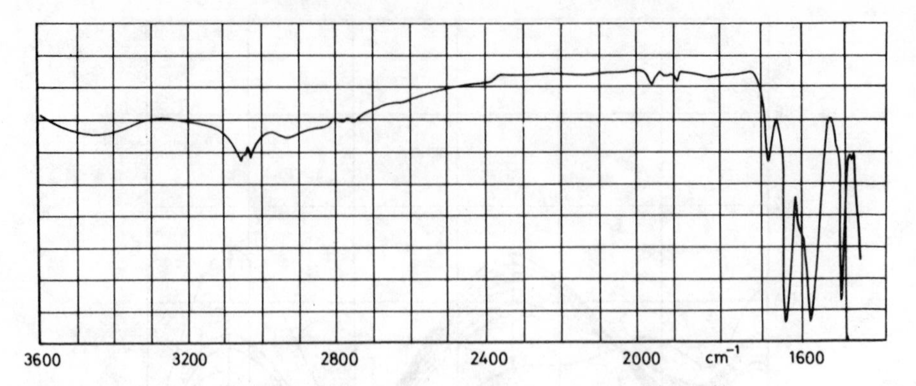

Fig. 7. Deuterization of the "chalcone sodium salt"

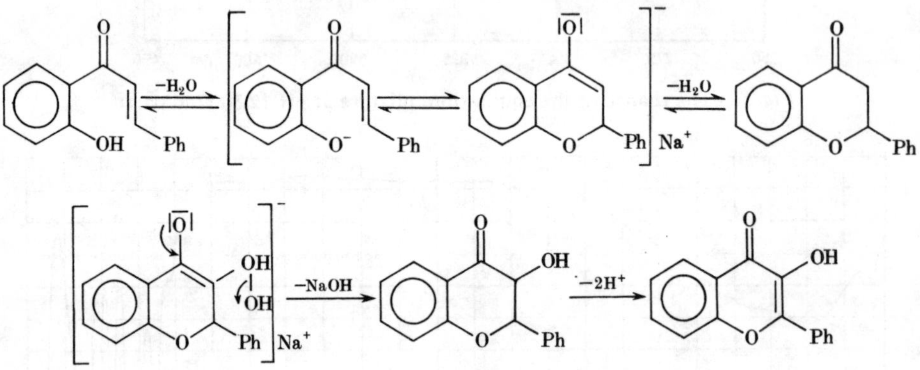

Fig. 8

is markedly shifted towards the phenolate ion; but the oxidation takes place with the enolized flavanone [9]. According to the literature [9–11] the rate-determining factor is the formation of an enolate anion.

On the basis of our studies the formation of chalcone epoxide is out of question, since chalcones present as phenolate ions, as shown by the experimental data, do not transform into epoxides because of Coulomb interaction [7]. At lower pH values substituted chalcones may transform into epoxides, and the rate of this reaction is of the same order as the rate of the chalcone ⇌ flavanone isomerization.

According to the results to be presented later, neither is there any probability of an oxidation taking place simultaneously with the cyclization, as proposed by Dean [7].

On the basis of the spectral investigations, 3-hydroxyflavanone is converted into 3-hydroxyflavone above pH 10.5. A new band appears at 410 nm,

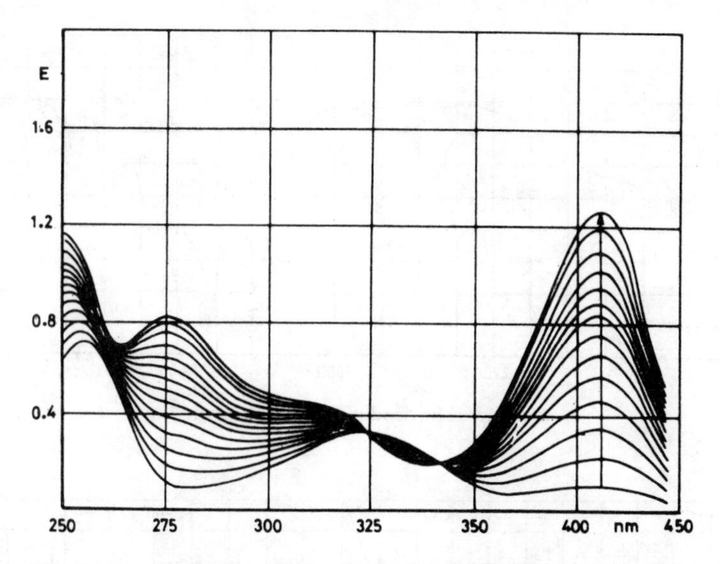

Fig. 9. Conversion of flavanon-3-ol to the "sodium salt" at pH 12.13

indicating a dissociation process similar to the previously explained case of 2′-hydroxychalcone (Fig. 9).

According to the experiments it seems that the course of the reaction at room temperature is determined by the pH of the medium, and not by the presence of hydrogen peroxide.

During the reaction an unstable intermediate product was isolated. Solely on the effect of dilution, this intense yellow material yields 3-hydroxyflavone.

The spectrum of this so-called sodium salt is diffuse, and differs from the characteristic spectrum of flavones (Fig. 10). The action of deuterium oxide on this salt results in deuterated 3-hydroxyflavone (Fig. 11). It seems that a metal-assisted enolization takes place during the reaction which may be followed by oxidative loss of the metal [12] to yield 3-hydroxyflavone.

Our experiments have also been extended to 2′-hydroxy-6′-methoxychalcone. According to the literature [3], this compound ought to have yielded 4-methoxy-aurone; instead, a reaction mixture containing five components was obtained in our experiments. From this two principal products were separated by fractional crystallization (Fig. 12).

In the course of identification the component with lower melting point was shown to be the aurone, while the product with higher melting point was found to be the corresponding flavanon-3-ol. The reaction was completely identical when 5-methoxyflavanone had been used as the starting material.

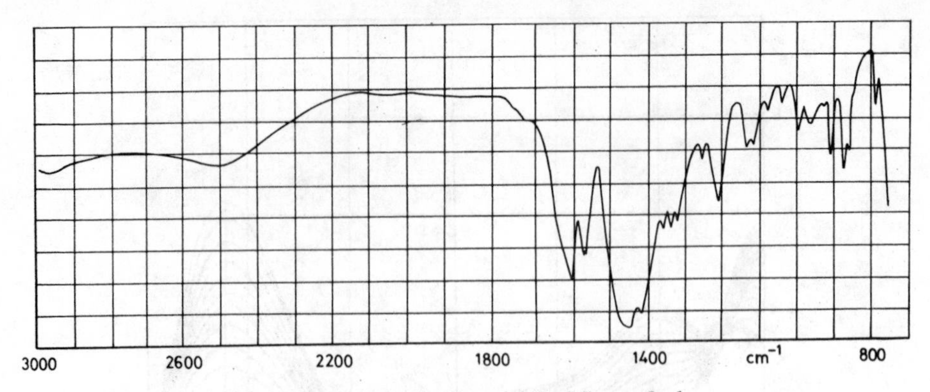

Fig. 10. The "sodium salt" of flavon-3-ol

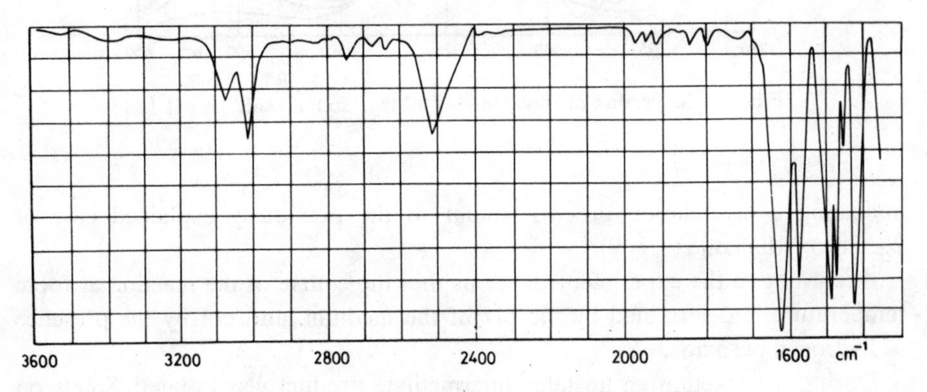

Fig. 11. Deuterization of the "sodium salt"

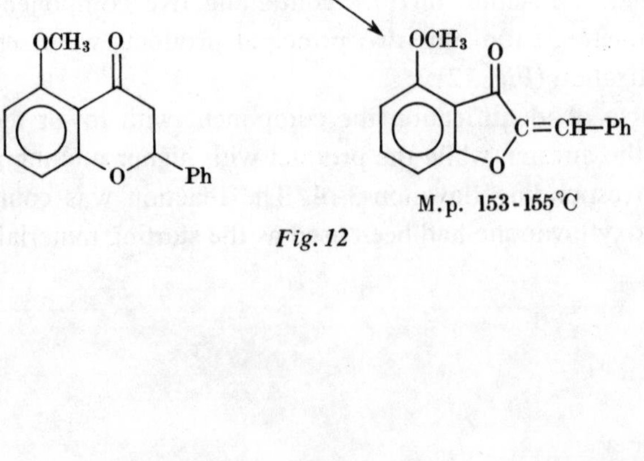

OCH_3 O

OH Ph

$NaOH/H_2O$

OCH_3 O

O Ph

OCH_3 O

OH

O Ph

M.p. 219-221°C

OCH_3

O

C=CH—Ph

O

M.p. 153-155°C

Fig. 12

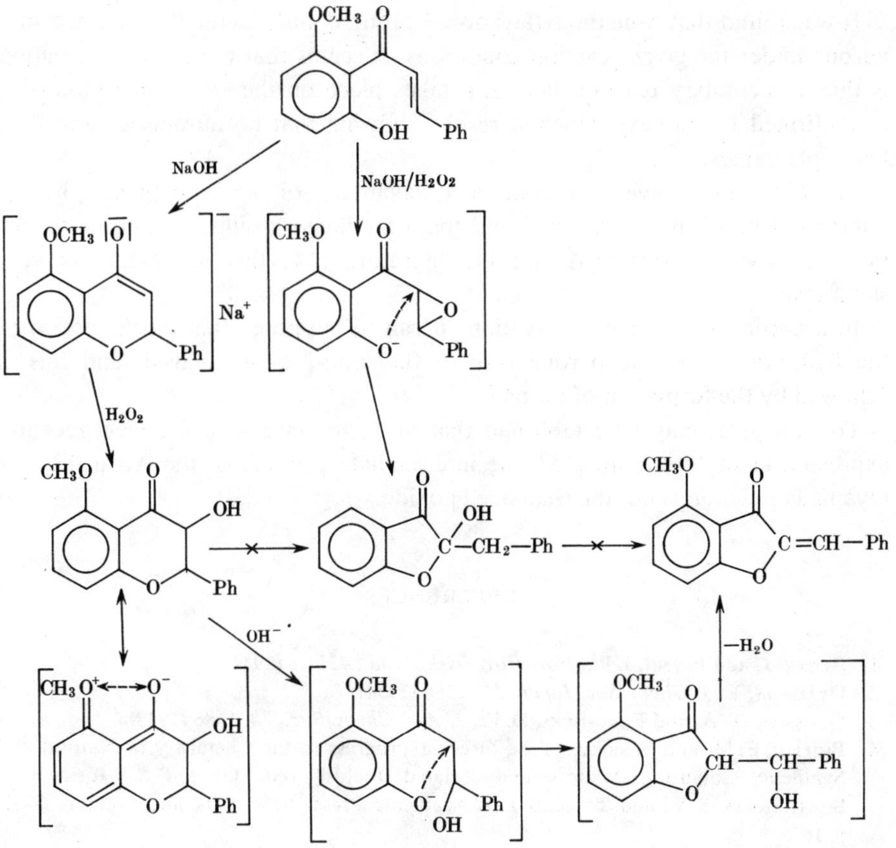

Fig. 13

A spectrophotometric study of the oxidation reaction of 2′-hydroxy-6′-methoxychalcone and the analogous flavanone has shown that the formation of the chalcone ⇌ flavanone equilibrium is accompanied here by structural changes different from those occurring in the case of 2′-hydroxychalcone. Namely, no new band characteristic of the phenolates appears, but a hypsochromic shift is observed, and the final compounds of the reaction mixture are formed in a fast and exothermic reaction. These facts indicate that the equilibrium may be shifted in the direction of enolized flavanone. This form may become stabilized by the methoxyl substituent.

In further studies we attempted to elucidate why aurone and flavanon-3-ol are formed in the course of the reaction. For the mechanism several propositions [3] have been made (Fig. 13).

It was found that 5-methoxyflavanon-3-ol in alkaline media is converted into aurone under the given reaction conditions. It seems that the aurone formation is thus a secondary reaction here and takes place on the effect of a base; this is confirmed by our experimental results showing that no aurone is formed at lower pH values.

The different behaviour of 5-methoxyflavanon-3-ol can be explained by the interaction between the carbonyl and the 5-methoxyl groups. According to our previous investigations [13] and the literature [14] this interaction is very significant.

In accordance with the observations it can be supposed that on the action of the hydroxyl anion the pyrone ring of flavanon-3-ol is cleaved, and this is followed by the formation of aurone.

To sum up it may be established that in accordance with the view recently expressed in the literature [15], the intermediate product of the Algar–Flynn–Oyamada reaction is not the chalcone epoxide.

REFERENCES

1. ALGAR, J. and FLYNN, J. P., *Proc. Roy. Irish Acad., 42B,* 1 (1934).
2. OYAMADA, T., *J. Chem. Soc. Japan, 55,* 1256 (1934).
3. GEISSMAN, T. A. and FUKUSHIMA, D. K., *J. Am. Chem. Soc., 70,* 1686 (1948).
4. PHILBIN, E. M. and WHEELER, T. S., "Recent Progress in the Chemistry of Natural and Synthetic Colouring Matters and Related Fields" (eds GORE, T. S., JOSHI, B. S., SUNTHANKAR, S. V. and TILAK, B. D.), Academic Press, New York and London, 1962, p. 167.
5. BOGNÁR, R. and STEFANOVSKI, J., *Tetrahedron, 18,* 143 (1962); *Magyar Kém. Foly., 68,* 296 (1962).
6. LITKEI, GY. and BOGNÁR, R., *Kémiai Közlemények, 34,* 249 (1970).
7. DEAN, F. M. and PODIMUANG, V., *J. Chem. Soc., 1965,* 3978.
8. ENEBÄCK, E., *Soc. Sci. Fennica Comm. Phys. Math., 28,* 1 (1963).
9. WIBERG, K. B. and GEER, R. D., *J. Am. Chem. Soc., 87,* 5202 (1965).
10. SPEAKMAN, T. P. and WATERS, W. A., *J. Chem. Soc., 1955,* 40.
11. MARSHALL, B. A. and WATERS, W. A., *J. Chem. Soc., 1960,* 2392.
12. PELTER, A., BRADSHAW, J. and WARREN, R. F., *Phytochemistry, 10,* 835 (1971).
13. DINYA, Z. and LITKEI, GY., *Acta Chim. (Budapest), 75,* 161 (1973).
14. SABATA, B. U. and ROUT, M. K., *J. Ind. Chem. Soc., 41,* 74 (1964).
15. GORMLEY, T. R. and O'SULLIVAN, W. J., *Tetrahedron, 29,* 369 (1973).

THE CONVERSION OF FLAVANONES INTO BENZOXAZEPINONES

by

A. LÉVAI and R. BOGNÁR

Institute of Organic Chemistry, Kossuth Lajos University
Debrecen, Hungary

In the recent past, in the course of our research on flavonoids, their conversion into seven-membered heterocyclic compounds with two hetero atoms has been studied. These investigations make possible the synthesis of numerous biologically active compounds starting from flavonoids. In our present paper we report on the conversion of flavanones into 2,3-dihydro-2-phenyl-1,4-benzoxazepin-5(4*H*)-ones and some reactions of the latter.

One of the possible methods for the conversion of flavanones into benzoxazepinones is the Schmidt reaction. This reaction of flavanone was described by Krapcho and Turk in 1966 [1] and according to them 2,3-dihydro-2-phenyl-1,5-benzoxazepin-4(5*H*)-one was obtained. In 1970 the reaction was reinvestigated by Misiti and Rimatori [2] and their studies proved that the major product was 2,3-dihydro-2-phenyl-1,4-benzoxazepin-5(4*H*)-one. In the reaction mixture they also managed to detect the presence of 2,3-dihydro-2-phenyl-1,5-benzoxazepin-4(5*H*)-one (3 %) and the tetrazole derivative of 2,3-dihydro-2-phenyl-1,4-benzoxazepine (5 %), as by-products.

Table I
Spectral Data

IR, cm^{-1} (KBr pellet)	NMR, δ ($CDCl_3$)			
	NH	Aromatic	$-\overset{|}{C}H-Ph$	$-CH_2-$
IV: ν_{CO} = 1660	8.64	7.38–8.30m	5.73q	3.76m
V: ν_{CO} = 1650	8.52	6.85–8.24m	5.68q	3.75m
VI: ν_{CO} = 1660	8.48	6.88–8.24m	5.66q	3.76m

The Schmidt reaction of substituted chromanones was studied by Lockhart *et al.* [3, 4] and those of 1-thiochromanones by Wünsch *et al.* [5]; it was found that the yield of the theoretically predicted 1,4- and 1,5-benzoxazepinone or 1,4- and 1,5-benzothiazepinone isomers was dependent on the substitution pattern of the benzene ring.

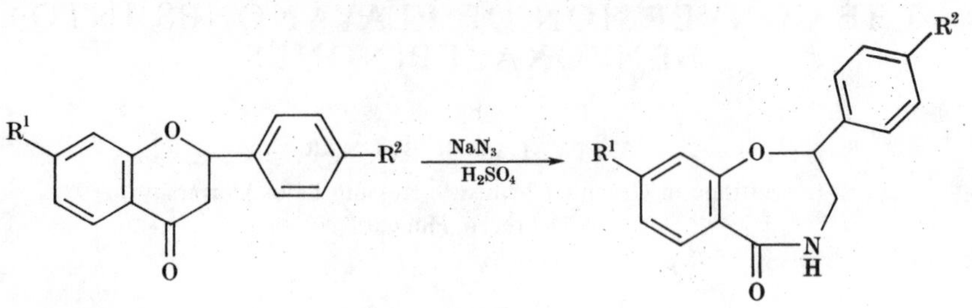

I: $R^1 = R^2 = H$
II: $R^1 = OCH_3$, $R^2 = H$
III: $R^1 = OCH_3$, $R^2 = Cl$

IV: $R^1 = R^2 = H$
V: $R^1 = OCH_3$, $R^2 = H$
VI: $R^1 = OCH_3$, $R^2 = Cl$

Fig. 1

Considering these results, we have started such investigations also on substituted flavanones. Compounds I, II and III were allowed to react with hydrazoic acid in acetic acid solution and the corresponding 2,3-dihydro-2-phenyl-1,4-benzoxazepin-5(4*H*)-ones (IV–VI) were obtained in all cases. The structures of the benzoxazepinones were elucidated by means of IR and NMR spectroscopy (Fig. 1).

The relationship between the substitution pattern of the flavanones and the structure of the benzoxazepinone isomers obtained from them is the subject of further studies which are in progress.

To our knowledge, thus far only few chemical conversions have been investigated in the field of benzoxazepinones. Krapcho and Turk [6] synthesized some N-substituted derivatives of these compounds. The reduction of benzoxazepinones was investigated by Misiti and Rimatori [2].

The object of our work in this field has been to study some novel reactions of benzoxazepinones and to synthesize new biologically active compounds in this way. Hitherto the following chemical transformations have been investigated.

Compounds IV, V and VI were allowed to react with phosphorus pentasulfide in pyridine solution, whereupon the corresponding 2,3-dihydro-2-phenyl-1,4-benzoxazepin-5(4*H*)-thiones (VII–IX) were obtained (Fig. 2).

Since N-substituted benzothiazepinones of similar structure were found to have a considerable activity on the central nervous system, the synthesis of N-substituted benzoxazepinones has been now studied. All active benzothiazepinones were N-aminoalkyl derivatives but, on the basis of pharmacological results obtained with flavonoids, it seemed expedient to synthesize some acid derivatives of benzoxazepinones as well.

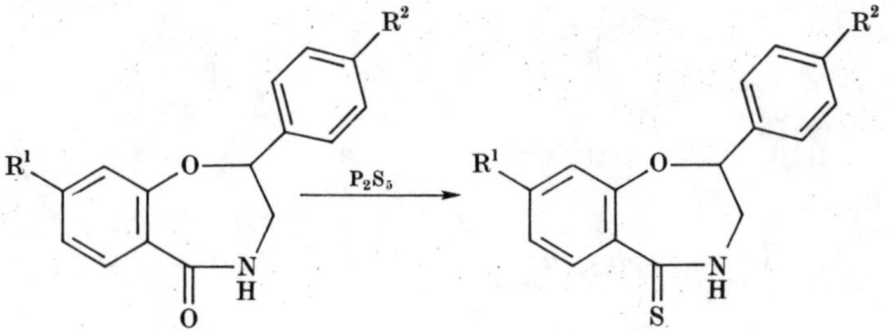

R^2 R^1 O N H O P_2S_5 R^2 R^1 O N H S

IV: $R^1=R^2=H$
V: $R^1=OCH_3$, $R^2=H$
VI: $R^1=OCH_3$, $R^2=Cl$

VII: $R^1=R^2=H$
VIII: $R^1=OCH_3$, $R^2=H$
IX: $R^1=OCH_3$, $R^2=Cl$

Fig. 2

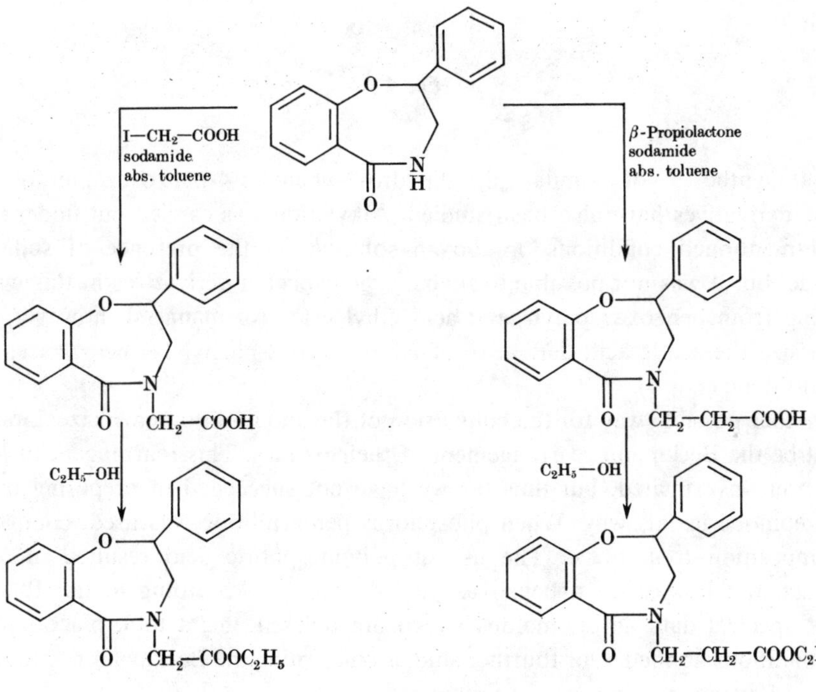

O N H O

$I-CH_2-COOH$ sodamide abs. toluene

β-Propiolactone sodamide abs. toluene

O N O CH_2-COOH

O N O CH_2-CH_2-COOH

C_2H_5-OH

C_2H_5-OH

O N O $CH_2-COOC_2H_5$

O N O $CH_2-CH_2-COOC_2H_5$

Fig. 3

2,3-Dihydro-2-phenyl-1,4-benzoxazepin-5(4*H*)-one was allowed to react with iodoacetic acid or β-propiolactone to obtain N-acetic acid or N-propionic acid derivatives, which yielded ethyl esters on esterification with ethanol. Starting with these esters, we intend to prepare some amide derivatives (Fig. 3).

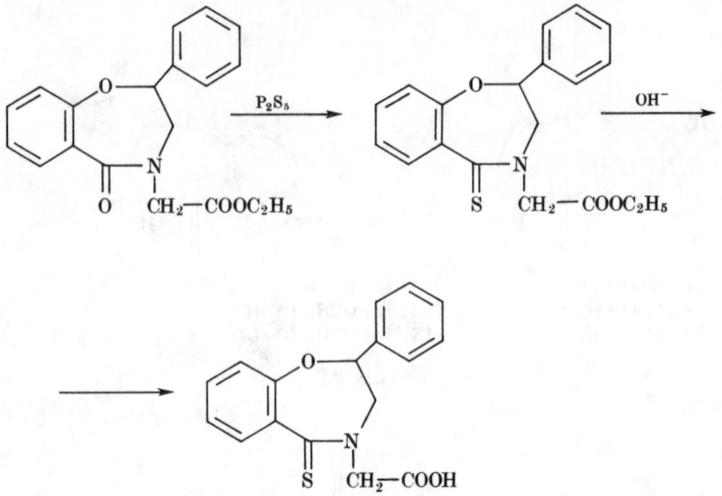

P_2S_5 → OH⁻ →

O, N, CH_2—$COOC_2H_5$; S, CH_2—$COOC_2H_5$; S, CH_2—COOH

Fig. 4

The syntheses of similar 2,3-dihydro-2-phenyl-1,4-benzoxazepin-5(4*H*)-thione derivatives have also been studied. Alkylation was carried out under the above-mentioned conditions, in dioxan solution, in the presence of sodium hydride, but it was not possible to prepare the expected derivatives in this way. Starting from benzoxazepinylacetic acid ethyl ester we managed, however, to synthesize the acetic acid derivative of 2,3-dihydro-2-phenyl-1,4-benzoxazepin-5(4*H*)-thione (Fig. 4).

Another possible way for the conversion of flavanones into benzoxazepinones might be the Beckmann rearrangement of their oximes. This rearrangement has also been investigated, but thus far we have not succeeded in preparing benzoxazepinone in this way. When phosphorus pentachloride was used, complete decomposition took place. The use of polyphosphoric acid resulted in one product, but it was not a benzoxazepine derivative. According to its IR and NMR spectral data an anomalous Beckmann rearrangement took place under the conditions applied. Our fourther studies concerning the Beckmann rearrangement of flavanone oximes are in progress.

*

The authors' thanks are due to Mrs. E. Hajnal for her careful assistance in the experiment work. The present study was sponsored by the Hungarian Academy of Sciences for which our gratitude is expressed.

REFERENCES

1. KRAPCHO, J. and TURK, CH. F., *J. Med. Chem., 9,* 191 (1966).
2. MISITI, D. and RIMATORI, V., *Tetrahedron Letters, 1970,* 947.
3. HUCKLE, D., LOCKHART, I. M. and WRIGHT, M., *J. Chem. Soc., 1965,* 1137.
4. EVANS, D. and LOCKHART, I. M., *J. Chem. Soc., 1965,* 4806.
5. WÜNSCH, K. H., STAHNKE, K. H. and EHLERS, A., *Chem. Ber., 103,* 2302 (1970).
6. KRAPCHO, J. and TURK, CH. F., U. S. Pat. 3,309,361; *C.A., 68,* 2930 (1968).

INTERCONVERSION OF THE NEOFLAVANOIDS AND OF THE ISOFLAVONOIDS

by

D. M. X. DONNELLY, Ph. J. KAVANAGH and P. J. M. GUNNING

Department of Chemistry, University College
Dublin, Ireland

I. Introduction

In the preceding century the heartwood of the Dalbergeae (Leguminosae-Lotoideae) was the principal source for the beautiful and durable rosewoods of commerce. The geographical distribution of the genus *Dalbergia* is influenced by the uniformly woody character of its species and by the extensive time lag before they attain the reproductive age. A few species are sufficiently maritime to be accidently carried across oceans (an example, *Dalbergia ecastophyllum*). In the last decade a phytochemical examination of some species indigenous to Africa, Asia and Brazil was completed. Many *Dalbergia* had the expected isoflavanoid-type compounds (C_6-C_2-C_6 skeleton) but the association of the C_6-C_1-C_6 skeleton (neoflavanoid class) with the *Dalbergia* and its close relative *Machaerium* and, as shown more recently, with some genera of Caesalpinioideae is now recognized and may have considerable taxonomic significance.

II. Neoflavanoids

Interestingly, the elucidation of a structure with a C_6-C_1-C_6 skeleton was first elaborated for the coumarin, calophyllolide (I), isolated from *Calophyllum inophyllium* (Guttiferae) [1]. To date, twenty-six 4-phenylcoumarins have been isolated from members of the genera *Calophyllum, Mesua* and *Mammea* [1, 2].

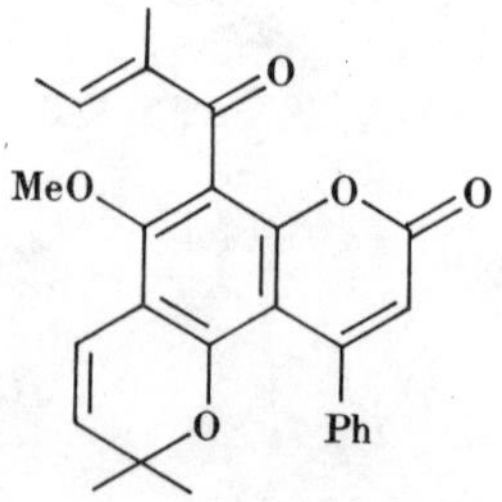

MeO O O O O Ph

I

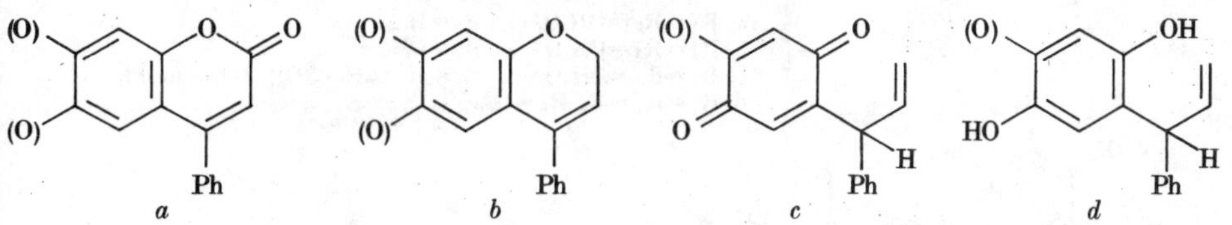

Fig. 1

The term neoflavanoid [3] was coined at a later date to describe the compounds isolated from the Leguminosae (Fig. 1) and includes, besides the 4-phenylcoumarins *(a)* and neoflavenes *(b)*, the open-chain quinones *(c)* and the 3,3-diarylpropenes *(d)*. The inclusion of the latter open-chain structures is in line with the assignment of 2′-hydroxychalcones and angolensin to the flavonoid and isoflavonoid classes, respectively.

The isolation from the *Dalbergia* heartwoods of closely related neoflavanoids in minor quantities and often of similar configuration emphasized the advantages of availability of a system of chemical interconversions.

1. Interconversions

High-yield biological-type conversions have been achieved in the neoflavanoid class [4]. The series includes 3,3-diarylpropenes (dalbergiquinols) (II), the dalbergiones (III), the neoflavenes (IV) and the corresponding coumarins (V). A schematic summary of interconversions and the reagents used is given in Fig. 2. The possible further oxidation beyond the 4-phenylcoumarin to the benzophenone (IX) has been proposed [5] and examples of its occurrence in the *Dalbergia* have been observed [5, 6].

Attempts to examine the proposed pathways (bio-oxidative or bio-reductive) of interconversion *in vivo* of these neoflavanoids and their biosynthesis by the technique of labelling has proved unsuccessful, as these compounds are apparently absent in the young plant.

Ollis and his associates [3] have demonstrated the reversible oxidation–reduction of a dalbergiquinol (IIa) to a dalbergione (IIIa) and subsequently reported [7] the isomerization of the quinone (IIIb) to the neoflavene, kuhlmannene (IVb, R′ = OH) by chromatography on neutral alumina. We have shown [4] that the dalbergiones could be obtained from the corresponding quinols under non-aqueous conditions with dichlorodicyanoquinone (DDQ) and that neutral alumina is unsatisfactory for the cyclization of 4-methoxydalbergione (IIIa) and 3′-acetoxy-4,4′-methoxydalbergione. The cyclization of the dalbergiones (IIIa–c) was achieved with N,N-dimethylaminopyridine in chloro-

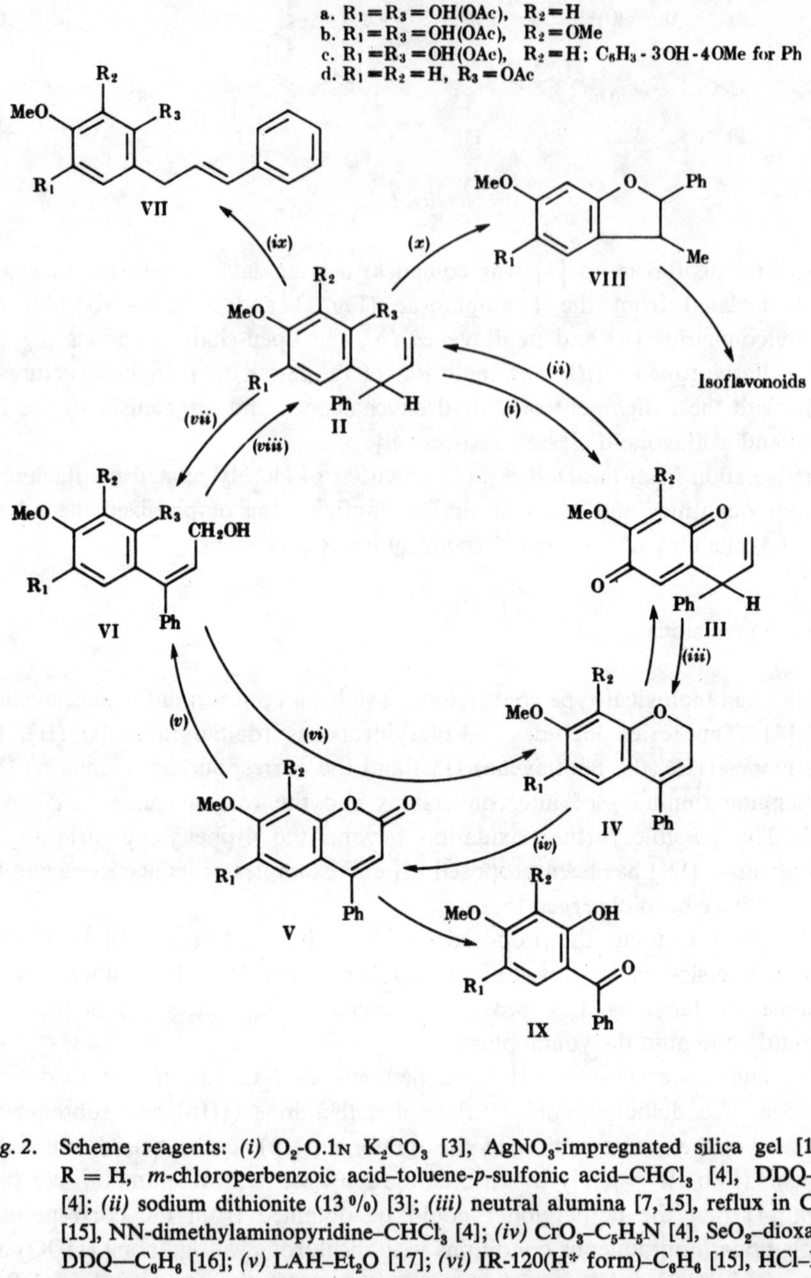

a. $R_1 = R_3 = OH(OAc)$, $R_2 = H$
b. $R_1 = R_3 = OH(OAc)$, $R_2 = OMe$
c. $R_1 = R_3 = OH(OAc)$, $R_2 = H$; C_6H_3 - 3OH - 4OMe for Ph
d. $R_1 = R_2 = H$, $R_3 = OAc$

Fig. 2. Scheme reagents: *(i)* O_2-O.1N K_2CO_3 [3], $AgNO_3$-impregnated silica gel [14], if R = H, *m*-chloroperbenzoic acid–toluene-*p*-sulfonic acid–$CHCl_3$ [4], DDQ–C_6H_6 [4]; *(ii)* sodium dithionite (13 %) [3]; *(iii)* neutral alumina [7, 15], reflux in C_5H_5N [15], NN-dimethylaminopyridine–$CHCl_3$ [4]; *(iv)* CrO_3–C_5H_5N [4], SeO_2–dioxan [4], DDQ—C_6H_6 [16]; *(v)* LAH–Et_2O [17]; *(vi)* IR-120(H^+ form)–C_6H_6 [15], HCl–EtOH [4]; *(vii)* LAH–$AlCl_3$–Et_2O [17], $SnCl_2$–HOAc [17], Zn–HCl–HOAc [17], DMF–Zn–HCl [17]; *(viii)* $Hg(OAc)_2$–HOAc [4], SeO_2–HOAc [4]; *(ix)* BF_3 [18], HCO_2H [13]; *(x)* 220 °C 30 min [13]

form at room temperature and gave quantitative yields of the neoflavenes, dalbergichromene (IVa), kuhlmannene (IVb) and melanene (IVc).

A selection of oxidizing agents was examined for the conversion of the chromenes (IVa–d) to the 4-arylcoumarins (Va–d). Selenium dioxide in refluxing dioxan gave 50 % yield and a similar result was achieved with CrO_3 in pyridine at 21 °C. The yields were 95 % in all cases with CrO_3 in pyridine at 55 °C for 5 hr. The neoflavene (IVa, R_1 = OAc) was also oxidized to the coumarin (Va, R_1 = OAc) by DDQ.

The 3,3-diarylpropenes (IIa–c) contain a hydroxyl group at C-5 essential for quinone formation and subsequent isomerization to the neoflavene. The latter reaction is mechanistically similar to the ubiquinone–ubichromenol transformation [8] (Fig. 3).

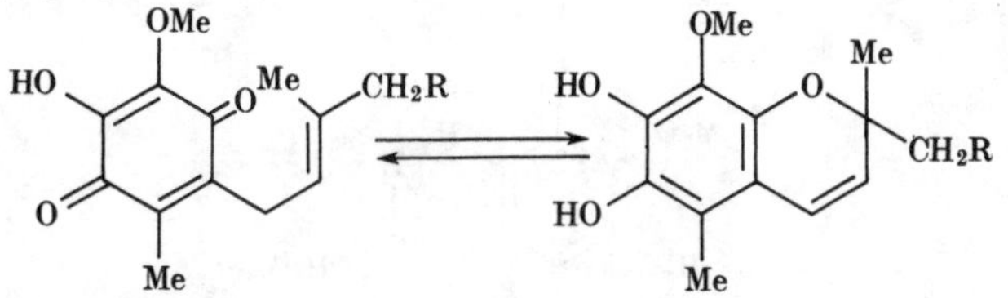

OMe, HO, O, Me, Me, CH_2R ⇌ OMe, Me, HO, O, CH_2R, HO, Me

Fig. 3

As a C-6 hydroxyl group is absent in the 4-phenylcoumarins which have been isolated so far from the genera of Guttiferae, a more general route to their synthesis *via* allylic alcohols, allylic acids or epoxides, was examined. The open-chain, α,β-unsaturated acid, for example, calophyllic acid (X), was demonstrated by labelling experiments to be the possible precursor of the 4-phenylcoumarin inophyllolide (XI) [9].

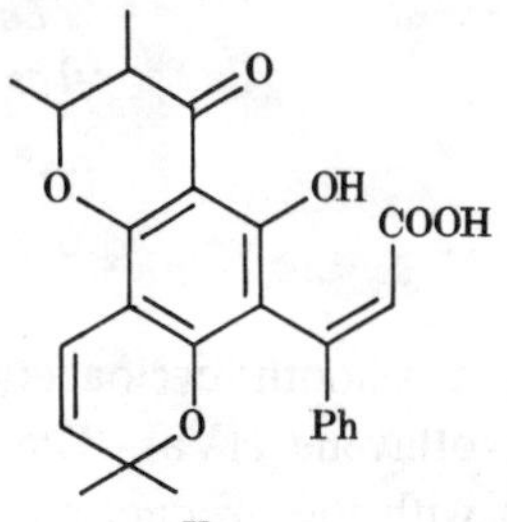

O, O, OH, COOH, O, Ph — X   O, O, O, O, O, Ph — XI

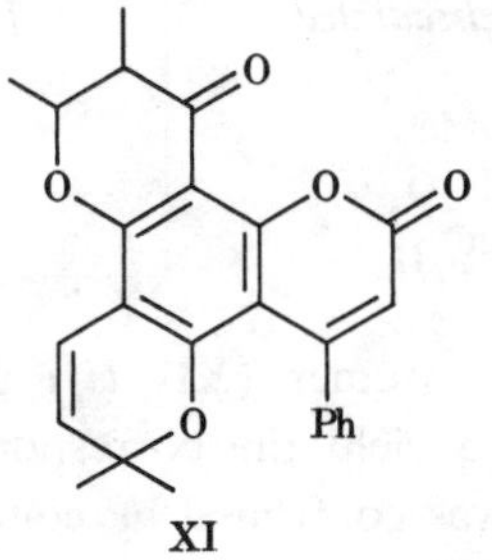

The oxidation of dalbergiquinol diacetates (II, $R_1 = R_3$ = OAc) by equimolar amounts of Hg $(OAc)_2$ in acetic acid at reflux temperature gave total conversion of 60 %, from which the allylic alcohols were identified (as their acetates). The formation of the allylic acetates (XII *(i), (ii)*) may be explained as proceeding *via* a concerted S_Ei' reaction mechanism [10] (Fig. 4).

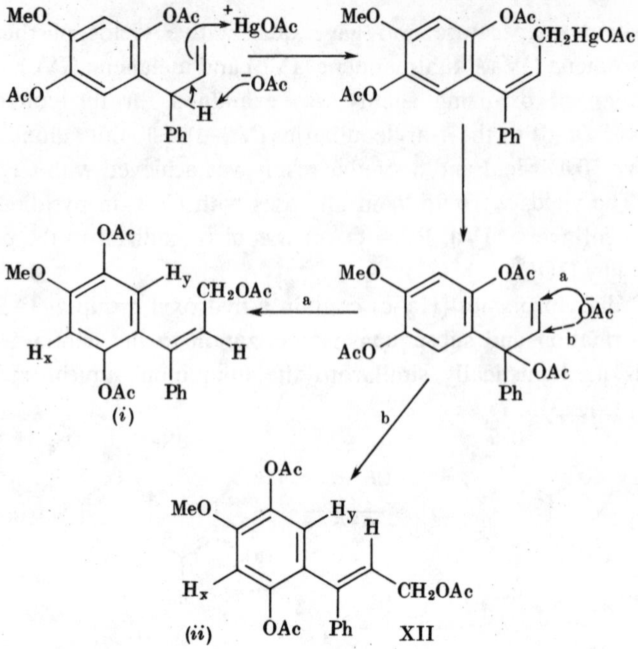

Fig. 4

The configurations *(i)* and *(ii)* were assigned by application of intramolecular nuclear Overhauser effects (NOE) (Table I).

Table I

NOE Measurements on 3–(2,5–Diacetoxy–4–methoxyphenyl)–3–phenylallyl Acetates (XII (*i*) and (*ii*)) [4].

Protons irradiated	*Proton observed*	*% Enhancement* (*i*)	(*ii*)
OMe	Hx	30	28
CH_2O	Hy	12	0
>C=C<H	Hy	0	18

The major isomer (XII *(i)*) underwent smooth cycloacetylation and re-acetylation to yield the corresponding neoflavene (IVa). The configurational assignment was confirmed by comparison with the product from the reduction (LAH) and subsequent acetylation of dalbergin (Va). The allylic alcohol VI was also obtained by SeO_2 oxidation of the quinols II.

The alternative method of oxidation of the 3,3-diarylpropenes *via* epoxides gave low yields. The major isomer (*(ii)* Fig. 5) is that formed by approach of the *m*-chloroperbenzoic acid to the less hindered side of the double bond. The epoxide did not undergo rearrangement to the allylic alcohol [11].

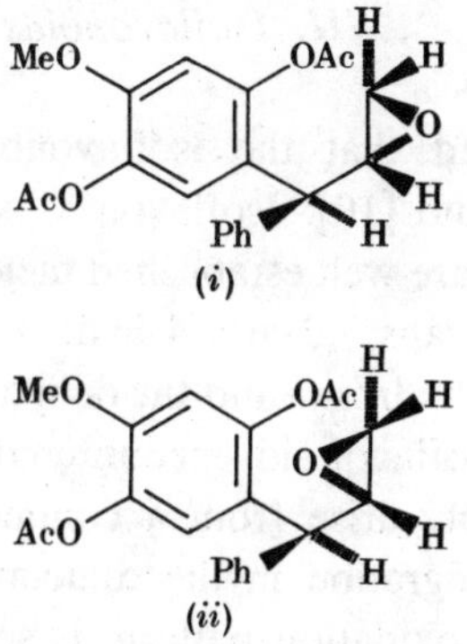

(i)

(ii)

Fig. 5

2. Rearrangements

A benzylstyrene (VII) and a dihydrobenzofuran (VIII) are depicted as offshoots of the main cycle in Fig. 2. The benzylstyrenes arise on acid rearrangement of 3,3-diarylpropenes, whilst a thermal rearrangement by a cyclic intramolecular process leads to a 2-phenyl-3-methyldihydrobenzofuran (Fig. 6). This

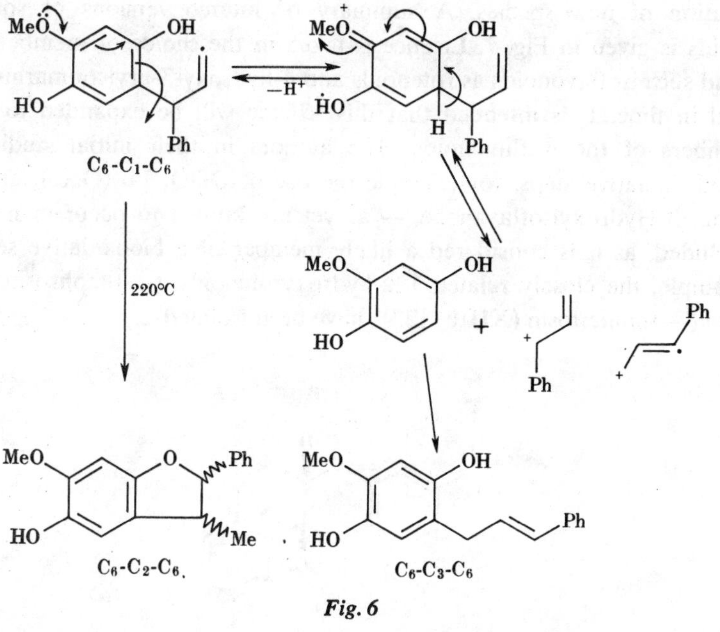

Fig. 6

dihydrobenzofuran can, in turn, be converted to give a coumestone — a member of the isoflavanoid class (Figs 2 and 7). Extensive work on these types of rearrangements has been carried out by Schmid *et al.* [12] and Jurd *et al.* [13].

III. *Isoflavonoids*

It is generally accepted that the isoflavonoids arise by the polyketide–chalcone–isoflavanone route [19]. Isoflavones, isoflavanones, rotenoids, pterocarpans and coumestones are well established members of this class. More recent additions include isoflavans, 3-aryl-4-hydroxycoumarins, 3-arylcoumarins, coumaranochromones, 6a-hydroxy- and the dehydro-pterocarpans. Undoubtedly, the different classes of isoflavonoid encountered in nature are closely related biogenetically, and possibly arise from a common precursor whose source is the result of a 1,2-aryl migration in the chalcone. In the *Dalbergia* a striking similarity, with regard to oxidation pattern, is seen in the isoflavonoids. However, an ordered pathway for their biosynthesis has not been established. The origin of the 2′-oxygenation function implicit in the ring systems of pterocarpans, coumestones and rotenoids is unproved. Two proposals suggested are *(i)* hydroxylation at C-2′ *via* attack of OH [20] and *(ii)* oxidative ring cyclization involving the carbonyl oxygen [21].

As in the case of the neoflavanoids, the availability of a system of chemical interconversion of the isoflavanoids was considered of use in a phytochemical examination of new species. A summary of interconversions of some isoflavonoids is given in Fig. 7. Licence is taken in the choice of members in the cycle and such isoflavonoids as rotenoids and 4-hydroxy-3-arylcoumarins will be included in time. It is intended that this scheme will be expanded to include all members of the isoflavonoids. The authors in their initial studies have examined oxidative steps, for example the use of DDQ, Pb $(OAc)_4$ and autooxidation. 3-Hydroxyisoflavanone — as yet not known to occur in nature — was included, as it is considered a likely member of a biooxidative sequence. For example, the closely related 12a-hydroxyrotenoids, (—)-tephrosin (XIIIa) [22] and (—)-millettosin (XIIIb) [23], have been isolated

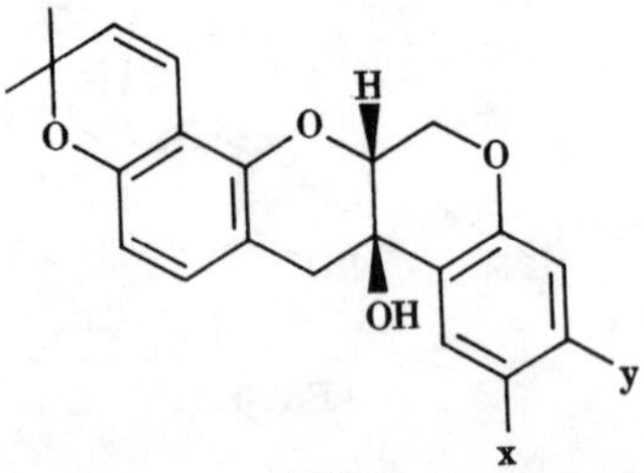

XIII

a. x=y=OMe : (–)-tephrosin
b. x, y=O·CH_2O : (–)-milletosin

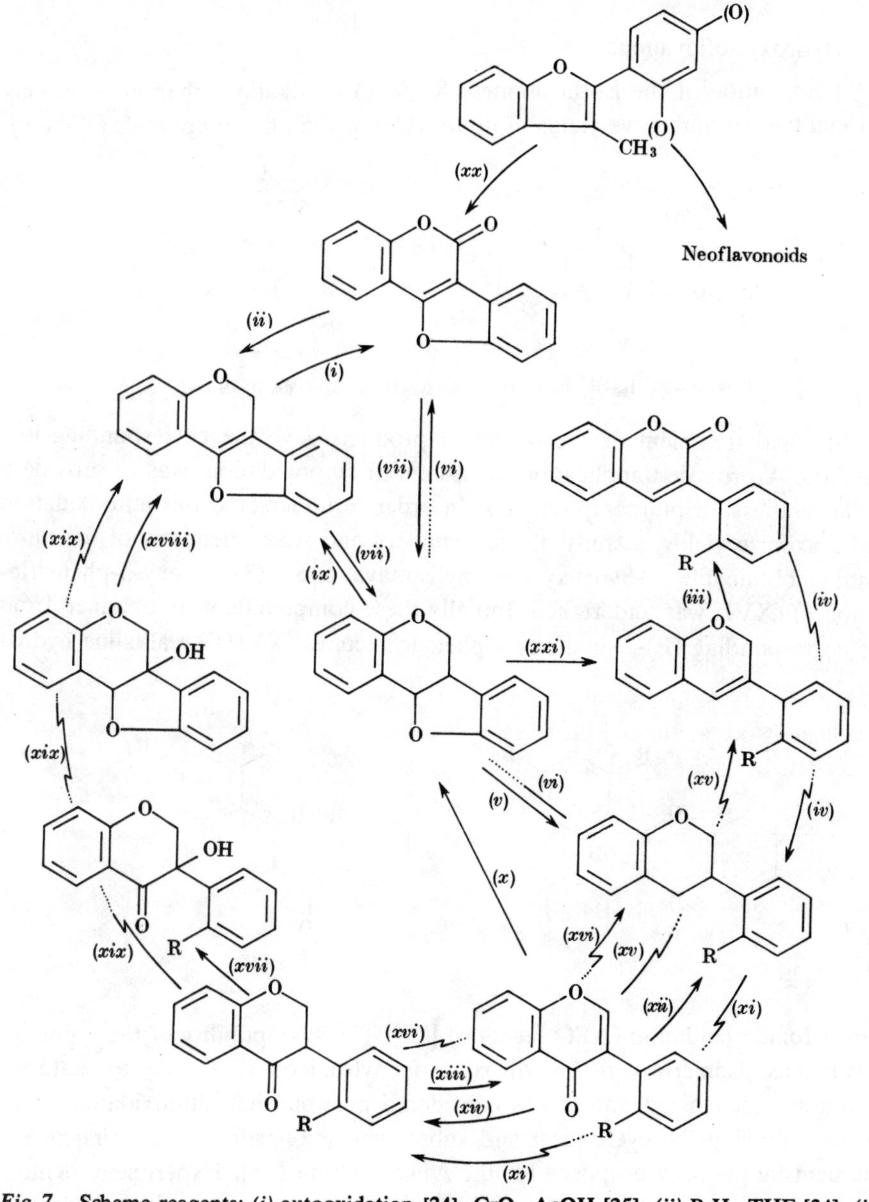

Neoflavonoids

Fig. 7. Scheme reagents: *(i)* autooxidation [24], CrO_3–AcOH [25]; *(ii)* B_2H_6–THF [24]; *(iii)* MnO_2–$CHCl_3$ [26]; *(iv)* LAH–BF_3 [27]; *(v)* H_2 [28]; *(vi)* if R = OH, DDQ–C_6H_6 [16]; *(vii)* Na–EtOH [29]; *(viii)* H_2 [30]; *(ix)* $Pb(OAc)_4$–C_6H_6 [31]; *(x)* $NaBH_4$–EtOH [32]; *(xi)* if R = H, DDQ–C_6H_6 [16], $KMnO_4$ [33], CrO_3 [34]; *(xii)* H_2 [35]; *(xiii)* DDQ–C_6H_6 [16], SeO_2, MnO_2–HOAc–H_2SO_4 [36]; *(xiv)* H_2 [34]; *(xv)* if R = OMe, KBH_4 [37]; *(xvi)* H_2 [28b, 34]; *(xvii)* O_2–NaOH [38]; *(xviii)* H* [39]; *(xix)* if R = OAc, OH, HCl–EtOH [40]; *(xx)* SeO_2–AcOH [16]; *(xxi)* H* [41]

1. 3-Hydroxyisoflavanone

Autooxidation of the isoflavanones (XIVa–c) in alkaline ethanolic solutions at room temperature gave yields of up to 61 % of the title compounds (XVa–c).

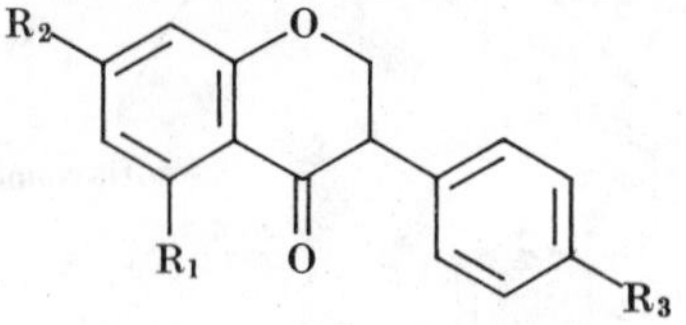

XIV 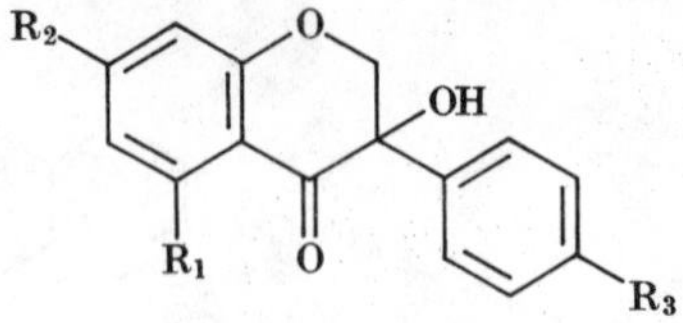 XV

a. $R_1=R_2=R_3=H$ b. $R_2=OMe$, $R_1=R_3=H$ c. $R_2=OMe$, $R_1=R_2=H$

Mild acid treatment of the oxidation products gave the corresponding isoflavones. A proposed application of this aerial autooxidation was to provide a facile synthesis of pterocarp-6a-enes. In order to understand this autooxidation and maximize yields, a study of the chemistry and stereochemistry of the more readily obtainable 3-hydroxy-2-phenylisoflavanones (3-hydroxy-3-phenylflavanones) (XVI) was undertaken. Initially these compounds were obtained from the corresponding (E)-2′-hydroxy-α-phenylchalcones (XVII) by alkaline hydro-

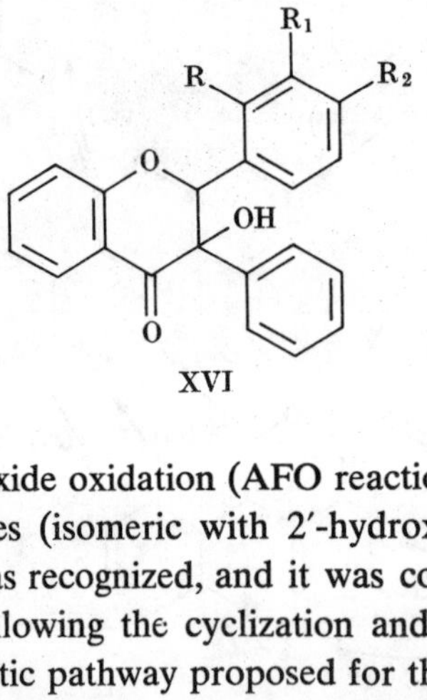

XVI 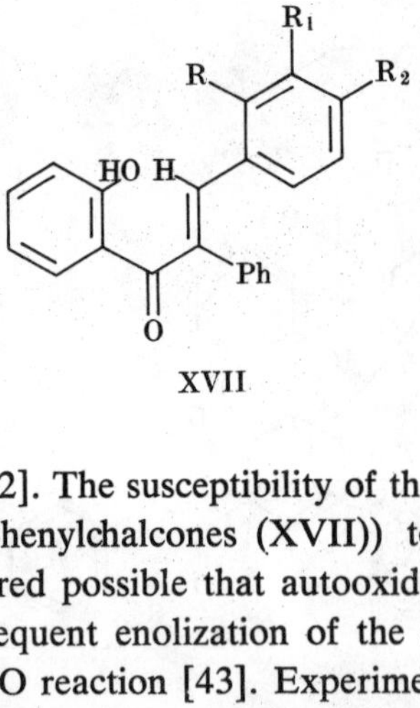 XVII

gen peroxide oxidation (AFO reaction) [42]. The susceptibility of the 3-phenylflavanones (isomeric with 2′-hydroxy-α-phenylchalcones (XVII)) to autooxidation was recognized, and it was considered possible that autooxidation could occur following the cyclization and subsequent enolization of the chalcone, a mechanistic pathway proposed for the AFO reaction [43]. Experiment demonstrated that direct treatment of the (E)-chalcone (XVII) with sodium hydroxide gave isomeric *trans*-3-phenylflavanone. However, with the addition of ethanol and oxygen the products were identical with that of the AFO reaction. Results of comparative studies (AFO, autooxidation) are given in Tables II and III.

Product dependence on pH was evident, and some results are tabulated in Table IV. The pH in the AFO reaction was 13.2.

Table II

% Yield of 3-Hydroxy-3-phenylflavanones

Chalcone XVII	AFO	Autooxidation (air)
$R=R_1=R_2=H$	68	53
$R=OMe; R_1=R_2=H$	20	42
$R_1=R_2=OMe; R=H$	14	51

Table III

% Yield of 3-Hydroxy-3-phenylflavanones on Autooxidation

XVII	Reaction time (hr)	O_2	Air	N_2
$R=R_1=R_2=H$	24	73	53	0
$R=OMe, R_1=R_2=H$	96	24.5	42	0
$R_1=R_2=OMe, R=H$	24	55	51	0

Table IV

Product Dependence on pH (% Yield)

pH	8	9	10	11
(E)-2-Hydroxy-α-phenylchalcone	84	60	0	14.5
2,3-*trans*-3-Phenylflavanone	16	40	100	53.6
3-Hydroxy-3-phenylflavanone	0	0	0	31.9

Autooxidation at a benzylic position adjacent to a carbonyl has been observed as a side reaction, e.g. in reactions on indan-1-one [44] and 2-acetylcoumaranones [45], and as the major product when (±)-deguelin is aerated in alkaline solution [23]. Autooxidation is considered to proceed *via* pathway *a* or *b* as shown in Fig. 8.

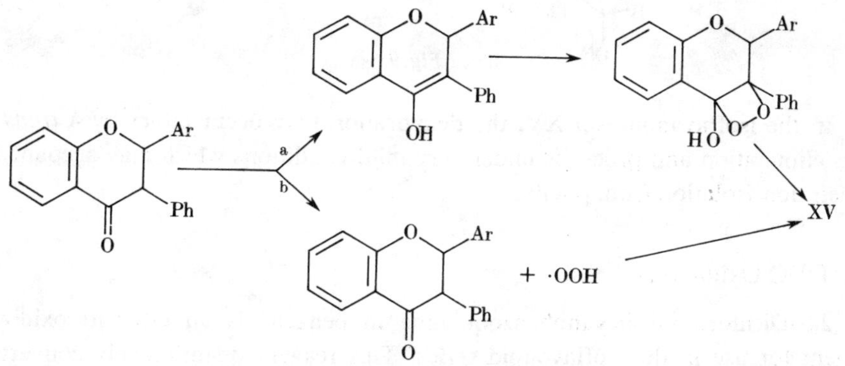

Fig. 8

The evidence used in the assignment of a stereochemistry to position 3 included analysis of the NMR spectra, dehydration experiments, and a synthesis of 3-hydroxy-3-phenylflavanone (XVI) by an unambiguous route. The compound XVI failed to dehydrate in experiments involving heat and acid, implying a 2-H, 3-OH *cis* relationship. *Cis*-dehydration has been demonstrated for the 12a-hydroxyrotenoids (XIIIa, b) but a Dreiding model of the isoflavanone (XVI) shows a steric inhibition to alignment of the 2-H and 3-OH groups. The synthesis of the isoflavanone XVI is summarized in Fig. 9 [38].

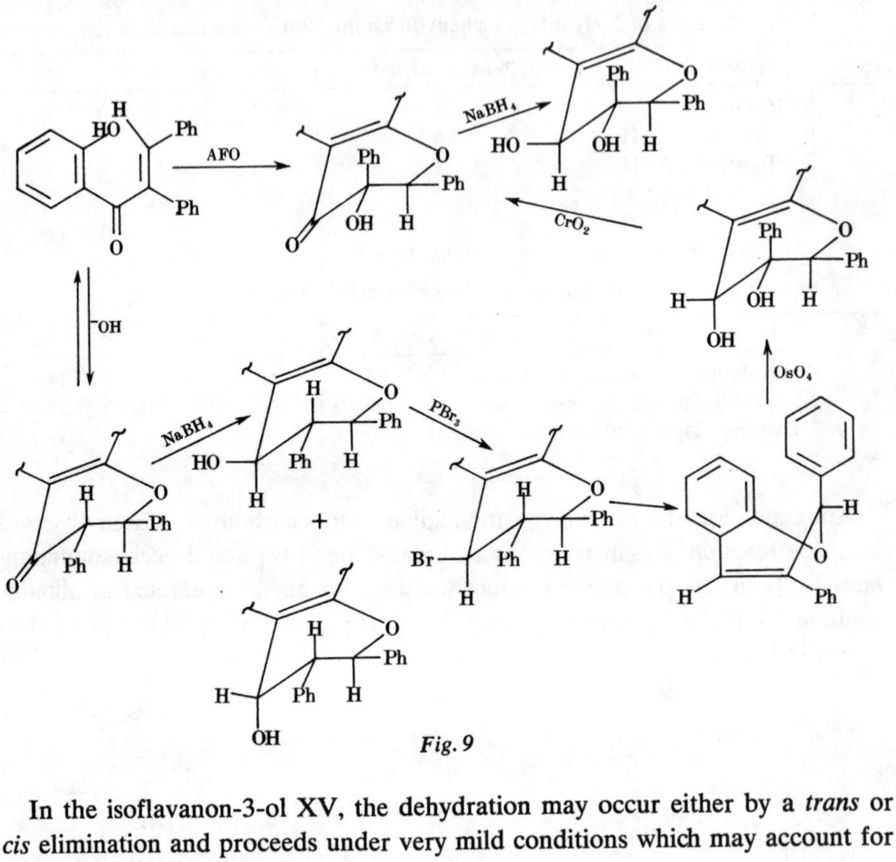

Fig. 9

In the isoflavanon-3-ol XV, the dehydration may occur either by a *trans* or *cis* elimination and proceeds under very mild conditions which may account for their non-isolation from plants.

2. DDQ Oxidation

2,3-Dichloro-5,6-dicyanobenzoquinone in benzene is an efficient oxidizing agent for use in the isoflavonoid series. This reagent quantitatively converts a 2′-hydroxyisoflavan to the corresponding coumestone *via* the pterocarpan; however, an isoflavan without a hydroxyl at the 2′-position affords an isoflavanone. An additional mole of DDQ can give rise to the isoflavone [16]. In an attempt to synthesize the pterocarp-6a-ene from a pterocarpan this reagent was unsuccessful as it led directly to the coumestone; the less active reagent chloranil gave no reaction, but success was achieved with Ph $(OAc)_4$ [31].

IV. Interconversion of Neoflavanoids and Isoflavonoids

The co-occurrence of neoflavanoids and isoflavonoids has been observed many times in *Dalbergia* species, but no views have been proposed to account for a possible inter-relationship. It is possible to postulate such a scheme (Fig. 10).

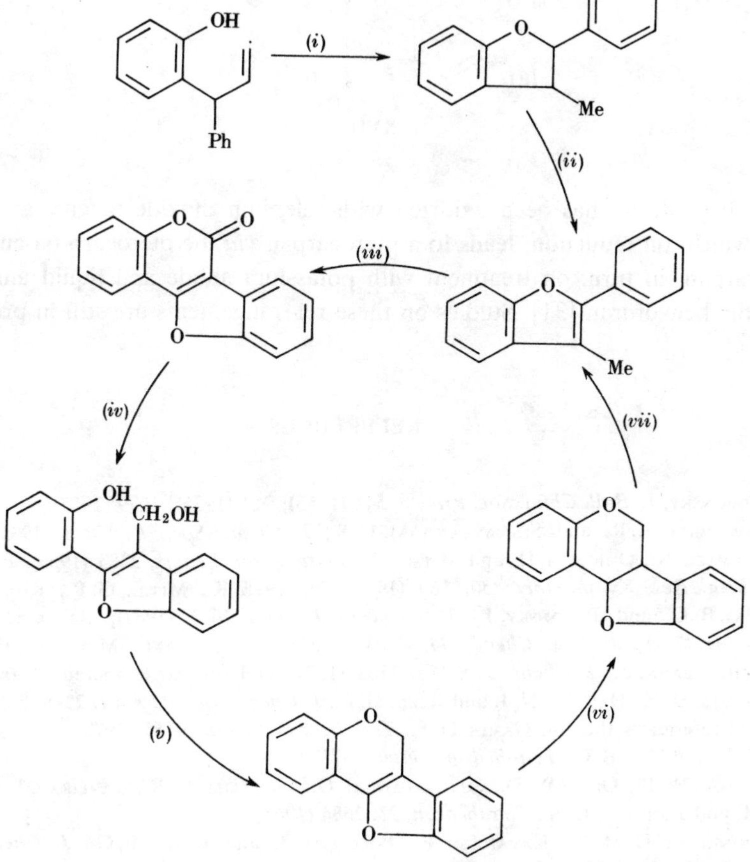

OH
Ph
(*i*)
O
Me
(*ii*)
O
O
O
(*iii*)
O
Me
(*iv*)
(*vii*)
OH
CH_2OH
O
O
O
(*v*)
O
O
(*vi*)

Fig. 10. Scheme reagents: *(i)* NN-Diethylaniline [12]; *(ii)* Pd/C–decalin [46]; *(iii)* SeO_2–HOAc [16]; *(iv)* LAH [25]; *(v)* diethylene glycol–heat [25]; *(vi)* H_2/Pd [30a]; *(vii)* KNH_2–liquid ammonia [41]

The thermal rearrangement of a dalbergiquinol (III) gives a dihydrobenzofuran (VIII) [12, 13] which may be dehydrogenated (Pd/C; DDQ) to afford the benzofuran, a key compound in the cycle linking isoflavonoids and neoflavanoids. An example of dehydrogenation was demonstrated for the structural proof of melanoxin (XVIII) a major component of the heartwood extractive of *Dalbergia melanoxylon*, Guill et Perr [46]. Conversion of the benzofuran to the allylic alcohol would represent the transformation of neoflavanoids to isoflavonoids, as the alcohol may be cyclized to the pterocarpene.

OH
OMe
H
MeO
O
HO
H
CH_3
XVIII

The benzofuran has been oxidized with selenium dioxide to give a coumestone which, on reduction, leads to a pterocarpan *via* the pterocarp-6a-ene. The pterocarpan, in turn, on treatment with potassium amide and liquid ammonia gives the benzofuran [31]. Studies on these rearrangements are still in progress.

REFERENCES

1. Polonsky, J., *Bull. Chim. Soc. France,* 541 (1955); 914 (1956); 1079 (1957).
2. Somanathan, R. and Sultanbawa, M. U. S., *J. Chem. Soc., Perkin I,* 1935 (1972); Kawazu, K., Ohigashi, H. and Mitsui, T., *Tetrahedron Letters,* 2383 (1968); *Bull. Inst. Chem. Res. Kyoto Univ., 50,* 160 (1972); Nigam, S. K., Mitra, G. R., Kunesch, G., Das, B. C. and Polonsky, J., *Tetrahedron Letters,* 2633 (1967); Breck, G. D. and Stout, G. H., *J. Org. Chem., 34,* 4203 (1969); Carpenter, I., McGarry, E. J. and Scheinmann, F., *J. Chem. Soc. (C),* 3783 (1971) and references therein; Crombie, L., Games, D. E., Haskins, N. J. and Reed, G. F., *J. Chem. Soc., Perkin I,* 2248, 2255 (1972) and references therein; Games, D. E., *Tetrahedron Letters,* 3187 (1972); Chakraborty, D. P. and Das, B. C., *Tetrahedron Letters,* 5727 (1966).
3. Eyton, W. B., Ollis, W. D., Sutherland, I. O., Gottlieb, O. R., Taveira Magalhães, M. and Jackman, L. N., *Tetrahedron, 21,* 2684 (1965).
4. Donnelly, D. M. X., Kavanagh, P., Polonsky, J. and Kunesch, G., *J. Chem. Soc., Perkin I,* 965 (1973).
5. Eyton, W. B., Ollis, W. D., Fineberg, M., Gottlieb, O. R., Salignac de Souza Guimarães, I. and Taveira Magalhães, M., *Tetrahedron, 21,* 2697 (1965).

6. Donnelly, D. M. X. and O'Reilly, J., Unpublished results on isolates from *Dalbergia melanoxylon* Guill et Perr.
7. Ollis, W. D., Redman, B. T., Roberts, R. J., Sutherland, I. O. and Gottlieb, O. R., *Chem. Comm.*, 1392 (1968).
8. Isler, O., Langemann, A., Mayer, H., Rüegg, R. and Schudel, P., *Bull. Nat. Inst. Sci. India*, 132 (1961).
9. Gautier, T., Cavé, A., Kunesch, G. and Polonsky, J., *Experientia, 28,* 759 (1972).
10. Rappoport, Z., Dyall, L. K., Winstein, S. and Young, W. G., *Tetrahedron Letters,* 3485 (1970).
11. Bohlmann, F. and Buchmann, U., *Chem. Ber., 105,* 863 (1972).
12. Schmid, E., Fráter, Gy., Hansen, H. J. and Schmid, H., *Helv. Chim. Acta, 55,* 1625 (1972).
13. Jurd, L., Stevens, K. and Manners, G., *Tetrahedron, 29,* 2347 (1973).
14. Mageswaran, S., Ollis, W. D., Roberts, R. J. and Sutherland, I. O., *Tetrahedron Letters,* 2897 (1969).
15. Mukerjee, S. K., Saroja, T. and Seshadri, T. R., *Tetrahedron, 27,* 799 (1971).
16. Kavanagh, Ph., Ph. D. Thesis (NUI) 1974.
17. Mukerjee, S. K., Saroja, T. and Seshadri, T. R., *Ind. J. Chem., 8,* 21 (1970).
18. Kumar, D., Mukerjee, S. K. and Seshadri, T. R., *Tetrahedron Letters,* 1153 (1967).
19. Grisebach, H., "Biosynthetic Patterns in Microorganisms and in Higher Plants", Wiley, New York, 1967, p. 15.
20. Ollis, W. D., in "Recent Advances in Phytochemistry", Vol. 1. Appleton-Century-Crofts, New York, 1968, pp. 329–378.
21. Wong, E., *Fortschritte der Chemie organischer Naturstoffe, 28,* 1 (1970).
22. Ollis, W. D., Rhodes, C. A. and Sutherland, I. O., *Tetrahedron, 23,* 4741 (1967).
23. Crombie, L. and Godin, P. J., *J. Chem. Soc.,* 2861 (1961).
24. Ferriera, D., Brink, C. V. d. M. and Roux, D. G., *Phytochemistry, 10,* 1141 (1971).
25. Bowyer, W. J., Chatterjea, J. W., Dhoubhadel, S. P., Handford, B. O. and Whalley, W. B., *J. Chem. Soc.* 4212 (1964).
26. Anirudhan, C. A., Whalley, W. B. and Badran, M. M. E., *J. Chem. Soc.,* 629 (1966).
27. Brink, C. V. d. M., Nel, W., Rall, G. J. H. and Weitz, J. C., *J. S. Afric. Chem. Inst., 19,* 24 (1966).
28. a. McGookin, A., Robertson, A. and Whalley, W. B., *J. Chem. Soc.,* 787 (1940);
    b. Späth, E. and Schläger, J. *Ber., 733,* 1 (1940);
    c. Suginome, H., *Experientia,* 162 (1962).
29. Krishnaswamy, N. R. and Seshadri, T. R., "Chemistry of Natural and Synthetic Colouring Matter", Academic Press, New York and London, 1962, p. 235.
30. a. Fukui, K. and Nakayama, M., *Tetrahedron Letters,* 1805 (1966);
    b. Adityachaudhury, N. and Gupta, P. K., *Phytochemistry, 12,* 425 (1973).
31. Gunning, P., M. Sc. Thesis (NUI), 1974.
32. a. Aghormurthy, K., Kukla, A. S. and Seshadri, T. R., *Curr. Sci., 30,* 218 (1961);
    b. Suginome, H. and Iwadare, T., *Experientia, 18,* 163 (1962).
33. Rall, G. J. H., Englebrecht, J. P. and Brink, A. J., *Tetrahedron, 26,* 5007 (1970).
34. Anderson, E. L. and Marrian, G. F., *J. Biol. Chem., 127,* 649 (1934).
35. Wessely, F. and Prilinger, F., *Monatsh., 72,* 197 (1938).
36. Robertson, A. and Whalley, W. B., *J. Chem. Soc.,* 1440 (1954).
37. Crombie, L. and Whiting, D. A., *J. Chem. Soc.,* 1569 (1963).
38. Keenan, P. J. J., Ph. D. Thesis (NUI), 1973.

39. PERRIN, D. and BOTTOMLY, W., *J. Am. Chem. Soc., 84,* 1919 (1962).
40. a. HARPER, S. H., KEMP, A. D., UNDERWOOD, W. G. E. and CAMPBELL, R. V., *J. Chem. Soc.* (C), 1109 (1969);
    b. SUGINOME, H. and IWADARE, T., *Bull. Chem. Soc. Japan, 39,* 1535 (1966);
    c. NÓGRÁDI, M., ANTUS, S., GOTTSEGEN, Á. and FARKAS, L., "Topics in Flavonoid Chemistry and Biochemistry", Proceedings of the Fourth Hungarian Bioflavonoid Symposium, Keszthely, 1973 (eds L. FARKAS, M. GÁBOR, F. KÁLLAY). Akadémiai Kiadó, Budapest, 1975.
41. BEVAN, C. W. L., BIRCH, A. J., MOORE, B. and MUKERJEE, S. K., *J. Chem. Soc.,* 5991 (1964).
42. WHEELER, T. S., *Record Chem. Progr., 18,* 133 (1957) and references therein.
43. DEAN, F. M. and PODIMUANG, V., *J. Chem. Soc.,* 3978 (1965).
44. BROWN, D. W., DENMAN, C. and O'DONNELL, H., *J. Chem. Soc.,* 3195 (1971).
45. DEAN, F. M. and MANUNAPICHU, K., *J. Chem. Soc.,* 3112 (1957).
46. DONNELLY, B. J., DONNELLY, D. M. X., O'SULLIVAN, A. M. and PRENDERGAST, J. P., *Tetrahedron, 25,* 4409 (1969).

THE REACTIVITY OF 4-HYDROXYISOFLAVAN. SYNTHESIS OF 4-(1-PYRIDINIUM)-ISOFLAVANS

by

V. SZABÓ and E. ANTAL

Institute of Applied Chemistry, Kossuth Lajos University
Debrecen, Hungary

In the course of our studies on the catalytic hydrogenation of isoflavonoids in lower states of oxidation, we have investigated the possibility of further reduction and the hydrogenolysis of the hetero ring in these compounds, comparing their behaviour with that of flavans; one of the compounds examined was 4-hydroxyisoflavan.

The experiments have shown that, in basic medium 4-hydroxyisoflavan cannot be reduced to isoflavan at all, and the rate is still extremely low if the reaction is effected in acid or neutral medium. Under the same reaction conditions, isoflavan can, however, be obtained at a rate several orders of magnitude higher if 4-acetoxyisoflavan is used instead of 4-hydroxyisoflavan as the starting material of the hydrogenation. Hydrolysis, as a side reaction, has been observed only in basic medium, when 4-hydroxyisoflavan is obtained in a yield of approximately 20–30 %. In acid or neutral medium pure isoflavan is formed as the result of a very fast hydrogenolytic reaction. The experimental rate curves are shown in Fig. 1.

Contrary to the corresponding reactions of 4-hydroxyflavan, heterolytic ring cleavage has not been observed in any one of the experiments.

Hydrogenolytic cleavage of the C_4–O bond of 4-hydroxyisoflavan and 4-acetoxyisoflavan is similar to the corresponding reaction of benzyl alcohols. This reactivity is due to the enhancement of the polarization of the C_4–O bond and can be explained by the internal mesomerism of the acetyl group (Fig. 2).

Accordingly, the polarization of the C_4–O bond, or its tendency to form a carbonium ion, can further be increased by introducing an efficient leaving ester group, such as the tosyloxy group. We considered this reaction as a possible route to a number of new C-4 substituted, and possibly biologically active, compounds (e.g. –CN, $-N_3$, Cl, F, Br, etc.), obtainable from 4-tosyloxyisoflavan by known S_N reactions.

The preparation of the tosyl esters of the unsubstituted and of 7-methoxy- and 4′methoxy-4-hydroxyisoflavans, respectively, was attempted in the usual

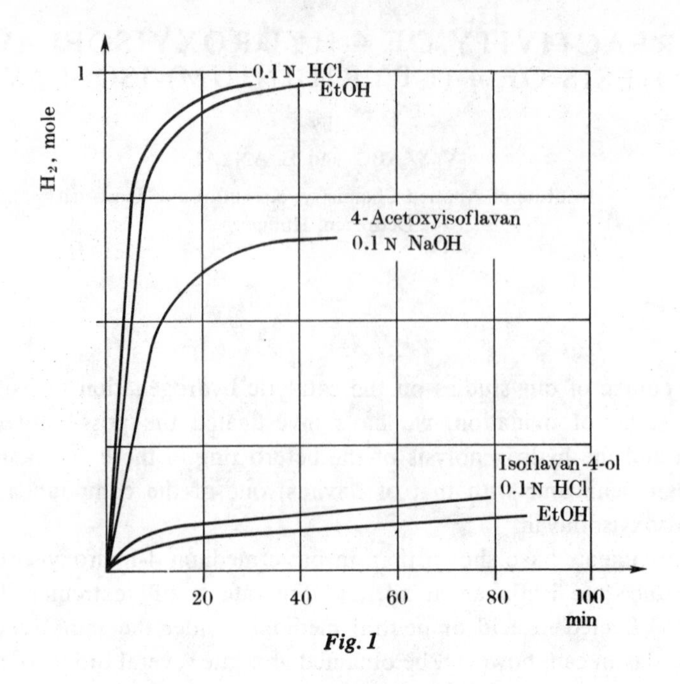

Fig. 1

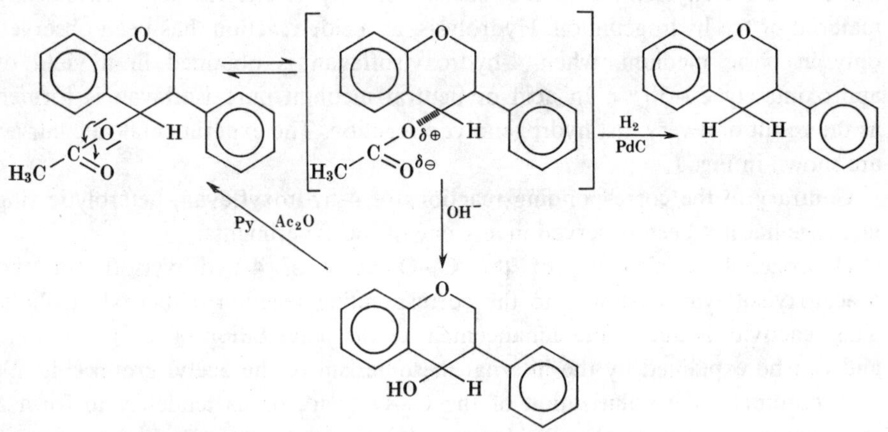

Fig. 2

manner, by allowing the compounds to react with tosyl chloride in pyridine solution.

After pouring the reaction mixture into water, a homogeneous solution was obtained, from the etheral extract of which only the unchanged 4-hydroxyisoflavans and isoflav-3-ene could be isolated. From the chloroform extract of the aqueous layer, however, well-crystallizable colourless compounds were precipitated by ethyl acetate. The elemental analyses of these compounds, however, were not consistent with the values calculated for the assumed tosyl esters.

Elemental analysis revealed the presence of one atom of nitrogen in addition to C, H, O and S. On the basis of this finding we concluded that salt-like pyridinium compounds had been formed instead of the expected tosylates (Fig. 3).

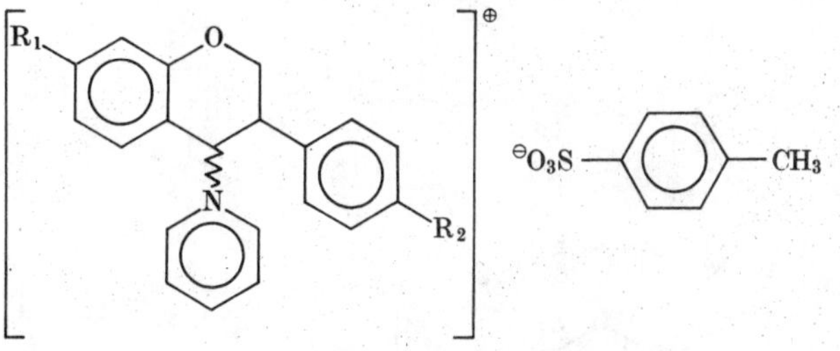

I; $R_1 = R_2 = H$
II; $R_1 = OCH_3$ $R_2 = H$
III; $R_1 = H$ $R_2 = OCH_3$

Fig. 3

Regarding the steps of formation of the pyridinium compounds, it is assumed that first the 4-hydroxyisoflavans react with tosyl chloride to give "true esters", which, as a result of the action of the strongly polarizing ester group, can dissociate in the manner shown in Fig. 4. The resulting carbonium ion will react with weak nucleophilic reagents in an S_N1-type reaction, and thus a stereoisomeric mixture is formed from the pure *cis*- or *trans*-4-hydroxyisoflavan.

The correctness of the above mechanism seems to be supported by the observation that either *cis*- or *trans*-7-methoxy-4-hydroxyisoflavan, as well as the mixture of the two isomers afford products (mixtures of isomers) having the same physical properties. Further evidence for our assumption is the fact that if the tosylation reaction is carried out in the presence of a stronger base (e.g. triethylamine) instead of pyridine at room temperature, isoflav-3-ene is obtained by elimination of the C-3 proton of the carbonium ion produced transiently.

The structures of the pyridinium compounds derived from the 4-hydroxyisoflavans are supported by their good solubility in water as well as their elemental analyses. The anion of the sulfonic acid salt can readily be replaced by chloride or iodide ions: the crystalline chloride and iodide salts are formed in a fast reaction in the cold.

The pyridinium compounds prepared exhibit extraordinary stability in acidic medium; they are not decomposed even on heating. In basic medium, however, decomposition to isoflav-3-ene occurs even at low temperatures (Fig. 5). Hydro-

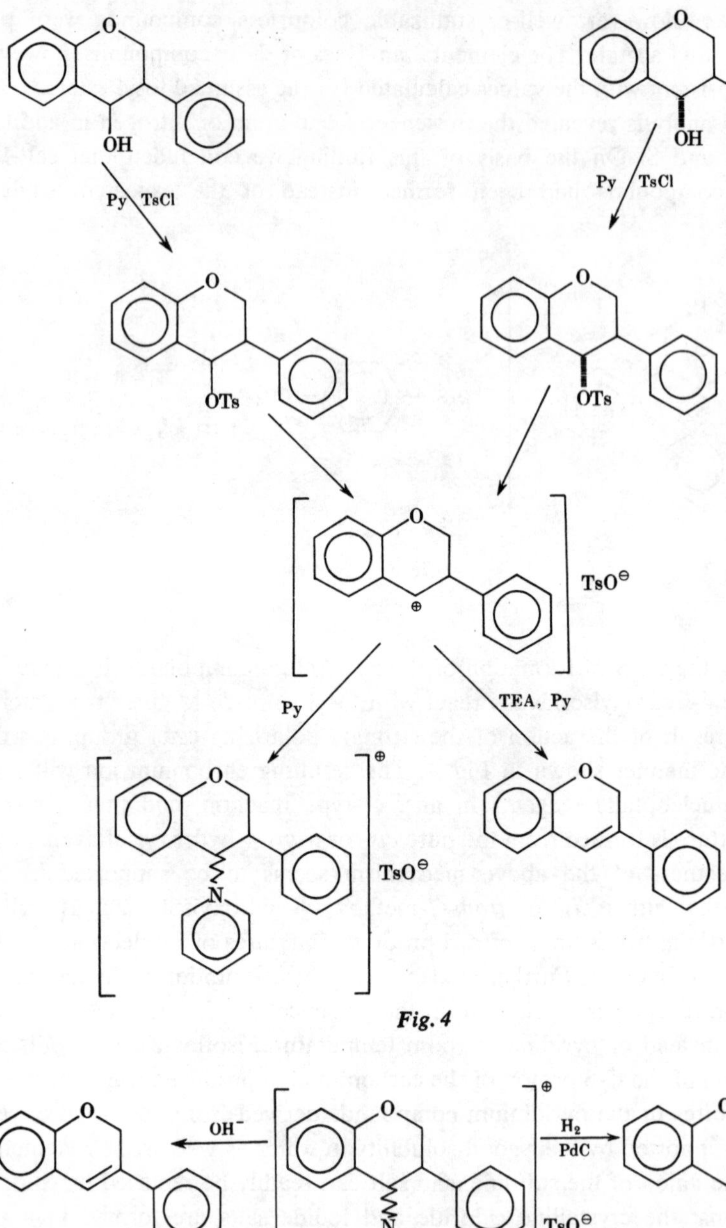

O OH Py TsCl O OTs O OH Py TsCl O OTs O ⊕ TsO$^{\ominus}$ Py TEA, Py O N TsO$^{\ominus}$ ⊕ O

Fig. 4

O OH^{-} O N TsO$^{\ominus}$ ⊕ H_2 PdC O

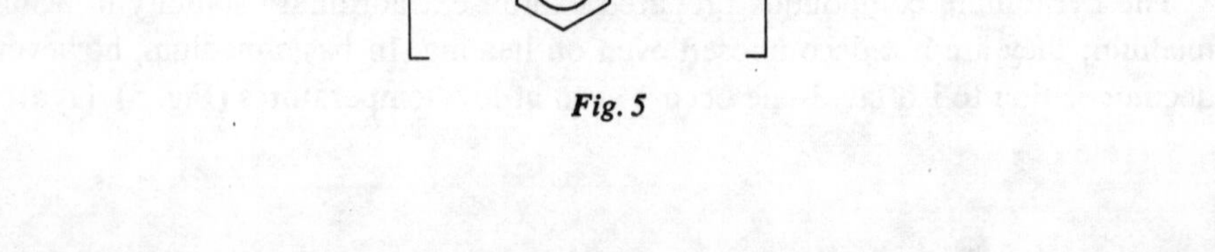

Fig. 5

genation of the pyridinium tosylate in the presence of Pd/C catalyst gives isoflavan.

These pyridinium compounds can be regarded as heteroaryl analogues of 4-arylisoflavans, which latter possess important physiological actions. It is hoped that also the new compounds may have similar properties.

The "pseudotosylation" reaction and the results of hydrogenation of 4-acetoxyisoflavans indicate that the C_4–O bond is loosened and therefore polarized by the internal mesomerism of the ester group. Therefore, these compounds may offer the possibility of other new and interesting conversions. The two reactions discussed above support the benzyl alcohol-like structure of 4-hydroxyisoflavans.

BROMINATED AND OTHER DERIVATIVES OF AZAFLAVANONES

by

G. JANZSÓ

Research Institute for Organic Chemical Industry
Budapest, Hungary

Connections between the big classes of natural products have always special interest. In spite of this, little has been done for the establishment of such connections between flavonoids and alkaloids. A known example speaking for the existence of a possible connection is the occurrence of melicopin in one species and the presence of meliternatin in other species of the same family, Rutaceae [1].

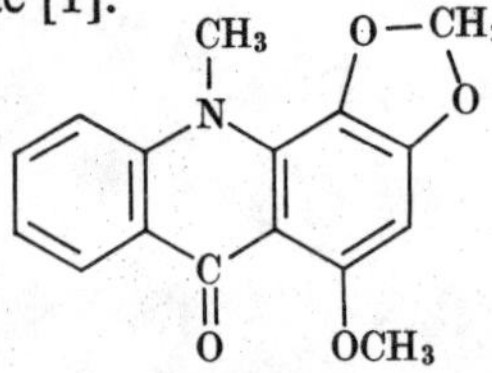

Melicopin 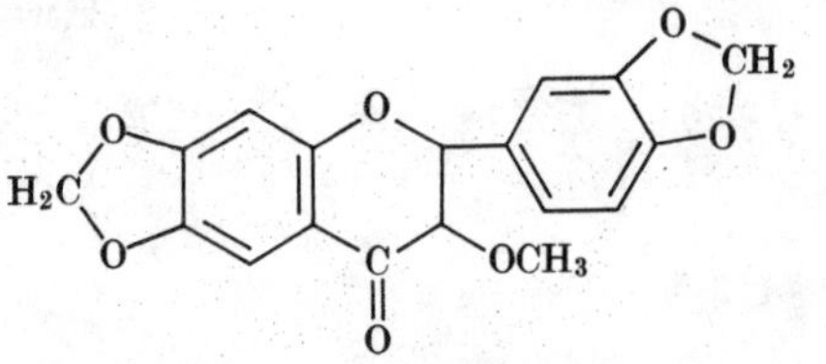   Meliternatin

The alkaloid edulein isolated from *Casimiroa edulis* [2] is a close analogue of 7-methoxyflavone:

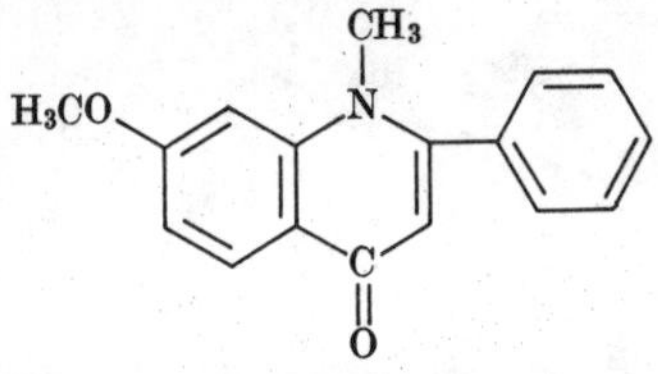

Edulein 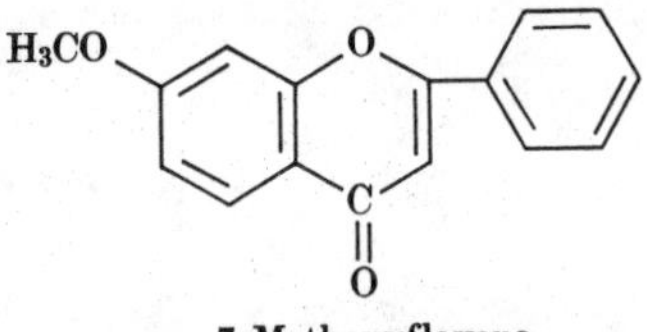   7-Methoxyflavone

Further, compounds which are both flavonoids and alkaloids at the same time, are represented by ficine and isoficine [3].

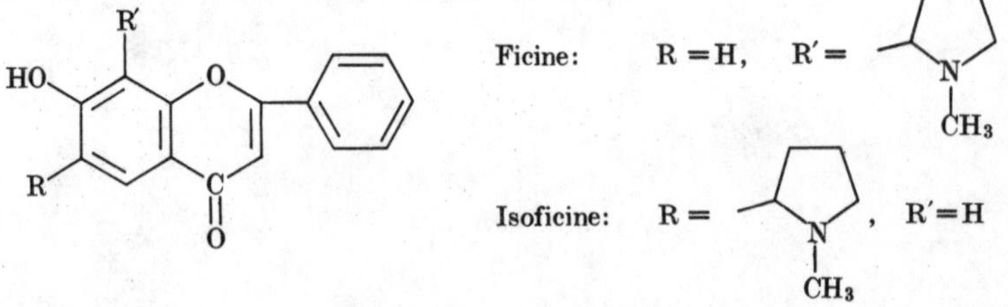

Ficine: R = H, R′ =

Isoficine: R = , R′ = H

We thought it would be interesting to start a study of flavonoids having the NH or an N–R group instead of the oxygen in the hetero ring; at first an investigation of the synthesis and reactions of the simple parent compounds was decided. These substances are, of course, strictly speaking phenylquinolones and tetrahydrophenylquinolones, but the very close structural relationship to flavanone may give us the right to call them azaflavonoids.

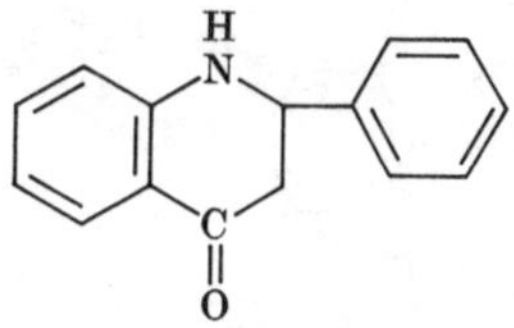

I

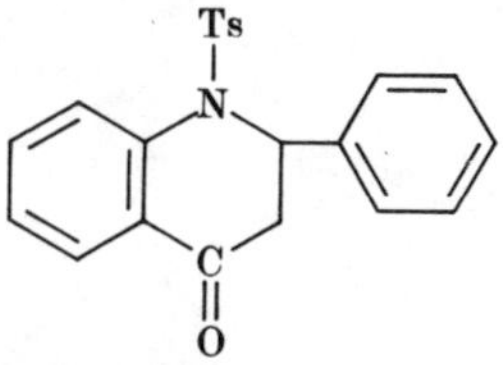

II

2-Phenyltetrahydroquinol-4-one (I) and its N-tosyl derivative (II) were synthesized as early as 1938 by Mannich [4] and in 1945 by Diesbach [5], respectively. On the analogy of these syntheses, we cyclized *o*-acetamido-chalcone (III) to 1-acetyl-2-phenyltetrahydroquinol-4-one (IV).

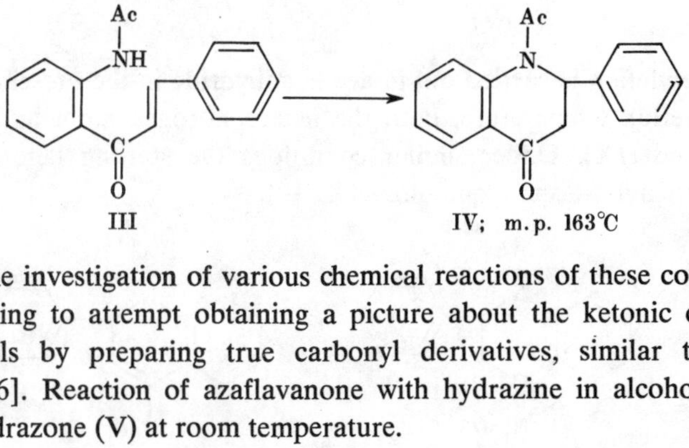

III

IV; m.p. 163°C

During the investigation of various chemical reactions of these compounds, it was interesting to attempt obtaining a picture about the ketonic character of these models by preparing true carbonyl derivatives, similar to those of flavanone [6]. Reaction of azaflavanone with hydrazine in alcoholic solution gave the hydrazone (V) at room temperature.

This product is always accompanied by the corresponding azine (VI), the formation of the latter becoming practically quantitative at reflux temperature.

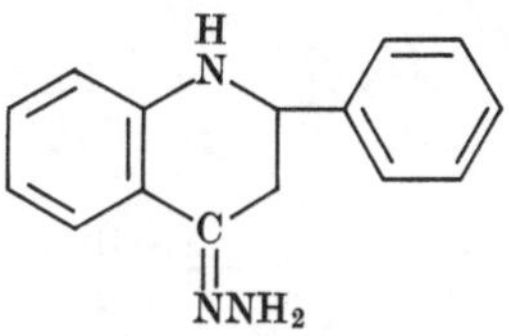

V; m.p. 73°C

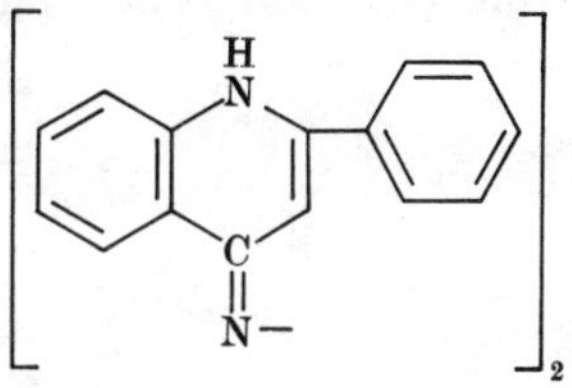

VI; m.p. 251°C

The high reactivity of the hydrazone amino group towards carbonyl compounds lends itself particularly well for proving chemically the hydrazone structure of V, by the preparation of a mixed azine (VII) with benzaldehyde beyond the spectroscopic evidence.

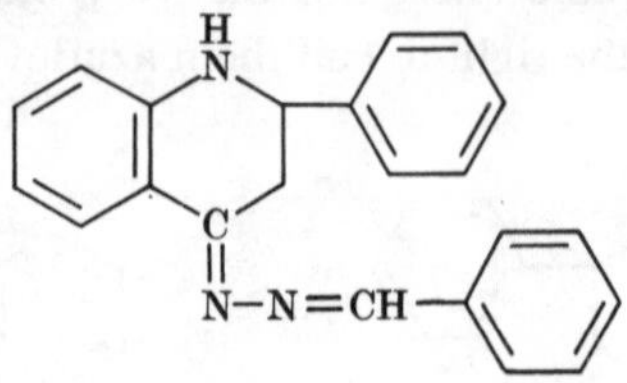

VII; m.p. 114°C

With hydroxylamine the oxime (VIIIa) is obtained; this can be readily acylated on the oxygen atom at room temperature.

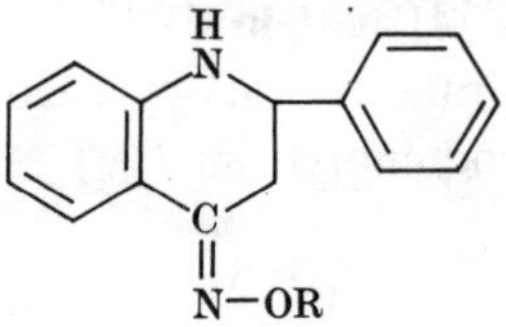

VIII; a. R=H; m.p. 164°C;
b. R=Ac; m.p. 145°C;
c. R=Ts; m.p. 135°C

If the acetylation is carried out in acetic anhydride in the presence of sodium acetate at reflux temperature, both the hetero nitrogen atom and the oxygen are acetylated (IX). Under similar conditions the starting ketone gives 1,2-dihydro-2-phenyl-4-acetoxyquinolone (X) [4].

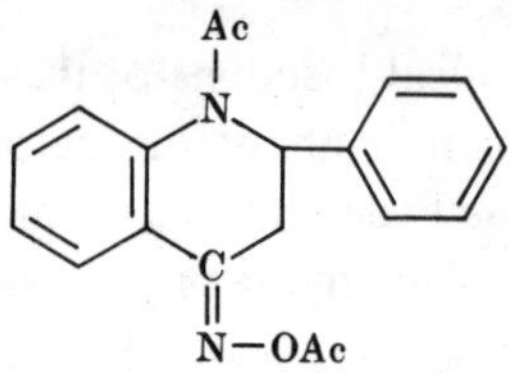

IX; m.p. 163°C

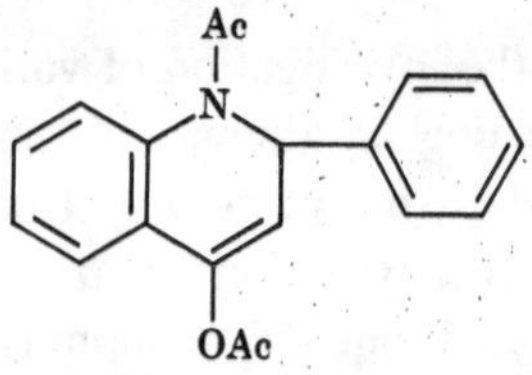

X

N-Tosylazaflavanone (II) shows similar behaviour towards hydroxylamine, giving the oxime (XI) and its derivatives.

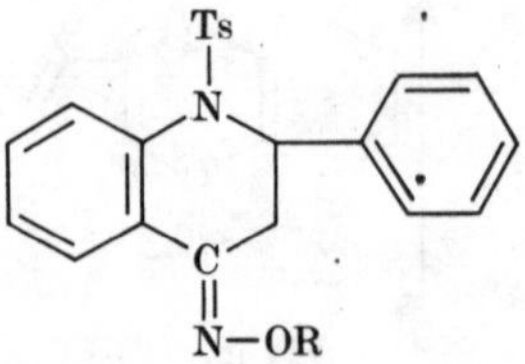

XI; R=H; m.p. 217°C
R=Ac; m.p. 179°C

As a summary of these results, one can say that these azaflavanones behave in carbonyl reactions similarly to each other and to the oxygen analogue, flavanone [6], when the C-4 atom is attacked with nitrogen-containing carbonyl reagents. However, they differ basically from flavanone in one respect, that is, their hetero ring is never opened, even under strongly alkaline conditions.

In the bromination reactions, azaflavanone and N-tosylazaflavanone reveal quite different features. When azaflavanone (I) is brominated in chloroform, the bromine enters firstly at the C-6 position [7] of the condensed aromatic ring (XII), independently of the molar ratio of bromine. Using two, three or more molecules of bromine, the compound is further brominated in the C-8 and C-3 positions with simultaneous hydrogen bromide elimination, affording two brominated azaflavones (XIII and XIV).

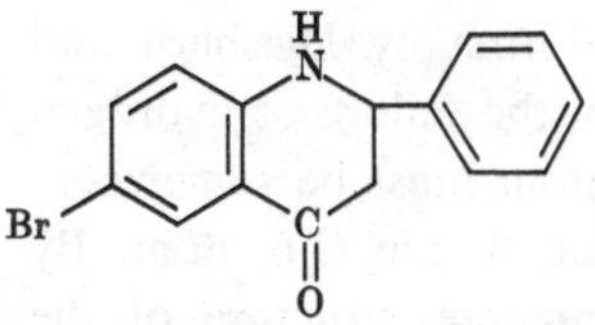

XII; m.p. 179°C 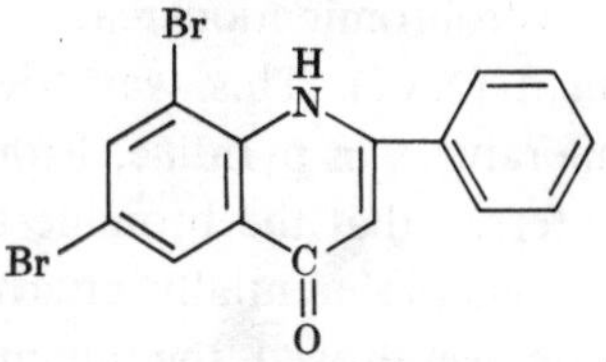 XIII; m.p. 213°C 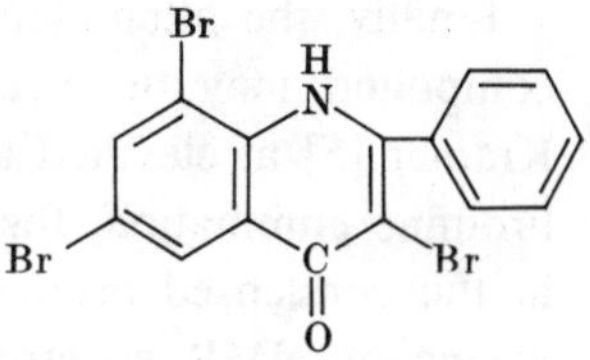 XIV; m.p. 239°C

With N-bromosuccinimide in carbon tetrachloride solution, azaflavanone gives 6-monobromo- or 6,8-dibromoazaflavanone (XII and XV, respectively), depending on the quantity of the brominating agent applied.

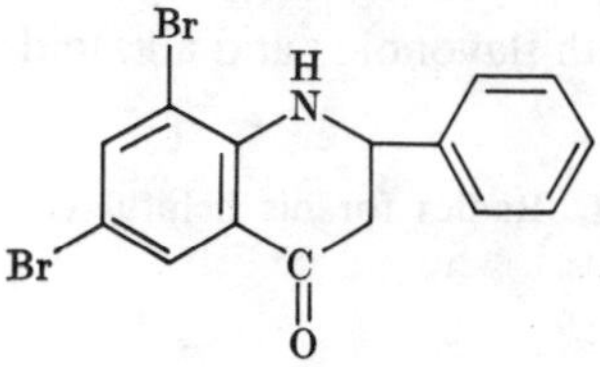

XV; m.p. 117°C

On the other hand, N-tosylazaflavanone (II) yields the 3-bromo derivative (XVI) with one molecule of bromine at reflux temperature in chloroform. Increasing the quantity of bromine, the 3,3-dibromo derivative (XVII) can be obtained under similar reaction circumstances. No further bromination can be observed even if the bromine is present in great excess, whether the N-tosyl compound or the 3,3-dibromo derivative is used as the starting material.

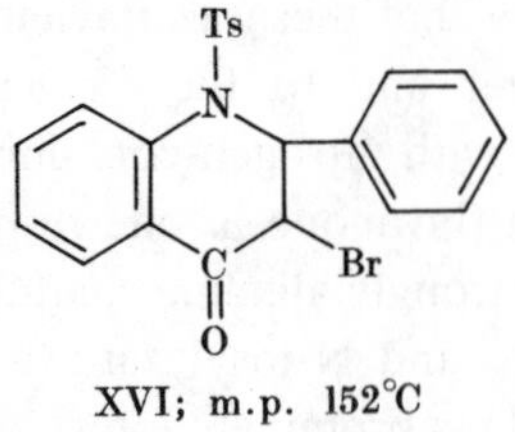

XVI; m.p. 152°C 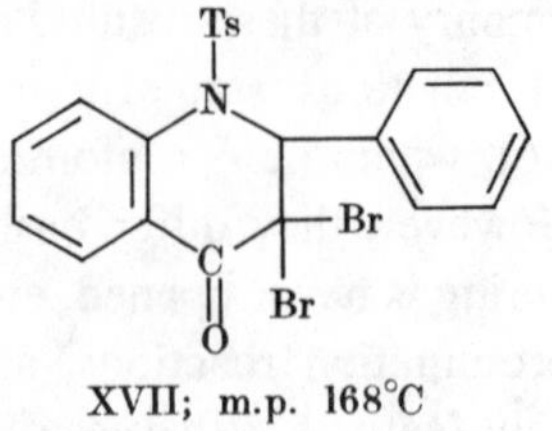 XVII; m.p. 168°C

An explanation for this different behaviour may come from a consideration of the character of the differently substituted nitrogen atoms in azaflavanone and the N-tosyl derivative. Regarding the NH group as a *para*-activating substituent, the C-6 atom as the point of attack of the bromonium ion becomes understandable. In the N-tosyl compound the former NH activating effect is absent, thus the α-carbon activity dominates in the electrophilic substitution.

Finally, the attempted dehydrobromination reaction of the 3-bromo-N-tosyl compound may be mentioned (XVI). This was tried first by Diesbach and Kramer [5] at elevated temperatures in pyridine. From the failure of hydrogen bromine elimination, they inferred that the bromine atom must be somewhere in the condensed aromatic ring, preferentially attached to the C-6 atom. By means of NMR spectroscopy we proved the unambiguous structure of the molecule to be the 3-monobromo derivative (XVI). Repeating the attempted dehydrobromination reaction in pyridine as well as in dimethyl sulfoxide, the product is unequivocally 3-bromoazaflavone, that is, toluenesulfinic acid is eliminated during the reaction. Work is now in progress to clarify the mechanism of this elimination reaction.

Later research may deal with the synthesis and examination of the chemical and perhaps interesting physiological properties of compounds uniting the structural elements of both flavonoids and alkaloids.

*

Thanks are due to Dr. L. Radics for his helpful co-operation in recording the NMR spectra of all new compounds.

REFERENCES

1. SCHLITTLER, E., in "Perspectives in Organic Chemistry". Alkaloids. Interscience, New York–London, 1956, p. 347.
2. SONDHEIMER, F. and MEISELS, A., *J. Org. Chem., 23,* 762 (1958).
3. JOHNS, S. R., RUSSEL, J. H. and HEFFERNAN, M. L., *Tetrahedron Letters, 1965,* 1987.
4. MANNICH, C. and DANNEHL, M., *Ber., 71,* 1899 (1938).
5. DIESBACH, H. and KRAMER, H., *Helv. Chim. Acta, 28,* 1399 (1945).
6. KÁLLAY, F., JANZSÓ, G. and KOCZOR, I., *Tetrahedron, 21,* 19 (1965).
7. JANZSÓ, G. and PHILBIN, E. M., *Tetrahedron Letters, 1971,* 3075.

SYNTHESIS OF TWO NEW FLAVONE GLUCOSIDES ISOLATED FROM *CIRSIUM LINEARE*

by

L. FARKAS and J. STRELISKY

Institute of Organic Chemistry, Technical University
Budapest, Hungary

Cirsium (corn thistle) is one of the commonest weeds which causes heavy losses in crops, therefore its investigation is not only of scientific, but also of economical importance. The broadest survey of various *Cirsium* species was carried out by Morita and his co-workers, who have examined twenty-seven species growing in Japan [1–5]. They identified some well known flavonoids like pectolinarin, luteolin 7-glucuronide, luteolin 7-glucoside, linarin, rhoifolin and rutin (Fig. 1).

In addition they discovered three new flavone glycosides and it was this field where our interest met with that of Morita, as all of the new aglycones had a 5-hydroxy-6,7-dimethoxy substitution pattern in ring A, and we had earlier synthesized several 5,6,7-trioxygenated flavones with various distribution of the hydroxyl and methoxyl groups. Some of these were 5-hydroxy-6,7-dimethoxyflavone [6] (as a model compound), scutellarein (4′-5,6,7-tetrahydroxyflavone) [7], eupatorin (3′,5-dihydroxy-4′,6,7-trimethoxyflavone) [8], eupatilin (5,7-dimethoxy-3′,4′,6-trimethoxyflavone) [9], and a few flavonols, which had been isolated from various *Eupatorium* species, such as eupatolitin (3,3′,4′,5-tetrahydroxy-6,7-dimethoxyflavone), eupalitin (3,4′,5-trihydroxy-6,7-dimethoxyflavone), eupatoretin (3,3′-dihydroxy-4′,5,6,7-tetramethoxyflavone) and eupatin (3,3′,5-trihydroxy-4′,6,7-trimethoxyflavone) [9].

The first of the above-mentioned new flavone glucosides was isolated in 1963 by Morita and Shimizu from *Cirsium maritimum* [3] and it was named cirsimarin, the aglycone being cirsimaritin (Fig. 2).

They assigned the structure 4′,5-dihydroxy-6,7-dimethoxyflavone 4′-O-β-D-glucopyranoside to cirsimarin, the correctness of which we proved bv total synthesis in 1967 [7]; the key steps of synthesizing the aglycone are shown in Fig. 3.

In 1973 Morita, Shimizu and Arisawa [10] reported the isolation of two new flavone glycosides of two new aglycones from *Cirsium lineare,* a Japanese species of corn thistle. One of the aglycones was stated to be 4′,5-dihydroxy-3′,6,7-trimethoxyflavone (cirsilineol). The other one, cirsiliol was identical with

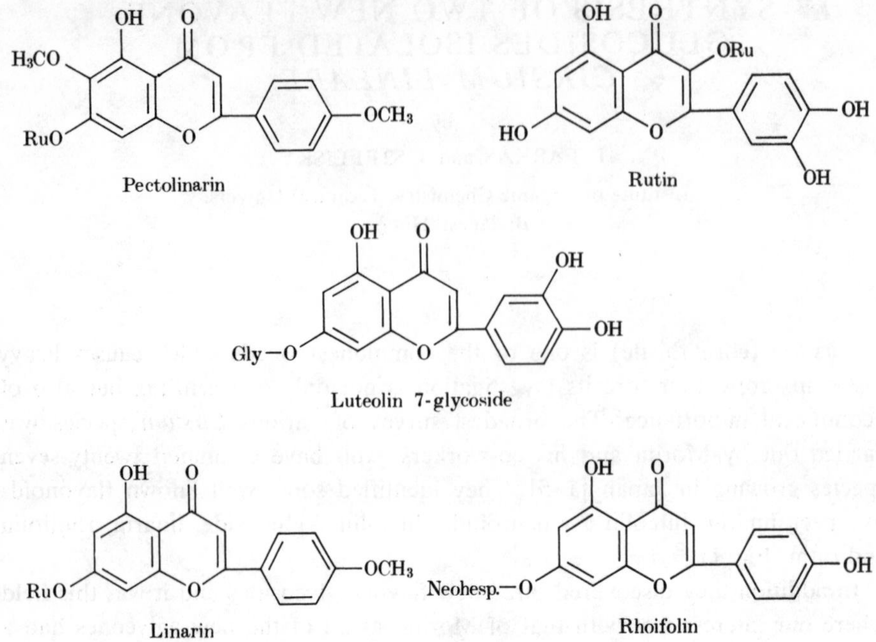
OH
O
H₃CO
RuO
O
OCH₃
Pectolinarin
OH
O
ORu
HO
O
OH
OH
Rutin
OH
O
OH
Gly—O
O
OH
Luteolin 7-glycoside
OH
O
RuO
O
OCH₃
Linarin
OH
O
Neohesp.—O
O
OH
Rhoifolin

Fig. 1

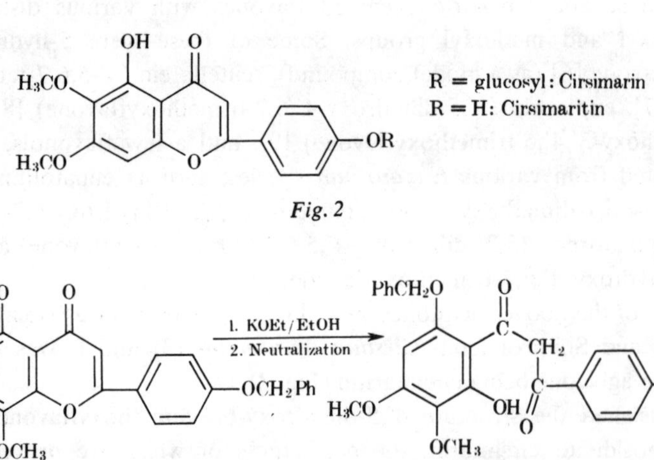
OH
O
H₃CO
H₃CO
O
OR
R = glucosyl: Cirsimarin
R = H: Cirsimaritin

Fig. 2

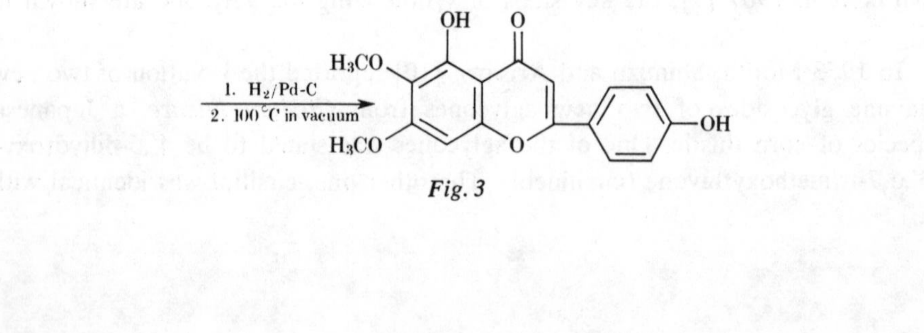
PhCH₂O
O
H₃CO
OCH₃
O
OCH₂Ph
1. KOEt/EtOH
2. Neutralization
PhCH₂O
O
C
CH₂
C
O
H₃CO
OH
OCH₃
OCH₂Ph
1. H₂/Pd-C
2. 100 °C in vacuum
OH
O
H₃CO
H₃CO
O
OH

Fig. 3

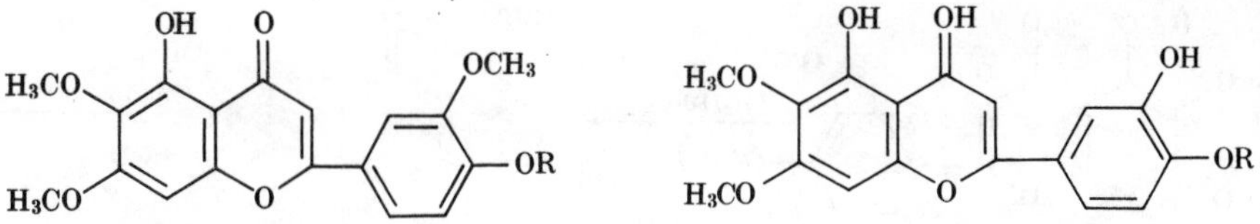

a. R=β-D-glucosyl
b. R=H: Cirsilineol

a. R=β-D-glucosyl
b. R=H: Cirsiliol

Fig. 4

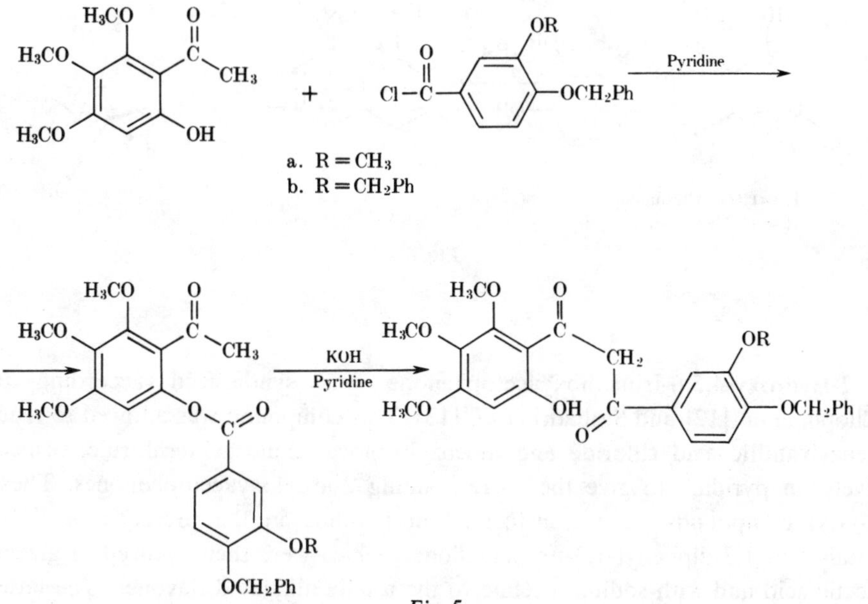

a. $R=CH_3$
b. $R=CH_2Ph$

Fig. 5

3′,4′,5-trihydroxy-6,7-dimethoxyflavone formerly isolated by Brieskorn and Biechele [11], as shown by the m.p.'s and spectra. Both aglycones were found by Morita in the form of their 4′-O-β-D-glucosides (Fig. 4).

For the synthesis of these compounds we did not follow the scheme used in the preparation of cirsimaritin [7] (base-catalyzed ring isomerization of the relatively readily accessible 5,7,8-substituted isomer), because of unfavourable experiences obtained in synthesizing eupatorin [8]; in the latter case a mixture of the two possible isomers, eupatorin (3′,5-dihydroxy-4′,6,7-trimethoxyflavone) and its 5,7,8-substituted analogue were formed, which were difficult to separate. The aglycones for glucosylation were now prepared by a more conventional but safer route to the flavones, making use of the Elbs persulfate oxidation to yield the acetophenone with the required substitution pattern (Fig. 5).

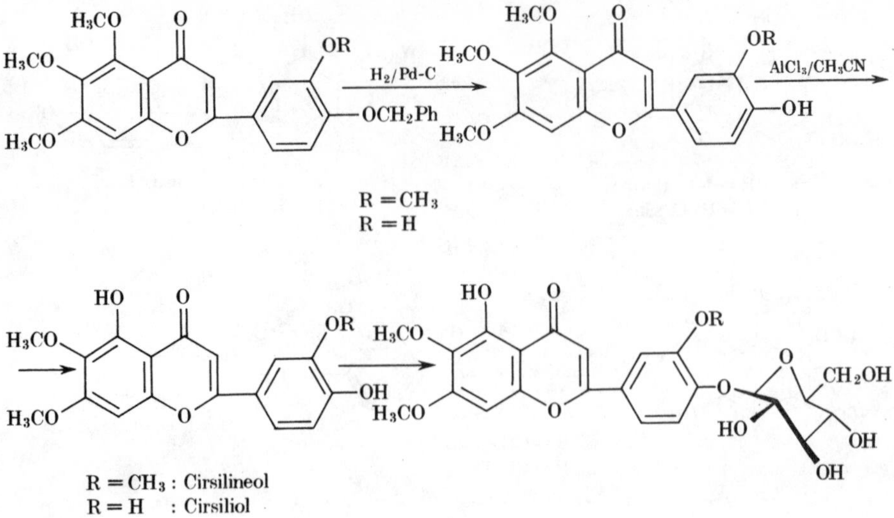

$R = CH_3$
$R = H$

$R = CH_3$: Cirsilineol
$R = H$: Cirsiliol

Fig. 6

2-Hydroxy-4,5,6-trimethoxyacetophenone was synthesized according to Chopin *et al.* [12] and Seshadri *et al.* [13]. This compound was allowed to react benzylvanillic acid chloride and dibenzylprotocatechuic acid chloride, respectively, in pyridine to give the corresponding 2-aroyloxyacetophenones. These O-acyl compounds were transformed in pyridine in the presence of KOH catalyst to 1,3-diphenyl-1,3-propanediones, which were then cyclized in glacial acetic acid and with sodium acetate to the totally alkylated flavones. The latter were debenzylated (Fig. 6) by hydrogen and palladium-on-charcoal to 4′-hydroxy-3′,5,6,7-tetramethoxyflavone and 3′4′-dihydroxy-5,6,7-trimethoxyflavone, respectively.

As a concluding step, selective demethylation was effected by aluminium chloride in boiling acetonitrile. Thus we finally obtained 4′,5-dihydroxy-3′,6,7-trimethoxyflavone in a fair overall yield; it was found identical with cirsilineol. Similarly, the properties of the synthetic 3′,4′,5-trihydroxy-6,7-dimethoxyflavone agreed with those of cirsiliol of plant origin. Identification was based on mixed melting point determination, chromatography, comparison of the infrared spectra and m.p. of the acetates (188 °C and 196 °C, respectively)*.

* In the meantime MATSUURA, S., KUNII, T. and MATSUURA, A. also synthesized the aglucones (*Chem. Pharm. Bull., 21,* 2757 (1973)), but by oxidation of the corresponding chalcones.

The attachment of the sugar moiety was accomplished by the Koenigs–Knorr reaction, by coupling the aglycones with α-acetobromoglucose in acetone in the presence of aqueous potassium hydroxide. The yields of both glucosylations were satisfactory. The tetraacetyl-4′-O-β-D-glucosides obtained in this way were saponified to the free glucosides. Synthetic 4′,5-dihydroxy-3′,6,7-trimethoxyflavone-4′-O-β-D-glucoside was identified with cirsiliol-4′-O-glucoside of plant origin, and 3′,4′,5-trihydroxy-6,7-dimethoxyflavone-4′-O-β-D-glucoside was found to be identical with natural cirsilineol-4′-O-glucoside, as shown by mixed m.p. determination, the UV spectra and chromatography.

REFERENCES

1. Nakaoki, T. and Morita, N., *J. Pharm. Soc. Jap., 79,* 1338 (1959).
2. Nakaoki, T. and Morita, N., *J. Pharm. Soc. Jap., 80,* 1296 (1960).
3. Morita, N. and Shimizu, M., *J. Pharm. Soc. Jap., 83,* 615 (1963).
4. Morita, N., Fukuta, M. and Shimizu, M., *Syoyakugaku Zasshi, 18,* 9 (1964).
5. Morita, N., Fukuta, M. and Shimizu, M., *Syoyakugaku Zasshi, 19,* 8 (1965).
6. Farkas, L., Major, Á. and Strelisky, J., *Chem. Ber., 96,* 1684 (1963).
7. Farkas, L., Strelisky, J. and Major, Á., *Acta Chim. Acad. Sci. Hung., 53,* 211 (1967).
8. Farkas, L., Strelisky, J. and Vermes, B., *Chem. Ber., 102,* 112 (1969).
9. Wagner, H., Flores, G., Hörhammer, L., Farkas, L. and Strelisky, J., *Tetrahedron Letters, 16,* 187 (1971); *Chem. Ber., 107,* 1049 (1974).
10. Morita, N., Shimizu, M. and Arisawa, M., *Phytochemistry, 12,* 421 (1973).
11. Brieskorn, C. H. and Biechele, W., *Arch. Pharmaz., 304,* 557 (1971).
12. Chopin, J., Molho, D., Pacheco, M. and Mentzer, C., *Bull. Soc. Chim. France, 1957,* 202.
13. Sastri, V. D. N. and Seshadri, T. R., *Proc. Ind. Acad. Sci., 23A,* 262 (1946).

2,5-DIHYDROXYFLAVANONES: A NEW TYPE OF NATURAL FLAVONOIDS

by

J. CHOPIN

Laboratoire de Chimie biologique, UER Chimie-Biochimie, Université de Lyon
Villeurbanne, France

From autumn to the beginning of spring, the buds of some trees are covered by a waxy material, containing flavonoids in non-glycosidic form.

From the acetone extract of *Populus nigra* buds, Wollenweber *et al.* [1] could isolate twenty flavonoids, covering the whole range of oxidation states: chalcones, flavanones, flavones, 3-hydroxyflavanones and flavonols, all possessing a phloroglucinol-type ring A.

One of these compounds, named P_4 by Wollenweber, was a colourless product, m.p. 170°C, molecular weight 286 (from the mass spectrum), the ultraviolet spectrum of which [λ_{max}^{EtOH} 287 nm (log ε : 3.99)] was shifted by $AlCl_3$ (λ_{max} 308 nm), but remained unchanged in the presence of sodium acetate.

By analogy with the other compounds identified, Wollenweber supposed P_4 to be 2′-hydroxy-4′,6′-dimethoxydihydrochalcone $C_{17}H_{18}O_4$ and sent the product to our laboratory for comparison with dihydrochalcones synthesized by Miss Chadenson and Miss Hauteville.

Direct comparison showed that Wollenweber's hypothesis was wrong, and further studies were needed on the very small amount of substance available.

Elemental analysis agreed with the formula $C_{16}H_{14}O_5$ and not with $C_{17}H_{18}O_4$. Alkali fusion led to phloroglucinol, but not to any hydroxybenzoic acid. The IR spectrum showed the presence of a chelated carbonyl group conjugated with an aromatic ring, ν 1,640 cm^{-1}, in agreement with the observed ultraviolet shift in the presence of $AlCl_3$.

The absence of a free hydroxyl group *para* to the carbonyl was evident from the lack of shift with NaOAc.

The following group could be deduced from these observations:

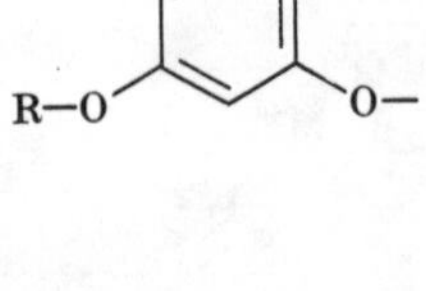

O—H····O
C—
R—O
O—

The ultraviolet spectrum, which showed in EtOH the absence of any other double bond conjugated with the carbonyl group, was quite different in NaOH and the absorption maximum shifted to 385 nm could be ascribed to a chalcone conjugated system.

These UV spectral properties were nearly identical with those of the 2,4,6-trihydroxydibenzoylmethane 4-glucoside isolated by Williams [2] from *Malus* leaves.

In agreement with a dibenzoylmethane structure of the compound P_4 was the presence in the mass spectrum of significant peaks at m/e 105 (C_6H_5-CO), 77 (C_6H_5), 167 ($CH_3O(OH)_2$ C_6H_2-CO) and 153 ($(OH)_3C_6H_2$-CO).

The NMR spectrum (60 MHz) of the trimethylsilyl derivative in CCl_4 corresponded to the enol ether.

OT

CH_3O—[ring]—CO—CH=C(OT)—[phenyl] T=Si$(CH_3)_3$

OT

δ_{TMS}ppm : 3.80 (s; 3p, OCH_3); 6.00 (s; 2p, H-3.5); 6.41 (s; 1p, H-α); 7.42 (m; 3p, H-3′, 4′, 5′); 7.92 (m; 2p, H-2′, 6′)

On the basis of these data compound P_4 could be regarded as 2,6-dihydroxy-4-methoxydibenzoylmethane (I). Indeed, it was easily and quantitatively converted into 5-hydroxy-7-methoxyflavone (tectochrysin) (II) by heating with a trace of sulfuric acid in acetic acid (Chadenson *et al.* [3]).

OH O ‖ C CH$_2$ C OH ‖ O CH_3O I

$-H_2O$ ⇄ $+H_2O$

OH O CH_3O O II

As we had previously noticed that the compound P_4 could be recovered unchanged after a 10-min boiling in ethanolic sodium hydroxide, it was tempting to synthesize P_4 by the ring opening of tectochrysin, because Farkas *et al.* [4] had previously observed a quantitative conversion of 5,7,8-trimethoxyflavone into the corresponding 2-hydroxy-3,4,6-trimethoxydibenzoylmethane by refluxing in ethanolic potassium ethoxide.

However, this reaction proved extremely difficult in our case, owing to the presence of a free 5-hydroxy group, and success was only obtained by refluxing tectochrysin with anhydrous potassium hydroxide in pyridine for 4 hours. The synthetic compound was identical with the natural one by m.p., mixed m.p., UV and IR spectra.

Total synthesis of the same product could also be realized by Baker–Venkataraman rearrangement of 4-O-methylphloracetophenone monobenzoate. Incidentally, we found a convenient procedure for the preparation of 4-O-methylphloracetophenone (IV) from phloroglucinol monomethylether (III) (Hauteville and Chadenson [5]).

OH, CH_3O, OH — III $\xrightarrow[AcOH-BF_3]{Ac_2O}$ OAc, CH_3CO, $COCH_3$, CH_3O, OAc $\xrightarrow{NaOMe}$ OH, $COCH_3$, CH_3O, OH — IV

Monobenzoylation of IV in aqueous sodium hydroxide gave an oil which was heated with anhydrous potassium hydroxide in pyridine at 50°C for 15 min, leading to the desired compound P_4.

OH, $COCH_3$, CH_3O, $OCOC_6H_5$ $\xrightarrow[Py]{KOH}$ OH, C(=O), CH_2, $C-C_6H_5$ (=O), CH_3O, OH

Monobenzoate rearrangement was chosen because it had been previously shown that Baker–Venkataraman rearrangement of 2,6-dibenzoyloxyacetophenone and phloracetophenone tribenzoate led to 3-benzoyl-5-hydroxyflavone and 3-benzoylchrysin, respectively.

Although these two synthetic ways confirmed the proposed dibenzoylmethane structure of the natural compound, it was surprising that the infrared spectrum in potassium bromide showed only one carbonyl band (ν 1,640 cm^{-1}) instead of the two (1,640 and 1,680 cm^{-1}) observed by Wagner *et al.* [6] for the colourless diketo forms of *o*-hydroxydibenzoylmethanes.

Moreover, when sufficient amounts of the synthetic compound became available for NMR studies, a new anomaly was found in the NMR spectrum of the product itself in deuterated chloroform.

Besides the methoxyl and ring protons a singlet of two protons at δ 3.05 ppm was present, instead of the δ 4.5 to 4.7 ppm observed for the diketo forms of *o*-hydroxydibenzoylmethanes by Wagner *et al.* [6].

This chemical shift was nearly the mean value of those given by the two protons in position 3 in flavanones.

This led us to the hypothesis of a rapid equilibrium between a small amount of the diketo form (Ia) and a large amount of the cyclic hemiketalic form (Ib).

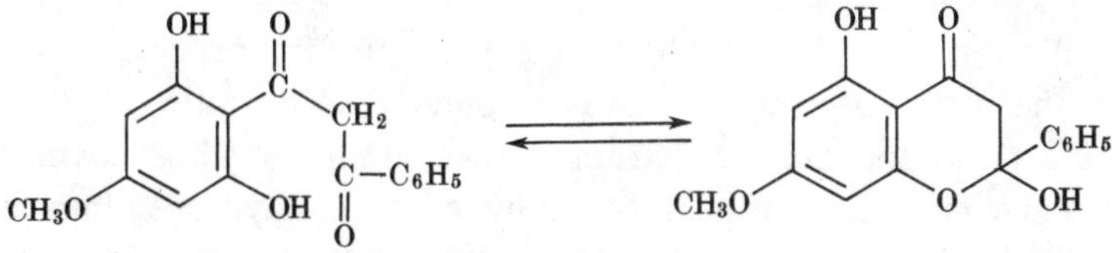

Ia ⇌ Ib

Indeed, the infrared spectrum in chloroform showed immediately two carbonyl bands, a strong one at ν 1,640 cm^{-1} and a weak one at 1,680 cm^{-1}, the spectrum remaining unchanged after 20 min. Thus, both forms are co-occurring in chloroformic solution.

We then made a study of the NMR spectrum at lower temperatures from 0 to — 60 °C in deuterated acetone, a better solvent than chloroform for our compound. In this solvent, the singlet of two protons at 3.05 ppm previously observed in $CDCl_3$ was replaced by a pair of doublets near 2.80 and 3.25 ppm, each doublet corresponding to one proton, *equatorial* and *axial,* respectively, in the 3-position of a flavanone, with a coupling constant of 17 Hz.

On the other side appeared a one-proton signal near 7 ppm. This signal was shifted to lower fields when the temperature decreased and, at — 60 °C, it became a sharp doublet at 7.26 ppm with a coupling constant of 2 Hz; the same coupling was then found in the *axial* proton at 3.30 ppm which became a quadruplet.

The proton at 7.26 being attributable to the 2-hydroxyl group in the cyclic form, this long range coupling with a proton in position 3 implies a *trans-diaxial* orientation of these substituents in the half-chair conformation of flavanones.

Such a coupling, about 1 Hz only, had been previously found in pyranoses and 2-hydroxy-2,3-dihydro-γ-pyrones.

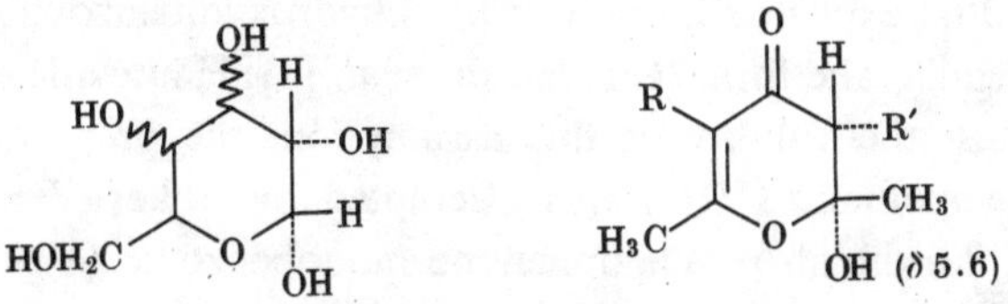

The other signals remained the same as in chloroform: 3.80 (s; 3p, OCH_3); 6.09 (s; 2p, H-6,8); 7.46 (m; 3p, H-3′,4′,5′); 7.74 (m; 2p, H-2′,6′); 12.40 (s; 1p, 5-OH).

The occurrence of only one signal for the protons in positions 6 and 8 had been already observed with 5-hydroxy-7,4′-dimethoxyflavanone and 5,7-dihydroxyflavanones.

No signal corresponding to the diketo form appeared between 4.5 and 4.7 ppm.

This cyclic structure (Ib) could also explain the importance of the peak M—17 in the mass spectrum of the compound P_4. The ratio $\frac{M-17}{M}$ is 30 % instead of 8 % in the mass spectrum of *o*-monohydroxydibenzoylmethanes. Flavanones give an important peak M—1 by loss of the hydrogen in position 2, and similarly, loss of the hydroxyl in position 2 can be expected for 2-hydroxyflavanones.

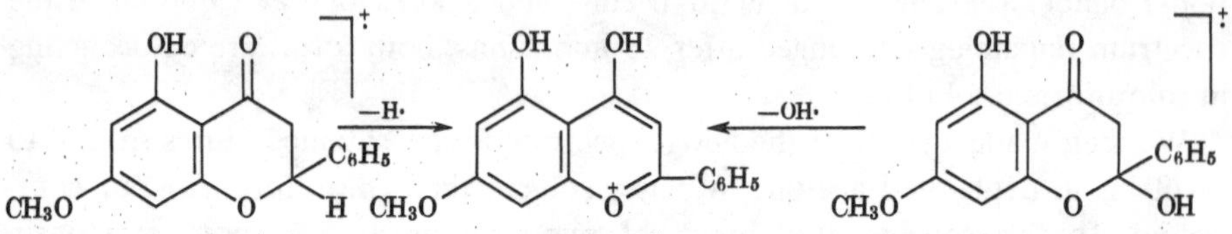

The 2-hydroxyflavanone structure had been considered for a long time as an intermediate step in the Baker–Venkataraman flavone synthesis. However, in their thorough study of the UV, IR and NMR spectra of more than ten *o*-monohydroxydibenzoylmethanes, Wagner *et al.* [6] found no trace of such a cyclic hemiketalic form.

This can be ascribed to chelation between the carbonyl group and the unique *ortho*-hydroxyl group, preventing participation of the latter in hemiketalization with the other carbonyl group.

In the compound P_4, two *ortho*-hydroxyl groups being present, one of them is chelated, but the other is available for hemiketalization.

Then, it could be expected that all *ortho*-dihydroxydibenzoylmethanes would exist in cyclic hemiketalic form (Chadenson *et al.* [7]; Hauteville *et al.* [8]).

In order to test the validity of this assumption, the preparation of 2,6-dihydroxydibenzoylmethane (VIII) was attempted by Baker–Venkataraman rearrangement of 2,6-dihydroxyacetophenone monobenzoate (V) under the conditions previously used for the synthesis of the compound P_4.

However, the reaction led to three products which could be separated in the order of increasing solubility: 3-benzoyl-5-hydroxyflavone (VI), 5-hydroxyflavone (VII) and the expected 2,6-dihydroxydibenzoylmethane (VIII).

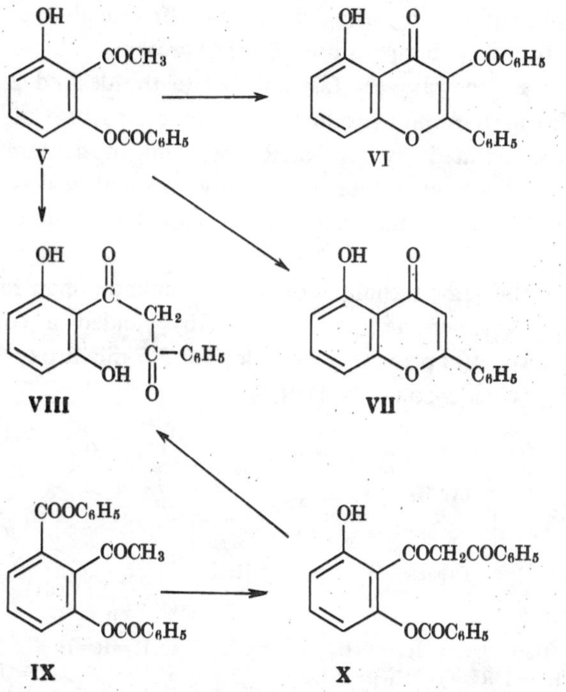

The m.p. 130°C and infrared spectrum ($\nu_{C=O}^{KBr}$ 1,645 cm^{-1}) of the latter did not agree with those (m.p. 155 °C; $\nu_{C=O}^{KBr}$ 1,632 and 1,663 cm^{-1}) described by Looker *et al.* [9] for a compound isolated in low yield from the rearrangement products of 2,6-dihydroxyacetophenone dibenzoate (IX) by powdered sodium hydroxide in pyridine at room temperature. Repeating their experiments, we could obtain (in agreement with Looker) 2-benzoyloxy-6-hydroxydibenzoylmethane (X). Mild saponification of the latter led us to the same 2,6-dihydroxydibenzoylmethane, m.p. 130 °C, as before.

The properties of this substance were quite parallel to those of the compound P_4 : λ_{max}^{EtOH} 270 nm (EtOH), 290 nm ($AlCl_3$), 385 nm (NaOH); $\nu C{=}O$ (KBr) 1,645 cm^{-1}. In the mass spectrum M—17/M = 42 % and the NMR spectrum in deuterated acetone showed the same characteristic signals of the cyclic hemiketalic form exclusively.

We then turned to the synthesis of 2-hydroxyflavanones related to widely spread flavones and which might be found in nature.

However, when applied to phloracetophenone tribenzoate (XIa), none of the preceding techniques of Baker–Venkataraman rearrangement gave satisfactory results, the flavone being the main product.

We could only succeed by using dimethyl sulfoxide instead of pyridine as a solvent. At ordinary temperature, in the presence of powdered sodium hydroxide, the reaction was very fast and led to the desired 2,5,7-trihydroxyflavanone (XIIa) as the main product. The cyclic structure of this compound was clearly demonstrated by its NMR spectrum in deuterated acetone at — 60 °C and at room temperature. In this case, a small peak at δ 4.57 may be ascribed to the $-CH_2-$ of the diketo form corresponding to less than 5 % of the mixture.

Application of the same technique of Baker–Venkataraman rearrangement to 4-O-methylphloracetophenone dianisoate (XIb) yielded a compound which showed again the spectral properties expected for 2,5-dihydroxy-7,4′-dimethoxyflavanone (XIIb) (Chadenson *et al.* [10]).

$COOR_2$, $COCH_3$, R_1O, $OCOR_2$ — NaOH / DMSO → OH, O, RO, O, OH, R′

XI — XII

a. $R_1 = C_6H_5-CO-$; $R_2 = C_6H_5$ — a. $R = R' = H$

b. $R_1 = CH_3-$; $R_2 = p\text{-}CH_3O-C_6H_4-$ — b. $R = CH_3$; $R' = OCH_3$

On the other hand, it was interesting to know whether these 2-hydroxyflavanones could be directly produced from the corresponding flavones by refluxing with potassium hydroxide in pyridine, as compound P_4 from tectochrysin.

Thin-layer chromatography of the reaction products showed that only 5-hydroxyflavone was completely converted into 2,5-dihydroxyflavanone.

Chrysin yielded only small amounts of 2,5,7-trihydroxyflavanone, along with phloracetophenone and phloroglucinol. With apigenin-7,4′-dimethyl ether, no trace of 2-hydroxyflavanone could be observed.

Chrysin 7-glucoside led to a complex mixture containing phloroglucinol, phloracetophenone, chrysin, chrysin 7-glucoside, 2,5,7-trihydroxyflavanone and a compound chromatographically identical with the natural dibenzoylmethane glucoside isolated by Williams.

Both compounds showed on silicagel the same bright fluorescence as all other 2-hydroxyflavanones studied. Moreover, the infrared spectrum of Williams' compound showed only one carbonyl band (ν 1,640 cm^{-1}) and it can be assumed that this compound, like the other 2,6-dihydroxydibenzoylmethanes studied, does exist in the cyclic hemiketalic form.

This assumption is further supported by the recent isolation by Williams from other *Malus* species of the isomeric 2,4,6-trihydroxydibenzoylmethane 2-glucoside, which showed the UV and chromatographic properties of a mixture of diketo and enol forms.

Thus, 2,5-dihydroxyflavanones appear as a new type of natural compounds, which can be found in the free state, as Wollenweber's compound, or in glycoside form, as Williams' compound (Hauteville *et al.* [11]).

Their place in flavonoid metabolism remains to be elucidated, since they could be equally considered as precursors or metabolites of flavones.

*

I must acknowledge the skill and patience of Miss Chadenson and Miss Hauteville throughout this work.

REFERENCES

1. Wollenweber, E. and Egger, K., *Phytochemistry, 10,* 225 (1971).
2. Williams, A. H., *Chem. and Ind., 1967,* 1526.
3. Chadenson, M., Hauteville, M., Chopin, J., Wollenweber, E., Tissut, M. and Egger, K., *Compt. rend. Acad. Sci., Ser. D, 273,* 2658 (1971).
4. Farkas, L., Major, A. and Strelisky, J., *Chem. Ber., 96,* 1684 (1963).
5. Hauteville, M. and Chadenson, M., *Bull. Soc. Chim. France, 1973,* 1780.
6. Wagner, H., Seligmann, O., Hörhammer, L., Nógrádi, M., Farkas, L., Strelisky, J. and Vermes, B., *Acta Chim. Acad. Sci. Hung., 57,* 169 (1968).
7. Chadenson, M., Hauteville, M. and Chopin, J., *J. C. S. Chem. Commun., 1972,* 107.
8. Hauteville, M., Chadenson, M. and Chopin, J., *Bull. Soc. Chim. France, 1973,* 1781.
9. Looker, J. H., Edman, J. R. and Dappen, J. I., *J. Heterocyclic Chem., 1,* 141 (1964).
10. Chadenson, M., Hauteville, M. and Chopin, J., *Compt. rend. Acad. Sci., Ser. C, 275,* 1291 (1972).
11. Hauteville, M., Chadenson, M. and Chopin, J., *Bull. Soc. Chim. France, 1973,* 1784.

SYNTHESIS OF NATURAL DIGLUCOSIDES OF QUERCETIN

by

B. VERMES, L. FARKAS and M. NÓGRÁDI

Institute of Organic Chemistry, Technical University

Budapest, Hungary

Introduction

The synthesis of flavonoid glycosides is a traditional field of research in our laboratory. During the past two years we have taken interest in the synthesis of quercetin glucosides.

Quercetin (I) is unique in the large family of natural products in the respect that this single aglucone has 36 known different mono- and diglycosides.

Methods for the structural elucidation of flavonoid glycosides are well documented, but there is relatively little information concerning the synthesis of these glycosides and until relatively recently the task of linking a sugar to a specific position of an aglucone moiety containing several hydroxyl groups seemed formidable and could only be solved in special cases. This problem is particularly acute with quercetin, which contains five phenolic hydroxyl groups.

In a previous work [1, 2] we undertook the synthesis of all possible quercetin monoglucosides: all of them are naturally occurring compounds (see Table I). The four known natural quercetin diglucosides [3] are also shown in Table I.

The selective synthesis of quercetin glycosides makes use of the fact that the C-7 and C-4′ hydroxyl groups of polyhydroxyflavones have preferential reactivity in nucleophilic substitutions, the next in this respect being then the C-3 hydroxyl. The C-3′ hydroxyl group is rather unreactive, and strong chelation with the carbonyl group usually prevents reaction at the C-5 hydroxyl. This gradation in reactivity is conditioned by the different acidity of the phenolic hydroxyl groups of quercetin. Enhanced acidity and reactivity in nucleophilic substitutions of the C-7 hydroxyl can be best rationalized by the electronegativity of the pyrone carbonyl group, affecting primarily the hydroxyl in *para* position (C-7) and to a lesser extent that attached to C-4′. The different acidities of the hydroxyl groups have been utilized many times before in selective benzylation, methylation and transacylation reactions.

Recently quantitative data were published by Tjukovina and Pogodeva [4] about the acidity of flavone hydroxyls. They measured the pK values of the hydroxyls by UV spectroscopy and, in accord with our observations, they found that the order of acidity was $7 > 4' > 3$.

Table I
Occurrence of Natural Quercetin Glucosides

I

Glucoside	Genus and species	Ref.
Quercetin 3 -glucoside	*Gossipium herbaceum*	[14]
Quercetin 5 -glucoside	*Lamium album*	[15]
Quercetin 7 -glucoside	*Gossipium herbaceum*	[14]
Quercetin 3′-glucoside	*Hyppocastaneum*	[16]
Quercetin 4′-glucoside	*Spirea ulmaria*	[17]
Quercetin 3,3′-diglucoside	*Hyppocastaneum*	[5]
Quercetin 3,4′-diglucoside	*Allium cepa*	[13]
Quercetin 7,4′-diglucoside	*Allium cepa*	[13]
Quercetin 3,7 -diglucoside	*Ulex europeus* L.	[8]

The Synthesis of Quercetin 3,3′-Diglucoside

Quercetin 3,3′-diglucoside (II) was isolated by J. Wagner [5] in 1961.

Taking into account what has been said above, it can be seen that blocking the C-7 and C-4′ hydroxyls of quercetin by selective benzylation will leave the C-3-OH as the most reactive one among the remaining three hydroxyl groups. In fact, coupling of 7,4′-dibenzylquercetin (III) [6] with α-acetobromoglucose afforded a single glucoside (IV) that gave quercetin 3-glucoside on catalytic debenzylation and saponification [7]. This compound has only two free hydroxyl groups; the C-3′-OH needed for coupling, and the other attached to C-5, having a low reactivity due to chelation with the carbonyl group.

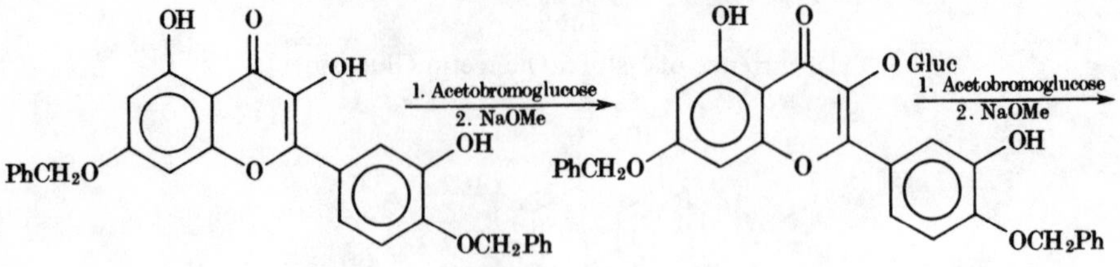

In a previous work with the similarly blocked 3,7,4′-tribenzylquercetin, we worked out suitable reaction conditions for the coupling of sugar to the rather unreactive C-3′ position and obtained, after some other steps, quercetin 3′-monoglucoside.

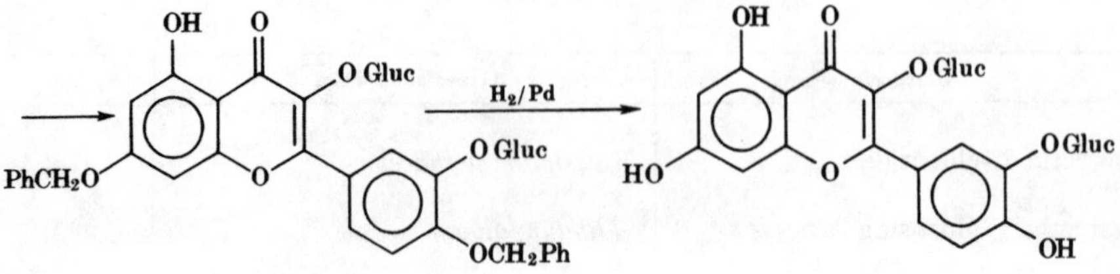

Now under the same conditions but starting with the 7,4′-dibenzyl-3-monoglucoside (IV) we obtained quercetin 3,3′-diglucoside dibenzyl ether (V) in a fair yield. After saponification and debenzylation the synthetic product was found identical with J. Wagner's [5] natural 3,3′-diglucoside (II) in every respect.

Quercetin 3,7-Diglucoside

For the synthesis of quercetin 3,7-diglucoside, quercetin 3-glucoside seemed to be a suitable starting material. Knowing the acidity gradient of the different hydroxyl groups of quercetin, it was expected that direct coupling of this compound with one mole of α-acetobromoglucose would give the 3,7-diglucoside [8].

This time we explored a new route for the synthesis of quercetin 3-glucoside that looked more simple than the previous one.

Naturally occurring rutin (VI), which is readily available, was benzoylated according to a patent of Jurd [9]. Subsequent hydrolysis of the sugar moiety with hot aqueous hydrochloric acid afforded the aglucone (VII) containing only one free hydroxyl group at the desired position (C-3). This compound was coupled with acetobromoglucose in the presence of silver oxide in quinoline, and the product was saponified.

In addition to the expected quercetin 3-glucoside (VIII) a new product was obtained which, as indicated by its solubility, spectra and R_f value, was evidently a diglucoside. When this diglucoside was compared with a sample of natural quercetin 3,7-diglucoside, it was found — as a pleasant surprise — that the two were identical (IX).

In order to establish firmly this identity, we effected total methylation and hydrolysis of the diglucoside to obtain quercetin 5,3′,4′-trimethyl ether (X)

[10]. In the same way also the structure of our quercetin 3-monoglucoside was checked, and from this we obtained 3-hydroxy-5,7,3′,4′-tetramethoxyflavone (XI) (cf. [11]). The same 3,7-diglucoside (IX) was also prepared from free quercetin 3-glucoside (VIII) by careful glucosylation.

In order to explain the formation of the diglucoside, we have to consider the transacylation reactions of the flavonoids first reported by us in 1968 [12]. We found that the migration of benzoyl groups in partially benzoylated flavonoids was a general reaction. It is caused and its course governed by differences in the acidity of the individual hydroxyl groups of polyhydroxyflavones. This transesterification is best catalyzed by bases, mainly by silver carbonate in combination with aromatic bases such as quinoline and pyridine. Under these conditions acyl groups are expected to shift from the hydroxyl of stronger acidity to another of weaker acid character. The interchange of acyl groups between hydroxyls of different acidity is obviously not restricted to those attached to the same flavone nucleus; for example, phenol itself is a suitable external acceptor for the leaving acyl group.

Previously we used this reaction in the case of quercetin pentabenzoate [2] (XII) to prepare quercetin 3,5,3′,4′-tetrabenzoate (XIV) as shown below. Considering these reactions we suppose that, under the circumstances of the glycosylation, i.e. in the presence of silver oxide in quinoline, an equilibrium between VII and quercetin-5,3′,4′-tribenzoate (XV) is established, the latter giving rise to the diglucoside IX.

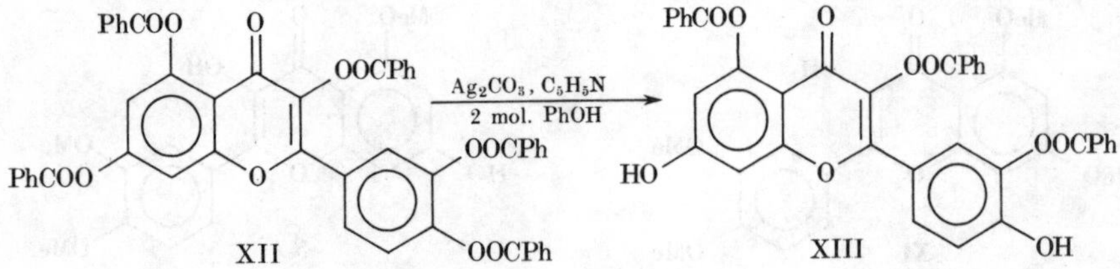

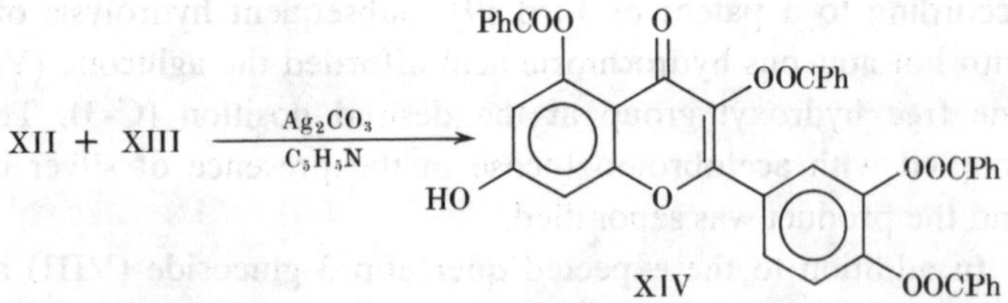

PhCOO O OH Ag_2O/quinoline PhCOO O OH + XII

PhCOO O OOCPh HO O OOCPh

OOCPh OOCPh

VII XV

Quercetin 3,7-diglucoside was first isolated by Harborne [8a] from *Ulex europeus*. Later the compound was also described by Birkofer and Kaiser [8b] as a product of the partial hydrolysis of quercetin 3-β-sophoroside-7-β-D-glucoside; they reported m.p. 246 °C. There was some confusion as to the m.p. of the diglucoside, because this value differed by about 30 °C from the m.p. of samples isolated by Olechnowicz-Stepien [8c], and synthesized by us earlier in another way [1]. The present product, obtained now by synthesis with a structure also checked by methylation, confirms the lower m.p.

Quercetin 3,4′-Diglucoside

Quercetin 3,4′-diglucoside (XVIII) and quercetin 7,4′-diglucoside (XXI) both were first isolated by Herrmann from *Allium cepa,* i.e. common onions [13].

The starting material of the synthesis of quercetin 3,4′-diglucoside was the known 7-benzylquercetin [6] (XVI). The coupling of this compound with α-acetobromoglucose and subsequent saponification afforded four glucosides. The chromatographic behaviour and relative amounts of these compounds were in agreement with our expectations regarding the relative reactivity of the phenolic hydroxyls of quercetin. The four products were fractionated by column chromatography and additional amounts of the diglucosides could be obtained by recycling the monoglucoside fraction into the glucosylation reaction. Both reactions gave predominantly the desired 3,4′-diglucoside-7-benzyl ether (XVII).

The free 3,4′-diglucoside (XVIII) was obtained by catalytic debenzylation of its 7-benzyl ether and was identical in every respect with the natural product.

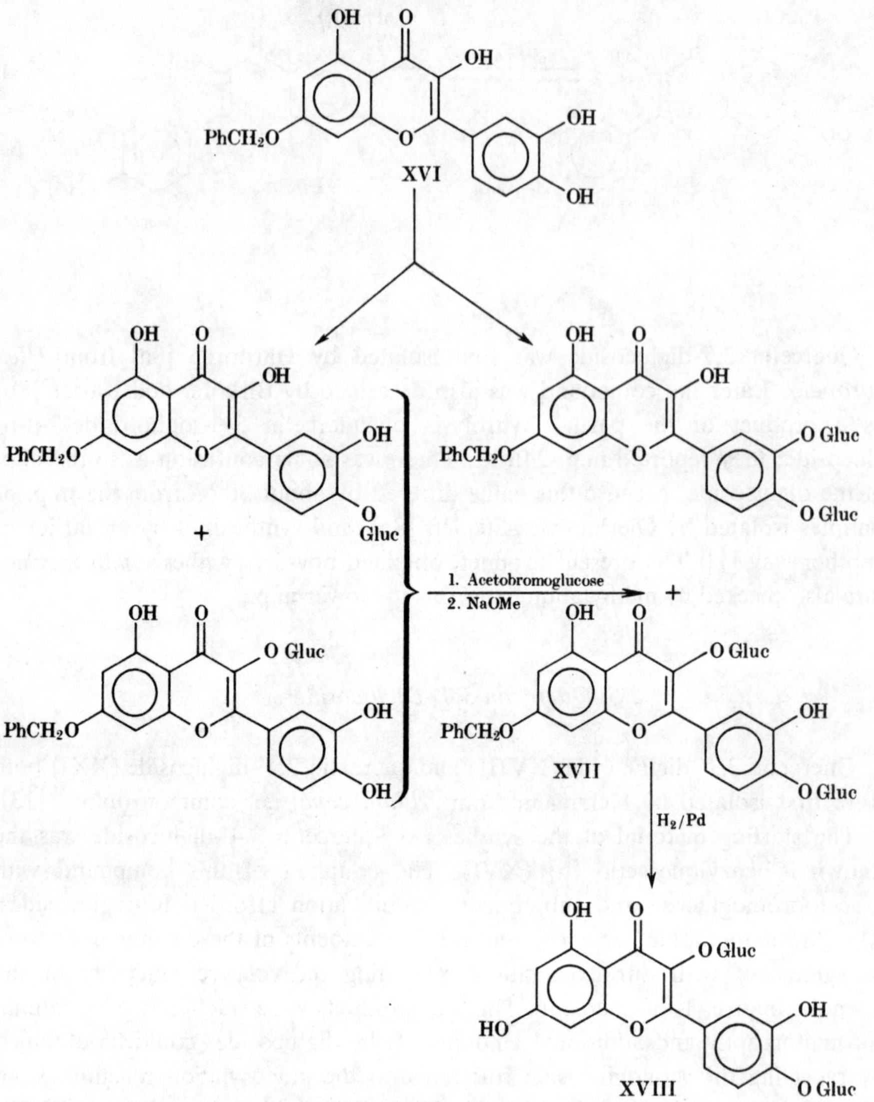

XVI

1. Acetobromoglucose
2. NaOMe

XVII

H_2/Pd

XVIII

Quercetin 7,4′-Diglucoside

Quercetin 7,4′-diglucoside has been synthesized from 7,4′-dibenzylquercetin (XIX) [6]. Benzoylation and catalytic debenzylation of this compound gave the dihydroxy derivative XX with free hydroxyls at the required positions. As it was

XIX

XX

XXI

expected, the coupling of this aglucone with a large excess of α-acetobromoglucose resulted in the 7,4′-diglucoside in one step. Catalytic saponification and debenzylation gave finally quercetin 7,4′-diglucoside (XXI).

REFERENCES

1. FARKAS, L., NÓGRÁDI, M., VERMES, B., WOLFNER, A., WAGNER, H., HÖRHAMMER, L. and KRÄMER, H., *Chem. Ber., 102,* 2583 (1969).
2. FARKAS, L., VERMES, B. and NÓGRÁDI, M., *Chem. Ber., 105,* 3505 (1969).
3. FARKAS, L., VERMES, B. and NÓGRÁDI, M., *Chem. Ber., 107,* 1518 (1974).
4. TJUKOVINA, N. A. and POGODEVA, N. N., *Khim. Prir., 7,* 11 (1971).
5. a. WAGNER, J., *Naturwissenschaften, 48,* 54 (1961).
   b. WAGNER, J., *Hoppe-Seylers Z. physiol. Chem., 335,* 232 (1969).
6. JURD, L., *J. Org. Chem., 27,* 1294 (1962).
7. HÖRHAMMER, L., WAGNER, H., ARNDT, H. G., HITZLER, G. and FARKAS, L., *Chem. Ber., 101,* 450 (1968).
8. a. HARBORNE, J. B., "Comparative Biochemistry of the Flavonoids", Academic Press, London, 1967, p. 67.
   b. BIRKOFER, L. and KAISER, C., *Z. Naturforsch., 17b,* 359 (1962).
   c. OLECHNOWICZ-STEPIEN, W., *Dissert. Pharm. Pharmacol., 22,* 16 (1970).
9. JURD, L., *U.S. Pat.* 3,661,890 (1972); *C.A., 77,* 75131 (1972).
10. KRISHNAMURTI, M., RAMANATHAN, J. D., SESHADRI, T. R. and SHENKARAN, P. R., *Indian J. Chem., 3,* 270 (1965).
11. RAHMAN, W. and ILYAS, M., *J. Org. Chem., 27,* 153 (1962).

12. Nógrádi, M., Farkas, L., Wagner, H. and Hörhammer, L., *Chem. Ber.*, *101*, 1630 (1968).
13. Herrmann, K., *Arch. Pharm. Berl.*, *291*, 238 (1967); Harborne, J. B., *Phytochemistry*, *4*, 107 (1965); Bandyukova, U. A. and Shinkorenko, G. L., *Farm. Zh.*, *22*, 54 (1967); *C.A.*, *67*, 57253 (1967).
14. Perkin, A. G., *J. Chem. Soc.*, *95*, 2183 (1908).
15. Harborne, J. B., *Phytochemistry*, *6*, 1569 (1967).
16. Wagner, J., *Naturwissenschaften*, *47*, 158 (1960).
17. Casparis, P. and Steinegger, E., *Pharm. Acta Helv.*, *21*, 341 (1947).

NATURALLY OCCURRING FLAVONE AND FLAVONOL BIOSIDES CONTAINING D-XYLOSE

by

V. M. CHARI and H. WAGNER

Institut für pharmazeutische Arzneimittellehre der Universität
München, FRG

Introduction

D-Xylose occurs relatively rarely as part of the disaccharide moiety in flavone and flavonol biosides [1]. To date there are only four such disaccharide combinations that have been reported to occur as the sugar part of such glycosides: sambubiose (2-*O*-β-D-xylopyranosyl-D-glucopyranose) (I), primeverose (6-*O*-β-D-xylopyranosyl-D-glucopyranose) (II), lathyrose (2-*O*-β-D-xylopyranosyl-D-galactopyranose) (III), and a xylosylglucuronide of unknown structure (Fig. 1). From the flowers of *Leucojum vernum* Linn. was isolated a kaempferol-xylosylglucoside whose disaccharide was assigned the structure 2-*O*-α-D-xylopyranosyl-D-glucopyranoside which is isomeric with sambubiose. Later work has, however, shown that the structure was incorrect.

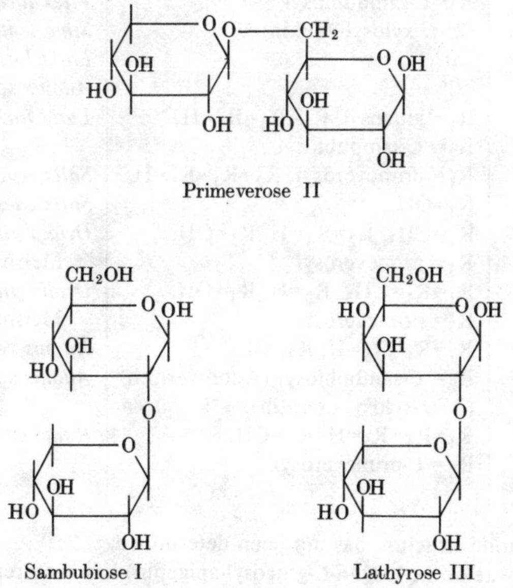

Primeverose II

Sambubiose I

Lathyrose III

Fig. 1

These disaccharides are mainly *O*-linked to the aglycones, though sambubiose also occurs as *C*-glycosylated with apigenin and luteolin. The other aglycones reported to be *O*-linked with these sugars are luteolin-7- and luteolin-7,4′-dimethyl ethers, kaempferol and quercetin. It can be seen from Table I that among such naturally occurring glycosides the *O*- and *C*-sambubiosyl derivatives seem to predominate.

Table I

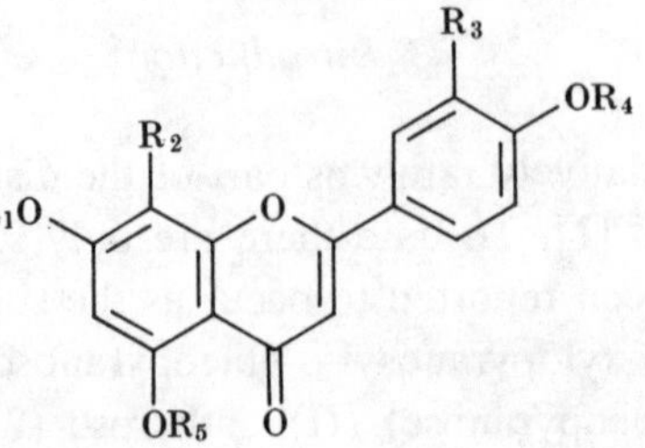

No.	Parent flavone	Substitution	Source	Reference
IV	Apigenin	$R_1=R_2=R_3=R_4=H$, $R_5=$ primeverosyl	*Ovidia pillo pillo* Meisner	[2]
V	Apigenin	$R_1=$ xylosyl-glucuronyl, * $R_2=R_3=R_4=R_5=H$	*Tanacetum niveum* (Lag) Schultz Bip.	[3]
VI	Apigenin	$R_1=R_3=R_4=R_5=H$, $R_2=C$-sambubiosyl (2″-*O*-xylosyl-vitexin) **	*Citrus sinensis* *Vitex lucens* T. Kirk *Stipa lemonii* Vasey *Larix laricina* Koch *Adonis* spp.	[4] [5] [6] [7] [8]
VII	Apigenin	$R_1=$ glucosyl, $R_3=R_4=R_5=H$, $R_2=$ *C*-sambubiosyl	*Larix laricina* Koch	[7]
VIII	Luteolin	$R_1=$ primeverosyl, $R_2=R_4=R_5=H$, $R_3=OH$	*Salix repens* L. *Salix caesia* Vill.	[9] [10]
IX	Luteolin	$R_1=CH_3$, $R_2=R_4=H$, $R_3=OH$, $R_5=$ primeverosyl	*Ovidia pillo pillo* Meisner	[2]
X	Luteolin	$R_1=R_4=CH_3$, $R_2=H$, $R_3=OH$, $R_5=$ primeverosyl	*Ovidia pillo pillo* Meisner	[2]
XI	Luteolin	$R_1=R_4=R_5=H$, $R_3=OH$, $R_2=$ *C*-sambubiosyl (Adonivernith) (2″-*O*-xylosyl-orientin)	*Adonis vernalis* L. *Adonis* spp.	[8]
XII	Luteolin	$R_1=R_4=R_5=H$, $R_3=OH$, $R_2=$ *C*-primeverosyl	*Ranunculus lingua* L.	[11]

* The disaccharide structure has not been determined.

** A xylosyl-isovitexin (xylosyl-6-*C*-glucosyl-apigenin) has been reported [12] to occur in *Tragopogon dubius;* the position of linkage of D-xylose to the glucose of isovitexin has not been established.

Table I (continued)

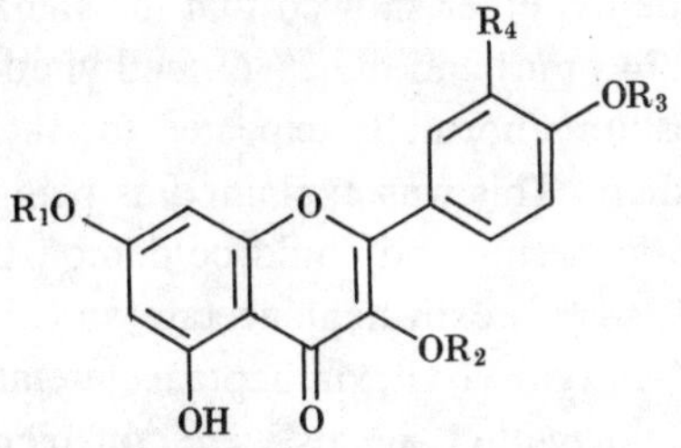

No.	Parent flavonol	Substitution	Source	Reference
			Phaseolus vulgaris L.	[13]
XIII	Kaempferol	R_1=R_3=R_4=H,	*Helleborus niger* L.	[14]
		R_2= sambubiosyl	*Leucojum vernum* L.	[15, 16]
XIV	Kaempferol	R_1= glucosyl, R_3=R_4=H,	*Helleborus niger* L.	[14]
		R_2= sambubiosyl	*Leucojum vernum* L.	[15, 16]
XV	Kaempferol	R_1= rhamnosyl, R_3=R_4=H,	*Lathyrus odoratus* L.	[17]
		R_2= lathyrosyl*		
XVI	Quercetin	R_1= glucosyl, R_3=H, R_4=OH,	*Helleborus foetidus* L.	[18]
		R_2= sambubiosyl		
XVII	Quercetin	R_1=R_3=H, R_2= sambubiosyl,	*Aesculus*	[19]
		R_4= glucosyloxy	*hippocastanum* L.	
XVIII	Quercetin	R_1=R_3=H, R_2= sambubiosyl,	*Leucojum vernum* L.	[16]
		R_4=OH		

* The disaccharide structure has not been established.

Past Work on Flavonoid Xylosyl-glucosides

Complete chemical structure analyses for these glycosides, prior to this work, have not yet been carried out. None of these glycosides has as yet been synthesized. The structures of the 5-*O*-primeverosides of apigenin, luteolin 7-*O*-methyl- and 7,4′-dimethyl ethers have been concluded from a consideration of their UV spectra, spectral shifts and mass spectral fragmentation of their perdeuteromethyl ethers [2]. The fragmentation observed for these compounds was that reported to be characteristic of a $1 \rightarrow 6$ interglycosidic linkage in flavonoid-*O*-biosides. Acid and enzymic hydrolysis of caesioside (VIII), isolated from *Salix caesia* Vill., and identification of the hydrolysates indicated the structure luteolin 7-primeveroside [10]. The luteolin xyloglucoside in *Salix repens* first reported [9] is probably identical with caesioside.

The structure of 2″-*O*-xylosyl-vitexin (8-*C*-sambubiosyl-apigenin) (VI) was elucidated [4] in the following manner. Acid hydrolysis yielded vitexin (8-*C*-glucosyl-apigenin), a known compound, and D-xylose. The PMR spectra of vitexin heptaacetate and the heptamethyl ether showed that the signal for the methyl protons of the 2″-*O*-acetate (1.74 ppm) and the 2″-*O*-methyl ether (3.0 ppm) are upfield compared with the positions normally expected for the methyl signals of sugar acetates and methyl ethers. This was explained as being due to the diamagnetic shielding of the aromatic A ring and could be demonstrated by an examination of the molecular models of vitexin heptaacetate and the heptamethyl ether. In the PMR spectra of 2″-*O*-xylosyl-vitexin heptaacetate and the heptamethyl ether the corresponding upfield methyl signals were conspicuously absent. Therefore it was concluded that the D-xylose residue was linked to the C-2″ of the glucose moiety in vitexin. Adonivernith (XI), initially isolated from *Adonis vernalis* and later from other *Adonis* spp. [8], on acid hydrolysis yielded orientin (8-*C*-glucosyl-luteolin) and D-xylose. The PMR spectrum of adonivernith decaacetate also showed the absence of a methyl singlet at 1.74 ppm, indicating again that D-xylose is linked to C-2″ of glucose in orientin.

The only flavonoid-*O*-sambubioside to be chemically characterized in the past is the anthocyanin sambucicyanin (cyanidin 3-*O*-β-sambubioside) from the berries of *Sambucus nigra* [20]. This glycoside was shown to be not identical with synthetic cyanidin 3-*O*-β-primeveroside and the 3-*O*-β-isoprimeveroside. Permethylated sambucicyanin on hydrolysis yielded the sugar partial methyl ethers identified as 2,3,4-tri-*O*-methyl-D-xylose and 3,4,6-tri-*O*-methyl-D-glucose. This led to the formulation of sambucicyanin as cyanidin 3-*O*-(2-*O*-β-D-xylopyranosyl-β-D-glucopyranose). This was the first report of such a disaccharide in the literature and was accordingly given the trivial name of sambubiose. No proof for the linkage configuration was, however, advanced. Similar xyloglucosides, tacitly assumed to be sambubiosides of delphinidin and pelargonidin have also been reported. Based on analogy and not on chemical proof, the xyloglucosides of the flavonols kaempferol and quercetin were assigned the sambubioside structure. The same applies to kaempferol 3-*O*-lathyroside-7-*O*-rhamnoside (XV). The structure elucidation of leucoside (m.p. 191 °C) and leucovernide (m.p. 206–207 °C), two kaempferol xyloglucosides from the flowers of *Leucojum vernum*, represents the first complete analysis of a flavonol *O*-sambubioside.

From the petals of the alpine flower *Leucojum vernum* Linn (Frühlingsknotenblume) were isolated two flavonoid glycosides and given the trivial names leucoside and leucovernide. The isolation was carried out by sequential solvent extraction of the aqueous alcohol concentrate followed by column chromatography on cellulose and polyamide. Acid hydrolysis of the two glycosides gave kaempferol, D-glucose and D-xylose. UV spectra and sodium acetate shifts indicated that leucoside was a 3-*O*-glycoside and leucovernide a 3,7-di-*O*-

glycoside. Partial hydrolysis of leucovernide yielded kaempferol 7-*O*-glucoside and a disaccharide containing D-xylose and D-glucose. Similar treatment of leucoside gave kaempferol and the same disaccharide. Enzymic hydrolysis of leucovernide gave rise to a glycoside that was identical with leucoside. This indicated that leucovernide was leucoside-7-*O*-glucoside. Similarly, it has been reported [18] that quercetin 7-*O*-β-glucoside-3-*O*-β-xyloglucoside, on treatment with β-glucosidase for two hours, was hydrolyzed to quercetin 3-*O*-β-xyloglucoside. Kaempferol 3-*O*-β-sophoroside-7-*O*-β-D-glucoside upon such treatment yields kaempferol 3-*O*-β-sophoroside.

Prolonged treatment of leucovernide with β-glucosidase gave kaempferol 3-*O*-β-glucoside (astragalin) along with kaempferol. Hence it was concluded that the disaccharide moiety was a xylosyl-glucose. A consideration of the electrophoretic mobility of the disaccharide in borate buffer (pH 8.25) and that of its reduced product xylosyl-glucitol in a sodium molybdate buffer (pH 5.0) led to the postulation of the structure of the disaccharide as being xylosyl (1 → 2) glucose. An α-interglycosidic linkage for the biose was suggested on the basis of its inertness towards β-glucosidase, whereas it was split by α-glucosidase. Leucoside and leucovernide were therefore assigned the structures kaempferol 3-*O*-(2-*O*-α-D-xylopyranosyl-β-D-glucopyranoside) (XXIIIa) and its 7-*O*-β-D-glucopyranoside (XIVa), respectively (Fig. 2).

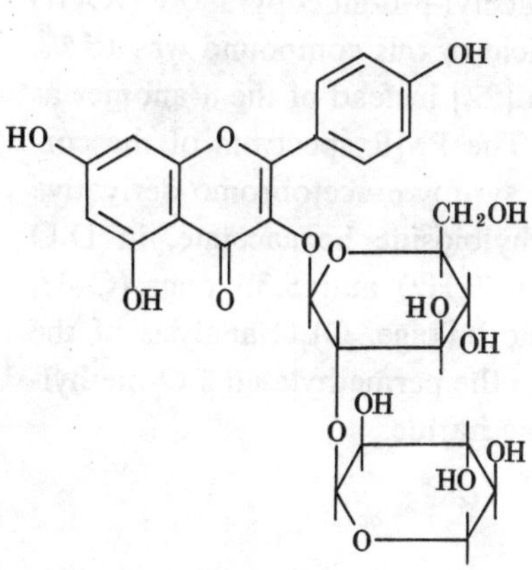

Leucoside
(XIIIa)

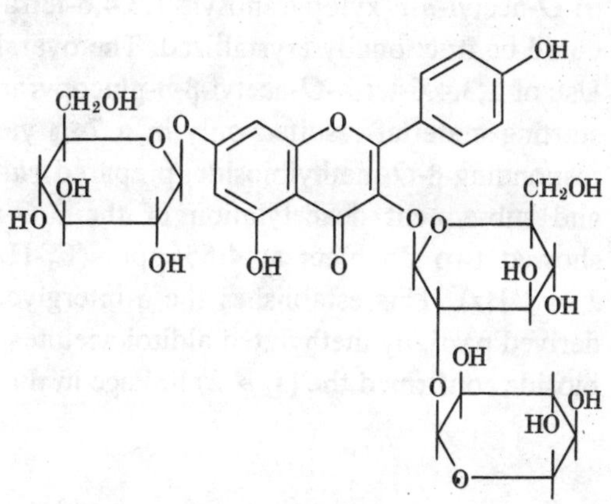

Leucovernide
(XIVa)

Fig. 2. Structures proposed for the *O*-xylosyl-glucosides of *Leucojum vernum*

Structure Investigation of the Leucojum Glycosides

(a) Synthetic work

The synthesis of the proposed structure of leucoside was undertaken in order to verify the constitution of the new disaccharide. The first step towards this objective was the synthesis of the hitherto unknown 2-*O*-(2,3,4-tri-*O*-acetyl-α-D-xylopyranosyl)-1,3,4,6-tetra-*O*-acetyl-β-D-glucopyranose (XXII). Condensation of 2,3,4-tri-*O*-acetyl-α-D-xylopyranosyl bromide (XX) with 1,3,4,6-tetra-*O*-acetyl-α-D-glucopyranose (XIX) in nitromethane solution in the presence of $Hg(CN)_2$ and $HgBr_2$ yielded a syrupy product (XXI) (Fig. 3). That this was a mixture of the α- and β-interlinked disaccharide 1-α-heptaacetates could be easily proved. The mixture (XXI) was deacetylated and reduced with $NaBH_4$ to the corresponding alditol-glycosides (XXIa). The PMR spectrum in D_2O of this product showed *inter alia* two doublets at 5.15 ppm ($J = 3$ Hz) and 4.6 ppm ($J = 8$ Hz) for the *equatorial* and *axial* interglycosidic protons in the alditol-glycoside. GLC of the trimethylsilyl ethers of the mixture (XXIa) at 220 °C showed the presence of two components. The formation of the α- and β-linked disaccharides under the above experimental conditions is well known [21]. From the syrupy reaction product (XXI) it was not possible to obtain a crystalline entity and the two disaccharide 1-α-heptaacetates had the same R_f values on silica gel in various solvent systems. As the 1-β-peracetates of disaccharides crystallize more easily, the 1-α-heptaacetates were converted to the corresponding 1-β-heptaacetates. From this mixture the required 2-*O*-(2,3,4-tri-*O*-acetyl-α-D-xylopyranosyl)-1,3,4,6-tetra-*O*-acetyl-β-D-glucopyranose (XXII) could be fractionally crystallized. The overall yield of this compound was 15 %. Use of 1,3,4,6-tetra-*O*-acetyl-β-D-glucopyranose [22] instead of the α-anomer as starting material resulted only in a 7 % yield. The PMR spectrum of the corresponding β-*O*-methylbioside, prepared *via* the syrupy α-acetobromo derivative and subsequent deacetylation of the β-*O*-methylbioside hexaacetate, in D_2O showed two doublets at 4.55 ppm (C_1-H, $J = 7$ Hz) and 5.33 ppm (C_1-H, $J = 3$ Hz). This establishes the α-interglycosidic linkage. GLC analysis of the derived partially methylated alditol acetates from the permethylated β-*O*-methylbioside confirmed the (1 → 2) linkage in the disaccharide.

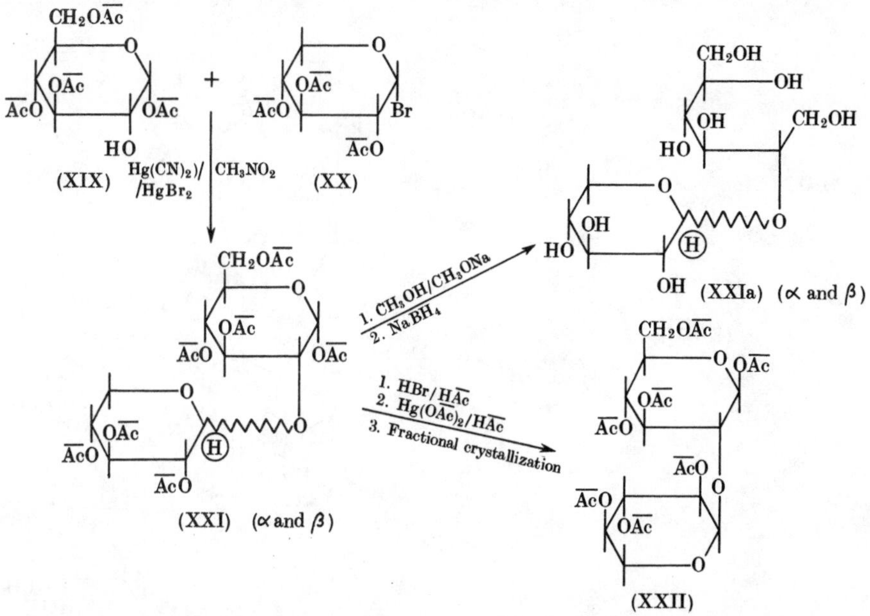

Fig. 3. Synthesis of 2-*O*-(2,3,4-tri-*O*-acetyl-α-D-xylopyranosyl) 1,3,4,6-tetra-*O*-acetyl-β-D-glucopyranose (XXII)

7,4′-Dibenzylkaempferol (XXIII), the synthesis of which was achieved through an Allan–Robinson reaction [23] (Fig. 4), was condensed with the syrupy α-acetobromodisaccharide (XXIV) in pyridine solution with silver car-

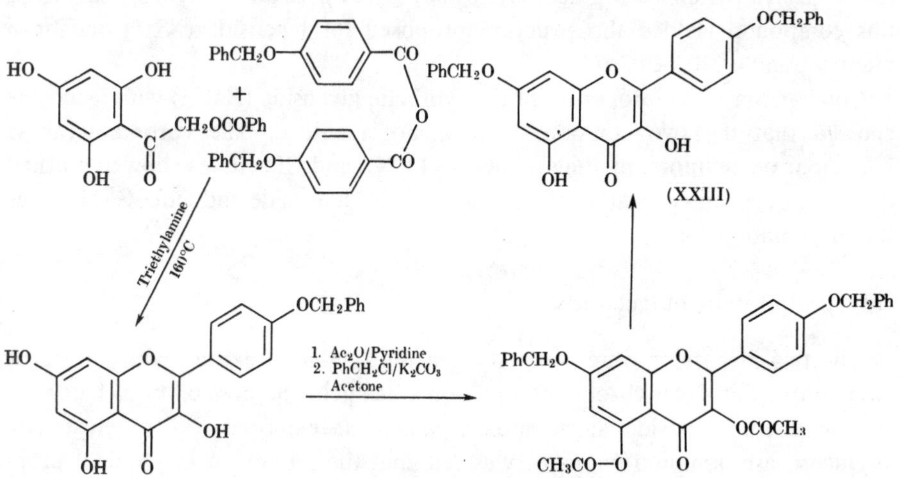

Fig. 4. Synthesis of 7,4′-Dibenzylkaempferol (XXIII)

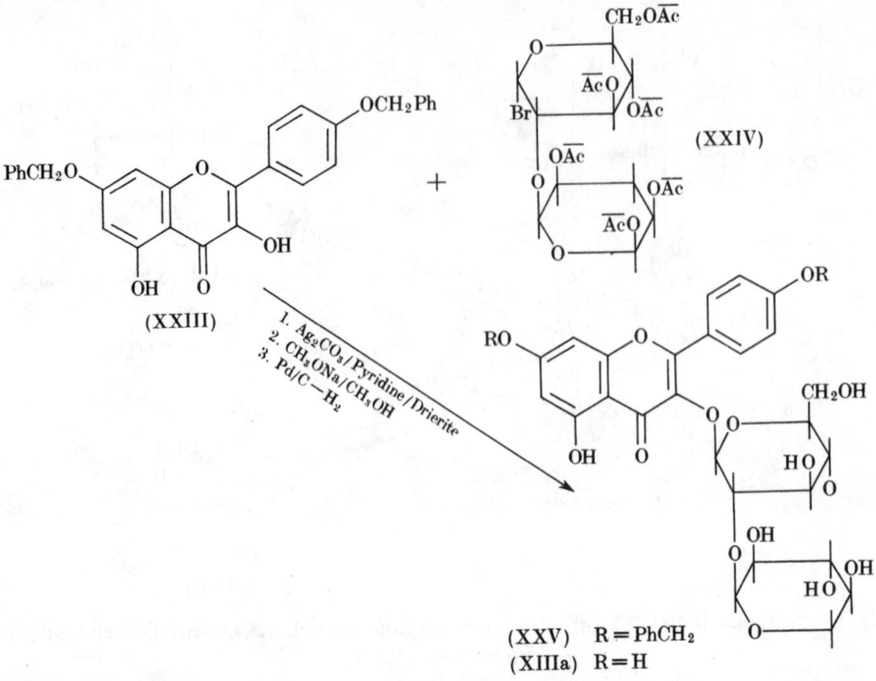

Fig. 5. Synthesis of the structure proposed for leucoside

bonate as acid acceptor (Fig. 5). The product was not isolated as such but was subjected to deacetylation to yield the crystalline 7,4′-dibenzylkaempferol 3-*O*-(2-*O*-α-D-xylopyranosyl-D-glucopyranoside) (XXV). Catalytic hydrogenolysis of this compound yielded the structure proposed for leucoside (XIII), having a melting point of 201–203 °C.

Comparison of the properties of the synthetic glycoside (XIIIa) with leucoside showed that the two compounds were not identical. The chromatographic behaviour on cellulose, melting points (≈11 °C) and CD curves showed marked differences. A reinvestigation of the structures of leucoside and leucovernide was therefore undertaken.

(b) Spectral study of leucoside

The two glycosides were isolated from the flavonoid extract concentrate by preparative thin-layer chromatography on silica gel. The possibility of leucoside having a sambubioside sugar moiety was next explored. As D-xylose and D-glucose are homomorphous, it was felt that the difference in physical properties between leucoside and the synthetic kaempferol glycoside should find correlation with those between kaempferol 3-*O*-β-sophoroside and kaempferol 3-*O*-kojibioside or even the corresponding glycosides of quercetin. A comparison of the R_f values of the glycosides on cellulose (solvent: 10 % and 15 % acetic acid) indicated that the β-interlinked bioside had the greater R_f value. The molar ellipticity at ca. 250 nm was much larger for the α-interlinked biosides.

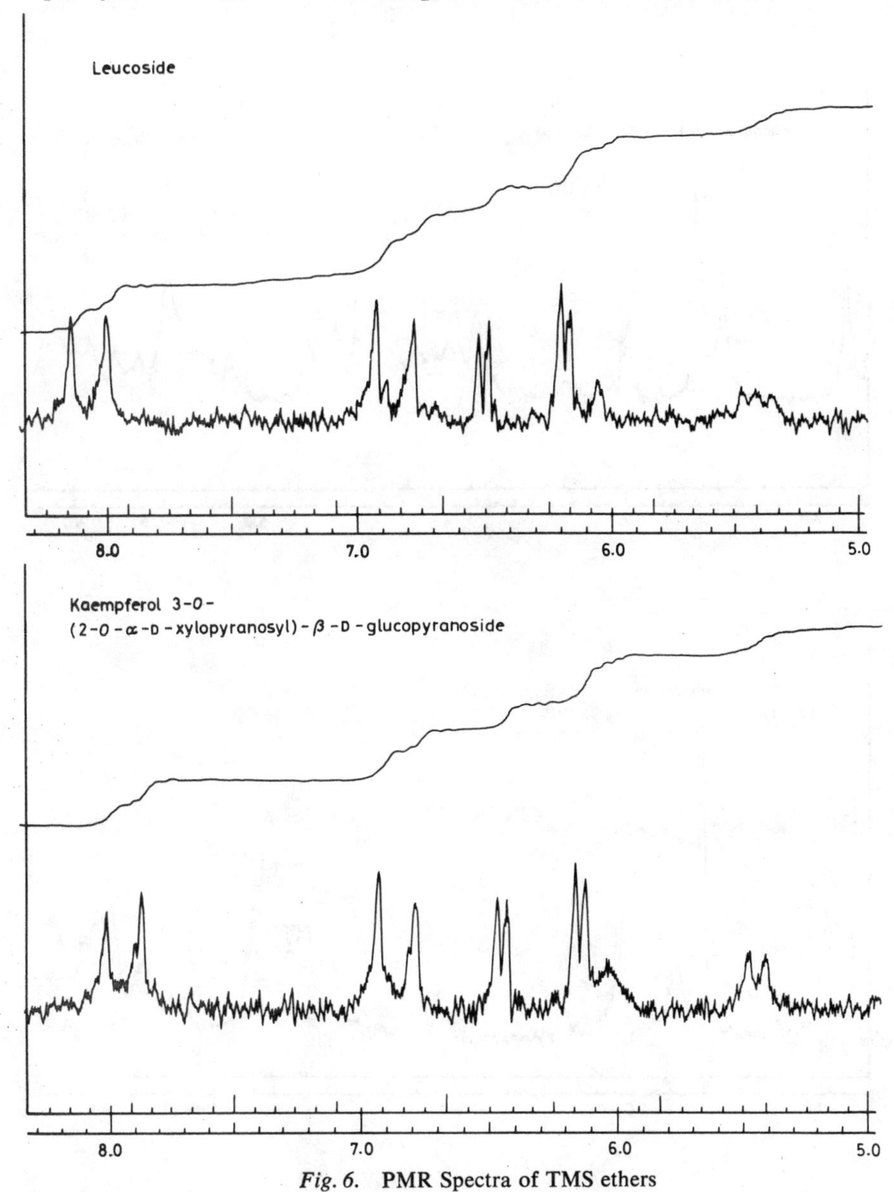

Fig. 6. PMR Spectra of TMS ethers

The PMR spectrum of the pertrimethylsilyl ether of leucoside showed *inter alia* a broad signal at 5.45 ppm (8 Hz) for the proton on C-1‴ of D-xylose and a doublet at 6.1 ppm (J = 8 Hz) for the proton on C-1″ of D-glucose (Fig. 6). The appearance of the interglycosidic anomeric proton on C-1‴ as a broad signal (ca. 7 – 9 Hz) at about 5.4 ppm in pertrimethylsilylated flavonoid bio-

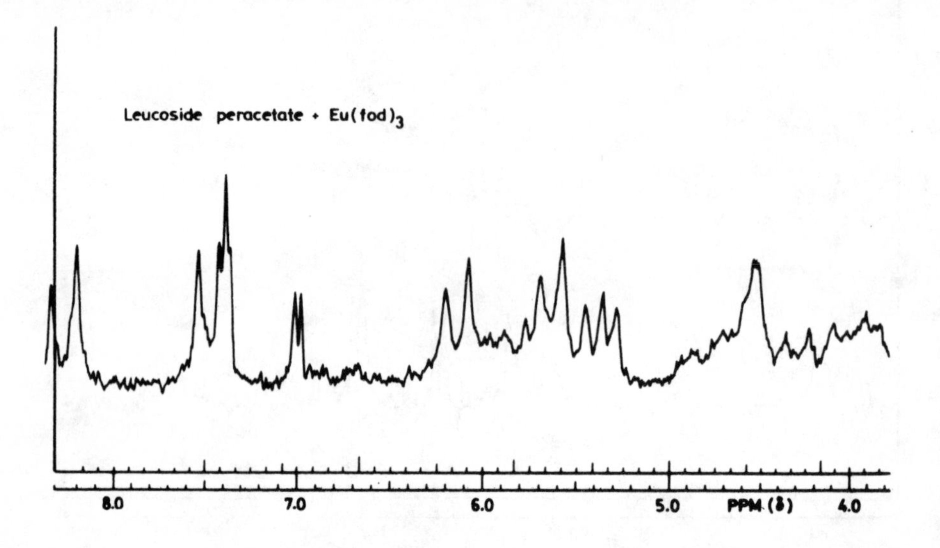

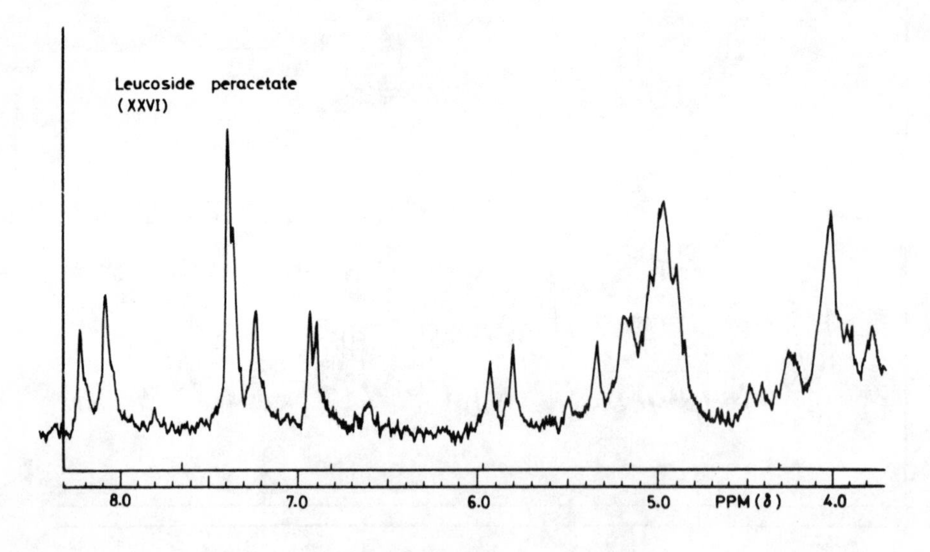

Fig. 7

sides with a β-linkage between the sugars is well documented [24]. The H-1″ resonances in the pertrimethylsilyl ethers of leucoside, the synthetic kaempferol glycoside, quercetin 3-*O*-β-sophoroside and quercetin 3-*O*-kojibioside occurred consistently at 6.1 ppm. This is more downfield than the corresponding H-1″ resonance in flavonol 3-*O*-β-glucosides (5.8 ppm). The deshielding is probably due to the proximity of the carbonyl group to the interglycosidic anomeric proton. The PMR spectrum of leucoside nonaacetate (XXVI) in the presence of $Eu(fod)_3$ (Fig. 7), however, provided conclusive evidence. In the absence of the shift reagent the H-1″ resonance appeared at 5.9 ppm (d; $J = 8$ Hz), whereas the H-1‴ signal was hidden amidst those of the other sugar protons. Addition of $Eu(fod)_3$ shifted the signals for the two anomeric protons downfield to 6.17 ppm (d; $J = 8$ Hz) and 5.65 (d; $J = 8$ Hz), respectively. The large values of the coupling constants for the two doublets establishes the β-configuration for both the flavonol-glucose and glucose-xylose linkages and a pyranose conformation for the two sugars.

Leucoside, permethylated according to the method of Hakomori [25], was subjected to mass spectrometry (Fig. 8). The fragmentation was in accord with that expected [26] for a kaempferol 3-*O*-bioside permethyl ether. The peak at m/e 328, which corresponded to a tri-*O*-methylkaempferol (A+H) was very intense. The methylated oligosaccharide fragment (OS) at m/e 379 gave rise to a relatively intense ($OS–CH_3OH$) peak at m/e 347. Such a fragmentation is considered to be a characteristic of a (1 → 2) bioside moiety [26].

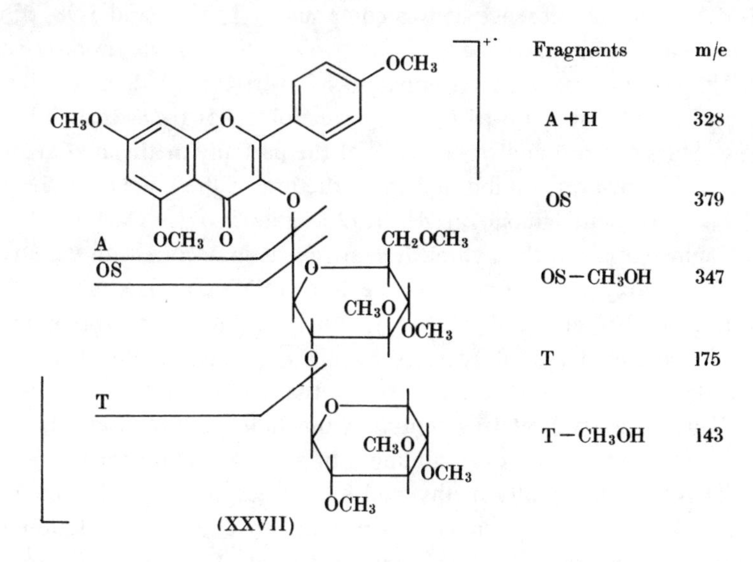

Fragments	m/e
A+H	328
OS	379
$OS-CH_3OH$	347
T	175
$T-CH_3OH$	143

Fig. 8. Mass spectrum of leucoside permethyl ether

The permethyl ether of the isomeric glycoside also exhibited the same fragmentation. An inspection of Table II indicates that the fragmentations (OS–CH_3OH) and T are relatively more intense for the β-interlinked biosides.

Table II *

Fragment m/e	A + H 328	A + H 358	OS 379	OS 423	OS–CH_3OH 347	OS–CH_3OH 391	T 175	T 218
Kaempferol 3–(2–*O*–α–D–xylosyl–β–D–glucoside)	100	–	2.4	–	17	–	40	–
Leucoside	100	–	7	–	56	–	112	–
Quercetin 3–*O*–kojibioside	–	100	–	0	–	4.6	–	5.8
Quercetin 3–*O*–sophoroside	–	100	–	0	–	14	–	14
Kaempferol 3–*O*–sophoroside	–	100	–	0	–	8.4	–	14

* The values give the intensities of the principal fragments relative to the (A+H) peak.

Permethylated leucoside was hydrolyzed and the sugar fragments were acetylated to yield the partially methylated alditol acetates. GLC of this mixture on ECNSS-M showed the presence of two compounds, T_G 0.68 and 1.98, of which the former was identified as being 2,3,4-tri-*O*-methylxylitol 1,5-di-*O*-acetate, while the latter could be ascribed either to 3,4,6-tri-*O*-methylglucitol 1,2,5-tri-*O*-acetate (T_G 1.98) or 2,4,6-tri-*O*-methylglucitol 1,3,5-tri-*O*-aceate (T_G 1.95) [27]. GLC-Mass spectral analysis (Fig. 9) of the partially methylated deuterated alditol acetates, however, established the structure of the glucitol derivative as being 3,4,6-tri-*O*-methylglucitol 1,2,5-tri-*O*-acetate-1 d (XXVIII). The mass spectral fragmentation of the xylitol derivative was in accord with the structure 2,3,4-tri-*O*-methylxylitol 1,5-di-*O*-acetate-1 d (XXVII). The absence of a primary fragment of m/e 118, always found in the mass spectra of partially methylated deuterated alditol acetates having an acetyl group at C-1 and a methoxyl at C-2, is unequivocal proof that the disaccharide linkage is $1 \rightarrow 2$ [27, 28]. This mass spectral fragmentation precludes a furanose structure for both the monoses in leucoside. Assuming a furanose structure for the D-glucose moiety, the resultant partially methylated alditol acetate would be 3,5,6-tri-*O*-methylglucitol 1,2,4-tri-*O*-acetate-1 d. The mass spectral fragmentation is different and is characterized by the fragment m/e 89, formed by fission of the bond between C-4 and C-5, as well as by the absence of the m/e 161 peak.

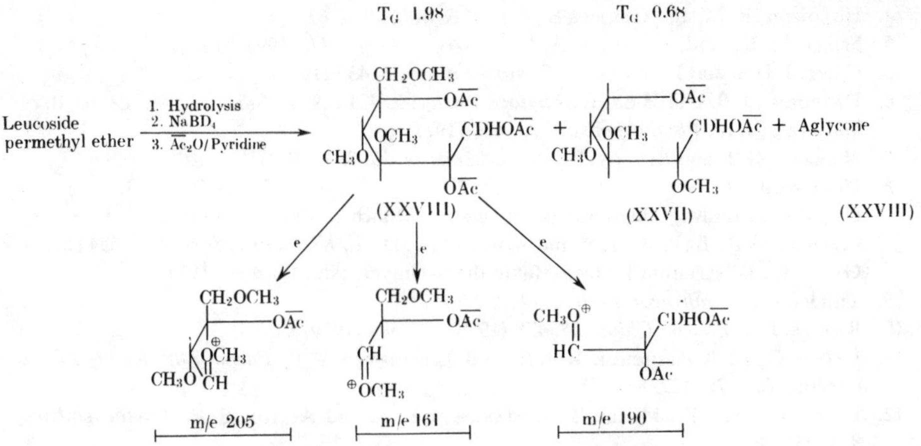

T_G 1.98 T_G 0.68

Leucoside permethyl ether → 1. Hydrolysis 2. $NaBD_4$ 3. Ac_2O/Pyridine → (XXVIII) + (XXVII) + Aglycone

(XXVIII) → m/e 205, m/e 161, m/e 190

Fig. 9. GC-MS analysis of leucoside permethyl ether

Permethylation and GLC analysis of leucovernide showed the presence of 2,3,4,6-tetra-*O*-methylglucitol 1,5-di-*O*-acetate (T_G 1.0) in addition to the two alditol acetate methyl ethers, as in the case of leucoside. This therefore definitely establishes the structures of leucoside and leucovernide as kaempferol 3-*O*-(2-*O*-β-D-xylopyranosyl-β-D-glucopyranoside) and its 7-*O*-β-D-glucopyranoside, respectively. The disaccharides sambubiose and lathyrose and their derivatives have been synthesized by us [29] and work is in progress towards the synthesis of leucoside and quercetin 3-*O*-β-sambubioside.

*

The award of an Alexander von Humboldt research fellowship (1968–1970) to V. M. Chari during the initial stages of this work is gratefully acknowledged by the authors.

REFERENCES

1. WAGNER, H., in: Zechmeister's "Fortschritte der Chemie organischer Naturstoffe" (eds. W. HERZ, H. GRISEBACH, G. W. KIRBY) Vol. 31, 1974, p. 153; HARBORNE, J. B., "Comparative Biochemistry of the Flavonoids", Academic Press, London and New York, 1967.
2. NUNEZ-ALARCON, J., RODRIGUEZ, E., SCHMID, R. and MABRY, T. J., *Phytochemistry, 12,* 1451 (1973).
3. HARBORNE, J. B., HEYWOOD, V. H. and SALEH, N. A. M., *Phytochemistry, 9,* 2011 (1970).

4. Horowitz, R. M. and Gentili, B., *Chem. & Ind., 1966,* 625.
5. Seikel, M. K. and Bushnell, A. J., *J. Org. Chem., 24,* 1995 (1959); Seikel, M. K., Chow, J. H.S. and Feldman, L., *Phytochemistry, 5,* 439 (1966).
6. Harborne, J. B. and Hall, E., *Phytochemistry, 3,* 421 (1964); Saleh, A. M., Bohm, B. A. and Maze, J. R., *Phytochemistry, 10,* 490 (1971).
7. Niemann, G. J. and Bekooy, R., *Phytochemistry, 10,* 893 (1971).
8. Hörhammer, L., Wagner, H. and Leeb, W., *Arch. Pharm., 65,* 264 (1960); Rosprim, L., Dissertation Ludwig-Maximilians-Universität, München, 1966; Chernobai, V. T., Komissarenko, N. F., Batyuk, V. S. and Kolesnikov, D. G., *Khim. Prir. Soedin., 5,* 634 (1970); Galle, K., Dissertation Ludwig-Maximilians-Universität, München, 1974.
9. Thieme, H., *Tetrahedron Letters, 1968,* 2781.
10. Rabate, J., *J. Pharm. Chim., 28,* 478 (1938); *29,* 584 (1939).
11. Drozd, G. A., Koreshchuk, K. J. E. and Litvinenko, V. I., *Farm. Zh. (Kiev), 24,* 56 (1969); *C. A., 70,* 112350.
12. Kroschewsky, J. R., Mabry, T. J., Markham, K. R. and Alston, R. E., *Phytochemistry, 8,* 1495 (1969).
13. Feenstra, W. J., *Meded. Landbouwhogeschool op zoekingstat Staat. Cent., 60,* 1 (1960).
14. Egger, K. and Keil, M., *Ber. dt. Bot. Ges., 78,* 153 (1965).
15. Hein, K., Dissertation Ludwig-Maximilians-Universität, München, 1965; Hörhammer, L., Wagner, H. and Beck, K., *Z. f. Naturforsch., 22b,* 988 (1967).
16. Chari, V. M. and Wagner, H., *Chem. Ber.,* In the press.
17. Harborne, J. B., *Nature, 187,* 140 (1960).
18. Harborne, J. B., *Phytochemistry, 4,* 107 (1965).
19. Wagner, J., *Z. Physiol. Chem., 335,* 232 (1964).
20. Reichel, L. and Reichwald, W., *Naturwiss., 47,* 41 (1960).
21. Helferich, B. and Zirner, J., *Chem. Ber., 95,* 2604 (1962).
22. Chari, V. M. and Wagner, H., *Chem. Ber.,* In the press.
23. Wagner, H., Danninger, H., Seligmann, O., Nógrádi, M., Farkas, L. and Farnsworth, N., *Chem. Ber., 103,* 3678 (1970).
24. Mabry, T. J., Markham, K. R and Thomas, M. B., "The Systematic Identification of Flavonoids", Springer Verlag, Berlin-Heidelberg-New York, 1970.
25. Hakomori, S., *J. Biochem. (Tokyo), 55,* 205 (1964).
26. Schmid, R. D., *Tetrahedron, 28,* 3259 (1972); Schmid, R. D., Varenne, P. and Paris, R., *Tetrahedron, 28,* 5037 (1972); Wagner, H. and Seligmann, O., *Tetrahedron, 29,* 3029 (1973).
27. Björndal, H., Hellerqvist, C. G., Lindberg, B. and Svensson, S., *Angew. Chem., 82,* 643 (1970).
28. Björndal, H., Lindberg, B. and Svensson, S., *Carbohydr. Res., 5,* 433 (1967).
29. Chari, V. M. and Wagner, H., *Chem. Ber.,* In the press.

MASS SPECTROMETRY OF FLAVONOID *O*-GLYCOSIDES

by

O. SELIGMANN and H. WAGNER
Institut für pharmazeutische Arzneimittellehre der Universität
München, FRG

Since 1963 detailed mass spectrometric studies have been made on flavonoids by a number of authors, notably by Reed and Wilson [1], Audier [2], Bowie and Cameron [3], Kingston [4], and Pelter *et al.* [5]. As aromatic compounds, flavones exhibit distinct molecular ion peaks. Typical fragmentation modes are the loss of CO and Retro Diels–Alder reaction. The mass spectra of flavone *C*-glycosides, investigated by Prox [6] and Chopin [7], show characteristic losses of water starting from the molecular ion. In contrast, flavone *O*-glycosides give essentially the spectra of the aglycones, i.e. the sugar is split off thermally; therefore, it is necessary to prepare derivatives. For this purpose the peracetates (PA), permethyl ethers (PME) and pertrimethylsilyl (PTMS) ethers may be advantageously used. This possibility of the structural analysis of flavone glycosides has been seldom used so far. It was only Weinges [8] who described the MS fragmentation of a few flavone glycoside acetates, and recently Schmid *et al.* [9–12] have shown that flavonoid di- and trisaccharides can be successfully investigated by mass spectrometry after perdeuteromethylation. In this way information is obtained about the structure of both the aglycone and the sugar moiety.

Our mass spectrometric investigations of flavonoid mono- and di-*O*-glycosides have been performed on the permethyl ethers [13]. First, a number of mono-glycosides were methylated on the micro scale, i.e. using about 0.1–1 mg of the glycosides. According to the method of Hakomori [14], the sodium methyl-sulfinylmethide procedure was used, which rapidly afforded the completely permethylated products. Somewhat less drastic is the variant using NaH and CH_3I in DMF as the solvent [15], and this was generally preferred in our experiments; thus the danger of C-methylation, especially in the case of flavonol glycosides, is reduced. The predominant fragmentation of such permethylated products is the loss of sugar with or without hydrogen transfer to the aglycone (Fig. 1). The flavonoid moiety promotes this fission so that the glycosyl cation is mainly produced. At the same time the fragmentation pattern of the sugar becomes simpler [16]. The aglycone appears with high intensity as the so-called

A+H peak, often as the base peak. Further fragmentation of the aglycone takes place in the well known way. The Retro Diels–Alder (RDA) fragments are here usually little pronounced, but — especially after perdeuteromethylation — they may afford indication of nuclear substitution. During the decomposition of the glycosyl cation the characteristic sequences of methanol elimination allow differentiation of the various sugars: e.g., hexoses afford the mass units 218, 187, 155 and 111, whereas pentoses such as arabinose give m/e 175, 143 and 111; a 6-deoxysugar (e.g. rhamnose) appears with the series 189, 157 and 125 mass units, and the m/e values of glucuronic acid are 201, 169 and 141 (Fig. 1). In

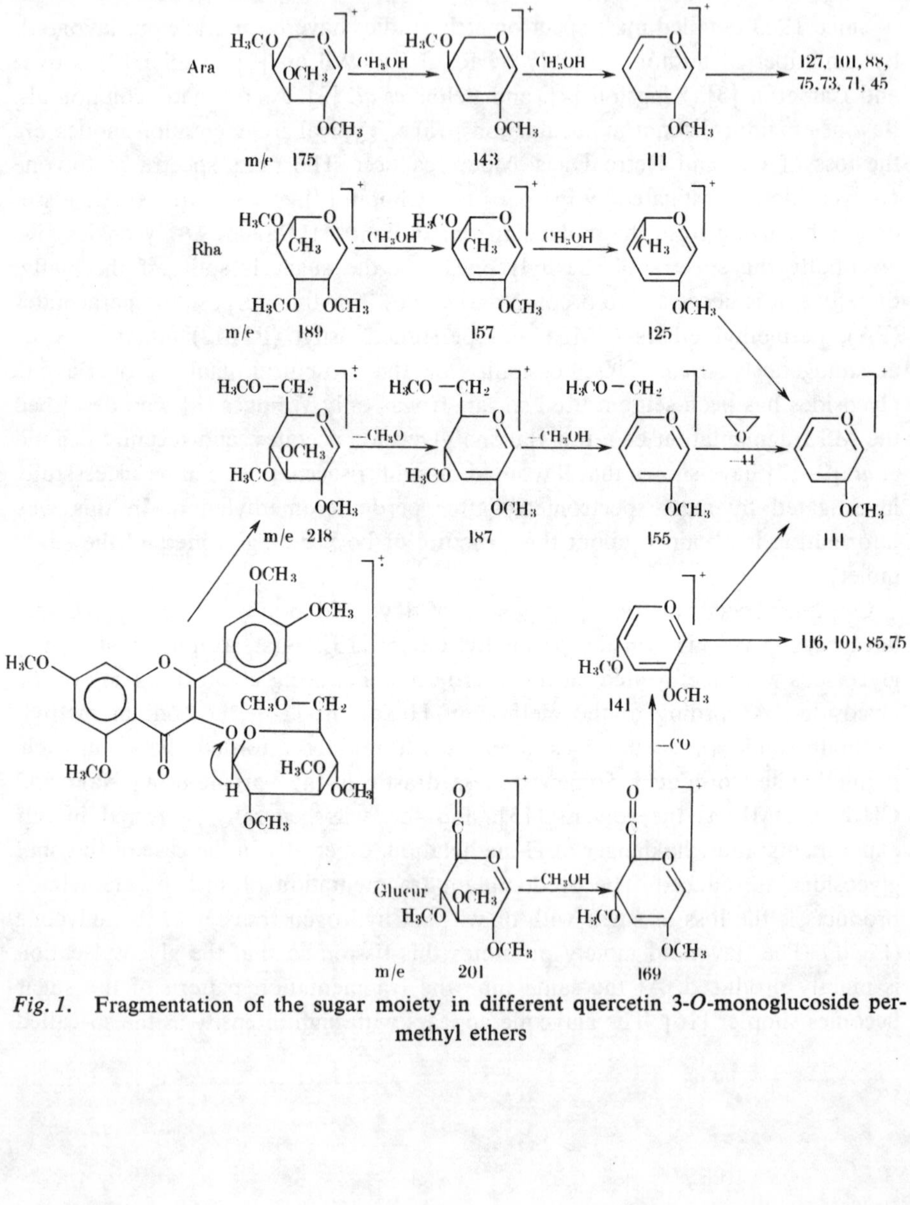

Fig. 1. Fragmentation of the sugar moiety in different quercetin 3-*O*-monoglucoside permethyl ethers

addition, there are smaller fragments, in general characteristic of methyl-sugars. The structures of these fragments are known since the studies of Kochetkov and Chizhov [16, 17] and Heyns and Scharmann [18, 19].

Our mass spectrometric studies of flavonoid monoglycoside permethyl ethers have shown that sugars attached to positions 5 and 3 are split off more readily than those at position 7. As shown in Fig. 2, the molecular ion peaks can be recognized with difficulty, or not at all, in the case of 3-*O*-monoglycosides, such as isoquercitrin. On the other hand, 7-*O*-glycosides, e.g. quercimeritrin, showed under our instrumental and experimental conditions a molecular ion peak of 50 % or higher, expressed as a percentage of the A+H peak, which was regarded as 100 %. The set of fragments with m/e 218, 187, 155 and 111 indicates the presence of glucose or galactose.

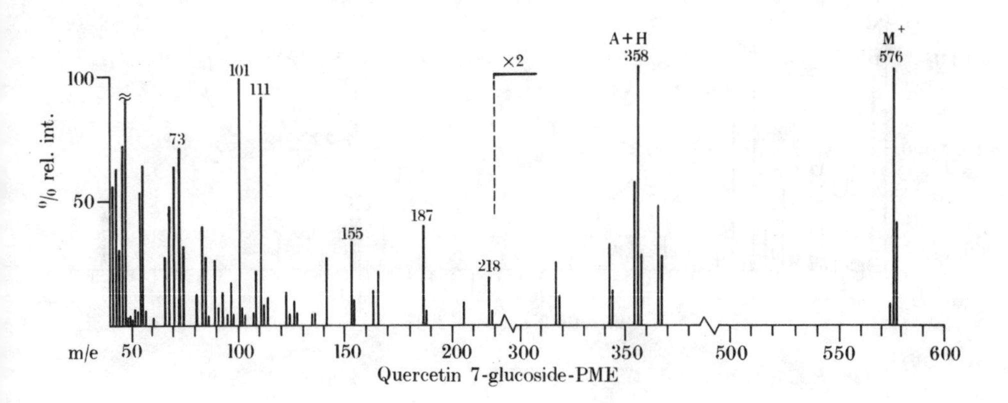

Quercetin 7-glucoside-PME

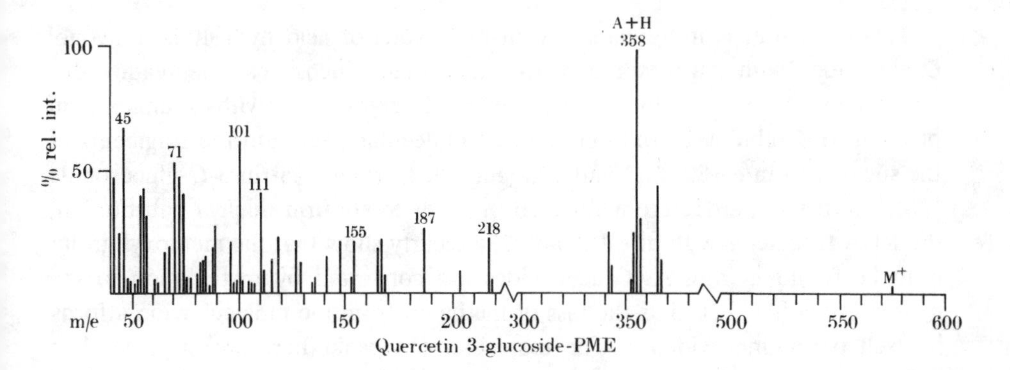

Quercetin 3-glucoside-PME

Fig. 2

The 4′- and 3′-*O*-glycosides represent an intermediate case, having small but distinct molecular ion peaks (3 % relative intensity for spireoside, and 8 % for quercetin-3′-glucoside) (Fig. 3).

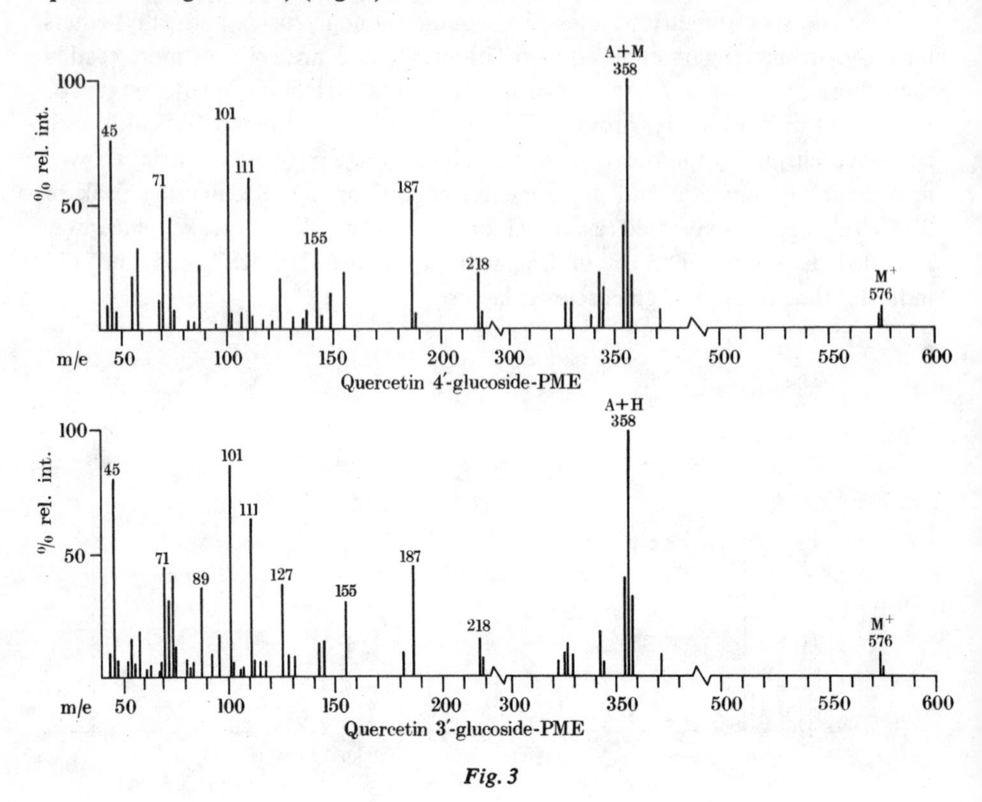

Fig. 3

This behaviour is in agreement with the results of acid hydrolysis. Flavonol *O*-glycosides with sugar components other than glucose can be readily differentiated, as shown by the example of myricitrin (with rhamnose in position 3) exhibiting an extremely small molecular peak and the fragments of the sugar with m/e 189, 157 and 125 (Fig. 4). In rhamnocitrin 3-*O*-glucuronide [20], which was perdeuteromethylated in order to confirm nuclear substitution, the RDA fragments with m/e 185 and 138 clearly show that the methoxyl group is to be found in ring A. Glucuronides are converted by methylation to the methyl esters [21, 22], thus the loss of methanol from the molecular ion affords in itself pertaining evidence. The molecular ion peaks here are, in general, a little more distinct. The fragments from the sugar moiety are shifted because of the CD_3O-groups to m/e 210, 175 and 147.

Table I lists other mono-*O*-glycosides, among them also biosides, with the typical intensity differences of the molecular ion peaks.

Table I

	A+H m/e	Rel. int. %		m/e	Rel. int. %
Isosakuranetin 7-glucoside-PME	314	100	M^+	544	48
= Isosakuranin	315	35			
Naringenin 7-neohesperidoside-PME	314	100	M^+	706	12
= Naringin	315	80	M-Rha	503	16
Luteolin 7-glucoside-PME	328	100	M^+	546	50
Luteolin 5-glucoside-PME	328	100	M^+	546	ø
= Galuteolin					
Chrysoeriol 7-glucuronide	328	100	M^+	560	45
methyl ester-PME			M-MeOH	528	40
Luteolin 7-neohesperidoside-	328	100	M^+	720	10
-PME	329	95	M-Rha	517	12
Diosmetin 7-rutinoside-PME	328	100	M^+	720	8
= Diosmin	329	40	M-Rha	516	14
Genistein 7-glucoside-PME	298	100	M^+	516	93
= Genistin					
Genistein 4'-glucoside-PME	298	100	M^+	516	5
= Sophoricoside					
Quercetin 5-glucoside-PME	358	100	M^+	576	0.2
Quercetin 3-glucuronide	358	100	M^+	590	2
methyl ester-PME			M-MeOH	558	1
Quercetin 3-galactoside-PME	358	100	M^+	576	0.2
= Hyperoside					
Quercetin 3-rhamnoside-PME	358	100	M^+	546	0.2
= Quercitrin					
Quercetin 3-rutinoside-PME	358	100	M^+	750	ø
= Rutin			M-Rha	545	8
Rhamnetin 3-dirhamno-	358	100	M^+	924	ø
galactoside-PME = Xanthorhamnin A			Disacch.	362	42
			Trisacch.	566	1
Rhamnazin 3-dirhamno-	358	100	M^+	924	ø
galactoside-PME = Xanthorhamnin B			Disacch.	362	10
			Trisacch.	566	3
Naringenin 7-galloyl-glucoside-PME	314	100	M^+	712	36
= Pruningallate	315	40	Gall.	195	1000
			Gall.+Glu.	399	120
Kaempferol 3-*p*- coumaroyl-	328	100	M^+	692	ø
glucoside-PME = Tiliroside			Coum.	161	125
			Coum.+Glu.	364	20

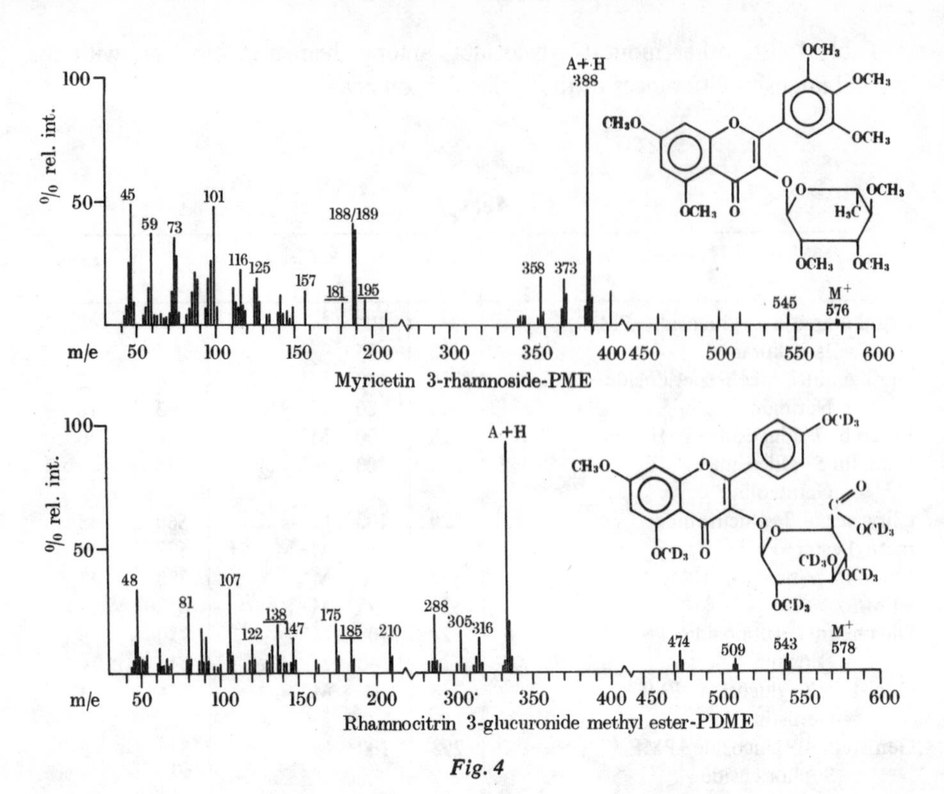

Fig. 4

It was obvious to extend this procedure to permethylated 3,7-*O*-diglycosides. In fact, it has been found that the signals corresponding to a fragment M^+—[3-*O*-sugar] appear with 40–50 % relative intensities, whereas those of a M^+—[7-*O*-sugar] fragment with intact 3-*O*-linkage have less than 1 % relative intensity. The conditions are shown in Fig. 5.

The loss of the 3-*O*-sugar is promoted by the formation of an intramolecular 5-membered ring chelate, leading to the A+H+[7-*O*-sugar] fragment, which appears with high intensity. Splitting at the C-7 oxygen atom is not sufficiently marked to yield a significant peak. After the loss of the second sugar moiety, the A+2H radical cation is obtained, which gives the reference peak.

This fission is elucidated in respect to two pairs of isomeric diglycosides. Isorhamnetin 3-gluco-7-rhamnoside permethyl ether exhibits a strong A+H+rhamnose peak at m/e 532, i.e. rhamnose must be in the 7-position. The A+H+glucose peak appears at m/e 562 with 0.5 % intensity only. The intensities for isorhamnetin 3-rhamno-7-glucoside permethyl ether are just reversed; there is a 42 % A+H+Glu peak against a 1.4 % peak of A+H+Rha. The same situation is found on comparing isorhamnetin 3-gluco-7-arabinoside per-

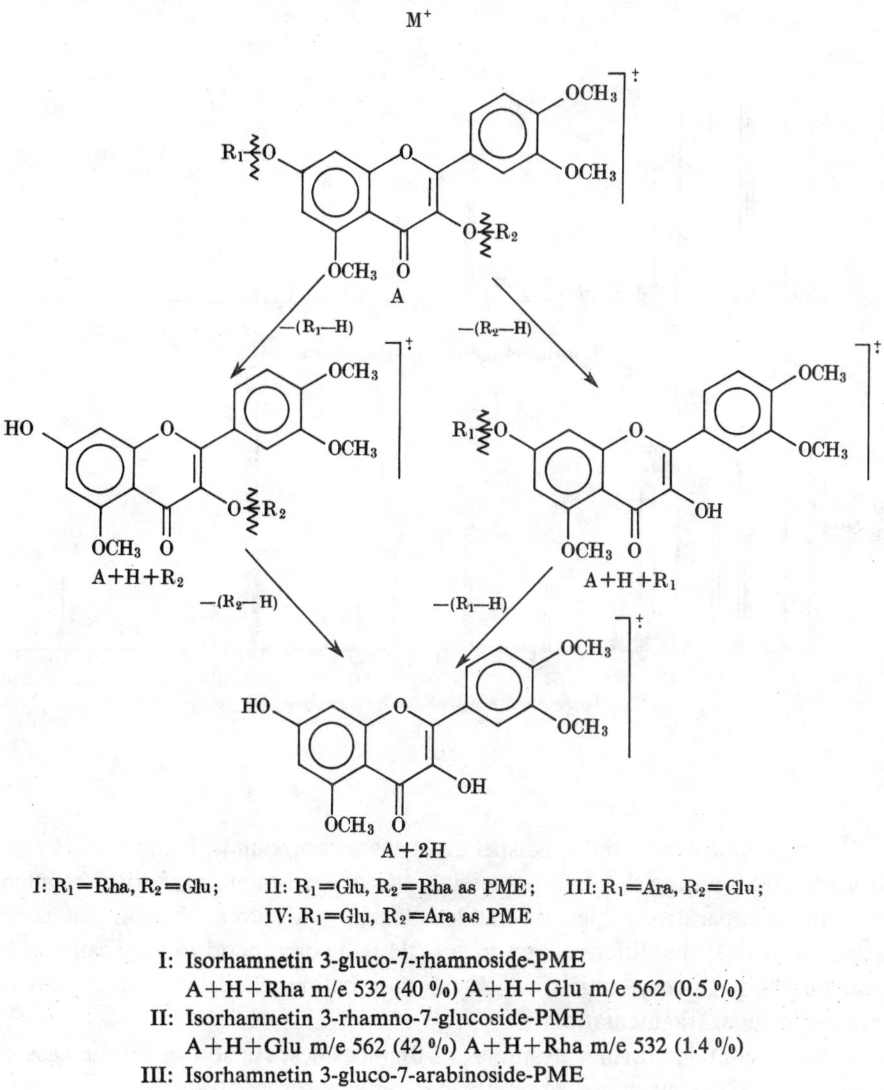

M^+

A

$-(R_1-H)$ $-(R_2-H)$

$A+H+R_2$ $A+H+R_1$

$-(R_2-H)$ $-(R_1-H)$

$A+2H$

I: R_1=Rha, R_2=Glu; II: R_1=Glu, R_2=Rha as PME; III: R_1=Ara, R_2=Glu;
IV: R_1=Glu, R_2=Ara as PME

I: Isorhamnetin 3-gluco-7-rhamnoside-PME
A+H+Rha m/e 532 (40 %) A+H+Glu m/e 562 (0.5 %)
II: Isorhamnetin 3-rhamno-7-glucoside-PME
A+H+Glu m/e 562 (42 %) A+H+Rha m/e 532 (1.4 %)
III: Isorhamnetin 3-gluco-7-arabinoside-PME
A+H+Ara m/e 518 (40 %) A+H+Glu m/e 562 (0.5 %)
IV: Isorhamnetin 3-arabino-7-glucoside-PME
A+H+Glu m/e 562 (45 %) A+H+Ara m/e 518 (1 %)

Fig. 5

methyl ether [23, 24] with its 3-arabino-7-glucoside isomer. The presence of an intense A+H+monosaccharide peak is always an indication that the sugar contained in this monoglycoside fragment is linked to the 7-position.

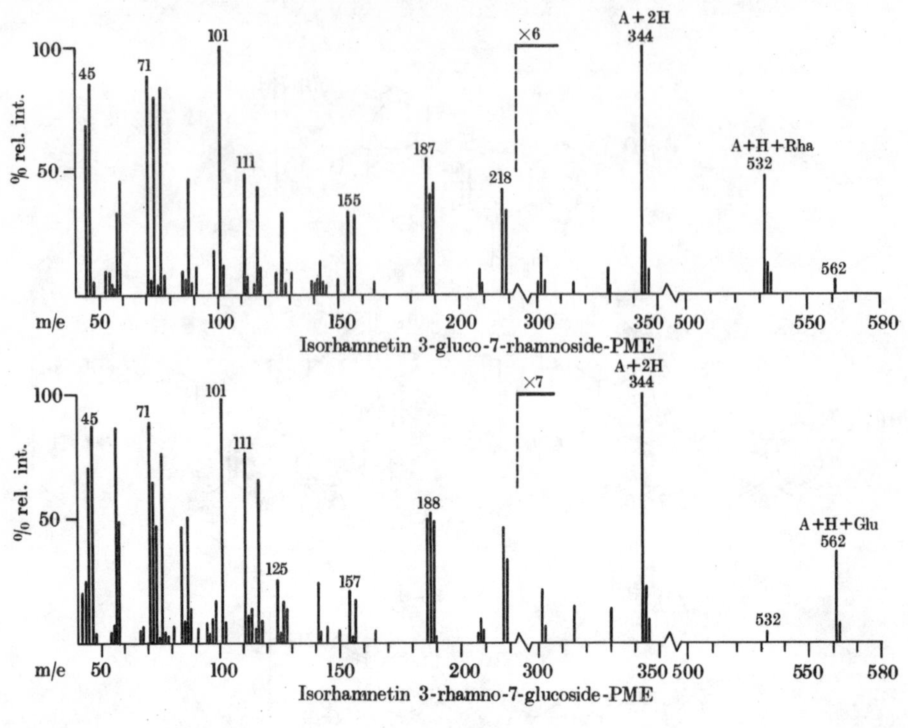

Fig. 6

Figures 6 and 7 show the mass spectra of these compounds. If the A+2H peak is made 100.0 during the data processing, the A+H+monosaccharide fragment with the comparatively higher relative intensity will reveal the kind of sugar attached to C-7; the difference up to A+2H will correspond to the rhamnosyl-, arabinosyl- or glucosyl cation, which can then be identified by its characteristic pattern of sugar fragmentation.

Table II contains further examples of diglucosides. As shown by the case of robinin, also the structure of glycosides with a disaccharide moiety can be elucidated in this way, and as Schmid *et al.* [9, 10] have earlier pointed out, indications can be obtained concerning the manner of attachment of the sugar. Even if the same sugar constituents occur in a molecule, mass spectrometric differentiation is possible e.g. between 7,3-, 7,4′- and 3,4′-*O*-diglucosides. In the case of flavanone *O*-glycosides an additional methyl group is found in the spectrum owing to ring cleavage to hydroxychalcone under the strongly alkaline conditions of methylation. An investigation of acylglycosides has shown that in the given circumstances of methylation acetyl groups are smoothly hydrolyzed,

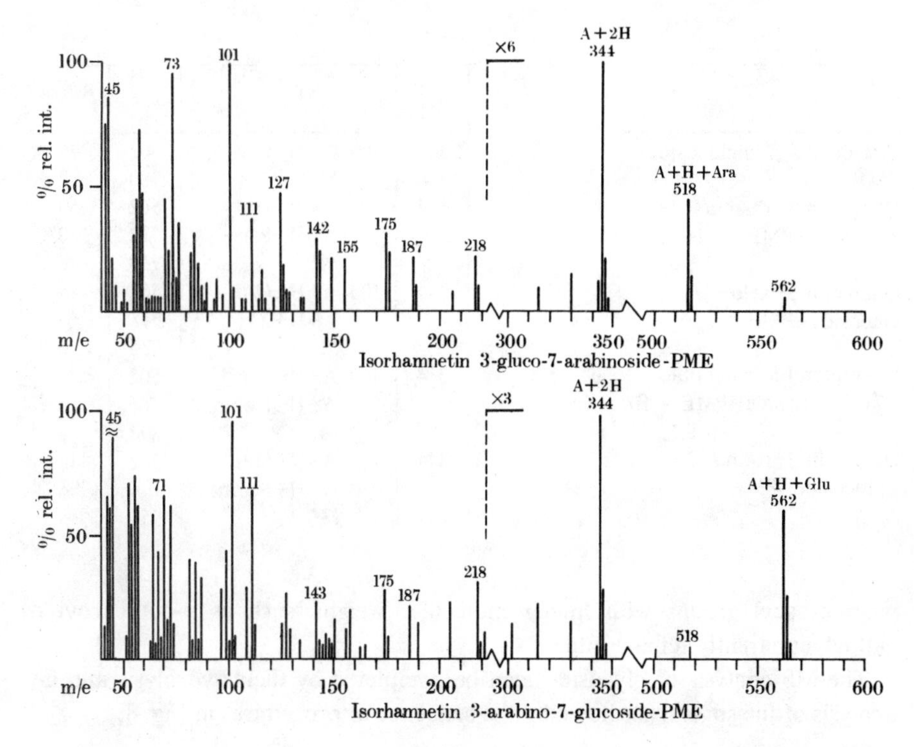

Fig. 7

Table II

	A+2H m/e	Rel. int. %		m/e	Rel. int. %
Eriodictyol 5,7–diglucoside–	330	100	A+H+Glu	548	48
PME			M^+	766	1
Apigenin 4′,7-diglucuronide	284	100	A+H+Glucur.	516	45
methyl ester–PME			A+H+(Glucur –MeOH)	484	50
			M^+	748	1.8
Kaempferol 3,7–dirhamnoside–	314	100	A+H+Rha	502	47
PME = Kaempferitrin			M^+	690	ø
Kaempferol 3–arabino–7–rhamnoside–	314	100	A+H+Rha	502	44
PME			A+H+Ara	488	0.5
			M^+	676	ø
Quercetin 3,7–diglucoside–	344	100	A+H+Glu	562	42
PME			M^+	780	ø
Quercetin 3,4′–diglucoside–	344	100	A+H+Glu	562	5
PME			M^+	780	ø

Table II (continued)

	A+2H m/e	Rel. int. %		m/e	Rel. int. %
Quercetin 4′,7-diglucoside-PME	344	100	A+H+Glu	562	46
			M^+	780	2
Quercetin 3-rhamno-7-glucoside-PME	344	100	A+H+Glu	562	50
			A+H+Rha	532	0.8
			M^+	750	ø
Quercetin 3-xylo-7-glucoside-PME	344	100	A+H+Glu	562	48
			A+H+Xyl	518	0.7
			M^+	736	ø
Kaempferol 3-robinobio--7-rhamnoside-PME = Robinin	314	100	A+H+Rha	502	46
			A+H+Robin.	706	ø
			M^+	894	ø
Quercetin 3-rutino-7--glucoside-PME	344	100	A+H+Glu	562	48
			A+H+Rutin.	736	ø
			M^+	954	ø

whereas acyl groups with higher molecular weight, such as *p*-coumaroyl or galloyl, apparently remain intact.

The MS analysis of glycosides can be completed by their hydrolysis and GC analysis of the sugars, as shown by the sequence of procedures in Fig. 8.

MS Analysis of Glycosides

Flavonoid *O*-glycoside
↓ Py/Ac_2O
Flavonoid *O*-glycoside-PA . . . . MS
DMF ↓ NaH/CH_3I
Flavonoid *O*-glycoside-PME . . . . MS
H^+ (or -PDME)
Monosacch.-ME Fl.-aglycone-ME . . . . MS
$NaBH_4$ ↓ Py/Ac_2O
Alditol acetate-ME . . . . GC–MS

Fig. 8

1–2 mg of an unknown flavone glycoside is first peracetylated and examined by mass spectrometry. Informative data can already be obtained in this step about the type of aglycone and the sugar. The peracetate is then directly permethylated (or deuteromethylated) with NaH and CH_3I (or CD_3I), and the product investigated again in the mass spectrometry. Subsequent acid hydrolysis gives the partly methylated aglycone, whose spectrum is also recorded. The partially methylated monosaccharide components can be isolated from the hydrolysate after reduction and acetylation, according to Björndal and Lindberg [25, 26], as the alditol acetates and their structures elucidated, where the combined use of GC and MS will be helpful [27].

The assignment of two different sugars to the several possible positions of glycosylation in a diglycoside is achieved according to Harborne's traditional method [28, 29] by selective and partial acid or enzymic hydrolysis. The rate of hydrolysis is followed in time and this markedly depends on the nature of the sugar and its position on the aglycone. Apart from the time consumption, the results of this method are often ambiguous, and the use of enzyme preparations of high substrate specificity is a prerequisite. As opposed to that, mass spectrometry affords the possibility of reliable differentiation on the micro scale and within very short time.

REFERENCES

1. Reed, R. I. and Wilson, J. M., *J. Chem. Soc. (C), 1963,* 5949.
2. Audier, H., *Bull. Soc. Chim. France, 1966,* 2892.
3. Bowie, J. H. and Cameron, D. W., *Australian J. Chem. 19,* 1627 (1966).
4. Kingston, D. G. J., *Tetrahedron, 27,* 2691 (1971).
5. Pelter, A., Stainton, P. and Barber, M., *J. Heterocyclic Chem., 2,* 262 (1965).
6. Prox, A., *Tetrahedron, 24,* 3697 (1968).
7. Chopin, J., in: "Synthesis of *C*-glycoflavonoids in Pharmacognosy and Phytochemistry" (eds. H. Wagner and L. Hörhammer), Springer Verlag, Berlin-Heidelberg-New York, 1971, pp. 111–128.
8. Weinges, K., Bähr, W. and Kloss, P., *Arzneimittel-Forsch., 18,* 537 (1968).
9. Schmid, R. D., *Tetrahedron, 28,* 3259 (1972).
10. Schmid, R. D., Varenne, P. and Paris, R., *Tetrahedron, 28,* 5037 (1972).
11. Schmid, R. D. and Harborne, J. B., *Phytochemistry, 12,* 2269 (1973).
12. Núnez-Alarcón, J., Rodríguez, E., Schmid, R. D. and Mabry, T. J., *Phytochemistry, 12,* 1451 (1973).
13. Wagner, H. and Seligmann, O., *Tetrahedron, 29,* 3029 (1973).
14. Hakomori, S., *J. Biochem., 55,* 205 (1964).
15. Brimacombe, J. S., Jones, B. D., Stacey, M. and Willard, J. J., *Carbohydr. Res., 2,* 167 (1966).
16. Kochetkov, N. K., Wulfson, N. S., Chizhov, O. S. and Zolötarev, B. M., *Tetrahedron, 19,* 2209 (1963).
17. Kotchetkov, N. K. and Chizhov, O. S., *Adv. Carbohydr. Chem., 21,* 39 (1966).
18. Heyns, K. and Scharmann, H., *Tetrahedron, 21,* 55 (1965).
19. Heyns, K. and Scharmann, H., *Tetrahedron, 21,* 507 (1965).
20. Wagner, H., Iyengar, M. A., Seligmann, O., Hörhammer, L. and Herz, W., *Phytochemistry, 11,* 2350 (1972).

21. Kováčik, V., Bauer, Š., Rosík. J. and Kováč, P., *Carbohydr. Res., 8,* 282 (1968).
22. Thompson, R. M., Gerber, N., Seibert, R. A. and Desiderio, D. M., *Drug Metabolism and Disposition, 1* (2), 489 (1973).
23. Rodríguez, E., Chin Sen, M., Mabry, T. J. and Dominguez, X. A., *Phytochemistry, 12,* 2069 (1973).
24. Wagner, H., Iyengar, M. A., Seligmann, O., Real, J. L. and Mitscher, L., *Lloydia, 36* (2), 166 (1973).
25. Björndal, H., Hellerqvist, C. G., Lindberg, B. and Svensson, S., *Angew. Chem., 82* (16), 643 (1970).
26. Björndal, H., Lindberg, B. and Svensson, S., *Carbohydr. Res., 5,* 433 (1967).
27. Wagner, H., Ertan, M. and Seligmann, O., *Phytochemistry, 13,* 857 (1974).
28. Harborne, J. B., *Phytochemistry, 4,* 107 (1965).
29. Harborne, J. B., *Phytochemistry, 3,* 151 (1964).

INVESTIGATIONS ON VITAMIN C_2 (SPARING FACTOR OF ASCORBIC ACID). A STUDY OF ITS METABOLISM

by

J.-M. GAZAVE, J.-L. PARROT, C. ROGER and M. ACHARD

Laboratoire de Physiologie Pathologique de l'Ecole-Pratique-des-Hautes-Etudes,
Faculté de Médecine Necker
Paris, France

Vitamin C_2 appears to be a secondary antiscorbutic factor: vitamin C (ascorbic acid) acting alone is unable to prevent or to cure scurvy, and vitamin C_2 acting alone is also unable to do so [1].

The chemical structure of vitamin C_2 is incompletely understood. The substance is found in various plants, mainly in *Citrus* fruit juices from which it can be extracted along with pectic substances [2]. It can also be obtained by extraction from various animal organs such as guinea pig liver, in which it is linked to lecithin [3]. However, it has not been possible to obtain vitamin C_2 in sufficiently pure state for structure determination by the usual chemical and physical methods, because it is very unstable and gives polymers even in the absence of oxygen and light [4].

We used an original biological method in order to determine the structure. This method consists in feeding a synthetic diet to animals and determining the urinary metabolites before and after oral administration of an impure vegetable or animal extract.

Our former investigations had shown that guinea pigs put on a basal synthetic diet, consisting of starch, cellulose, casein, mineral salts, vitamins B_1, B_2, B_6, PP, PAB, A, D and E, fish oil and vegetable oils, developed scurvy within 15 days and the illness caused death within about 21 days.

The administration of vitamin C (ascorbic acid) in the theoretically sufficient quantity (1 mg per 100 g of body weight per day) had no effect at all; but we just had to add a small quantity of a monomeric flavanol to put an end to the deficiency [1].

It is precisely the same method by which we have checked the value of our animal and vegetable extracts. From the results it seemed reasonable to infer that the C_2-factor belonged to the group of flavanols.

Another preliminary study, using Tayeau's and Masquelier's technique showed that this substance was identical with neither their "leucoanthocyanidol", nor one of its derivatives [5, 6, 7].

Given to guinea pig, flavanols follow two different metabolic pathways, both involving the cleavage of the heterocycle but in different ways, as shown in Figs 1 and 2 [8, 9].

First metabolic pathway

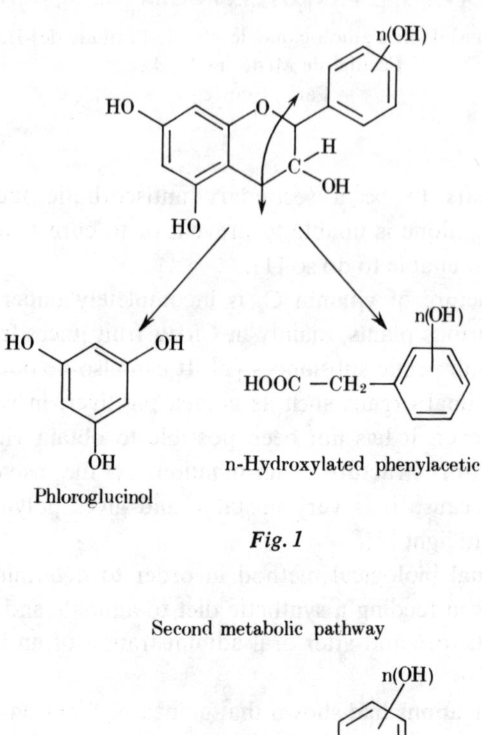

Phloroglucinol

n-Hydroxylated phenylacetic acid

Fig. 1

Second metabolic pathway

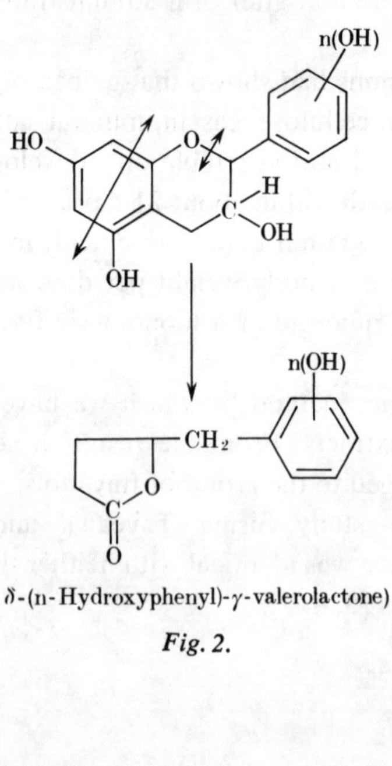

δ-(n-Hydroxyphenyl)-γ-valerolactone)

Fig. 2.

In either case, one of the metabolic products has a phenyl group carrying the same number of hydroxyl groups as the original compound.

Our recent experiments for determining the products of metabolism were carried out as described below.

Guinea pigs were put on the basal diet mentioned before, to which ascorbic acid was also added. After 8 days, the urine of the animals was mixed with 2 N HCl, then neutralized with sodium hydrogen carbonate, and the solution was extracted with ethyl ether which removed neutral phenols and urinary pigments. The aqueous phase was treated with 2 N sulfuric acid adjusting pH 2 acidity. The solution was then shaken with a mixture of 25 % ethanol and 75 % ethyl ether. The ethereal phase was evaporated to dryness, the residue taken up in ethanol (dissolution of free phenolic acids) and finally submitted to chromatographic analysis. The aqueous phase was refluxed for two hours in order to liberate the glycosylated phenolic acids. The solution was extracted with a mixture of 25 % ethanol and 75 % ethyl ether. The ethereal phase was evaporated to dryness, the residue taken up in ethanol, and then submitted to chromatographic analysis.

The same process was repeated 24 hours later, after having added 1 g of the animal or vegetable extract, rich in the C_2-factor, to the deficiency diet.

The chromatographic analyses were made on silicagel layer. Three developing systems were used:

(1) Benzene–methanol–acetic acid (90:16:6) (BMA)
(2) Butanol–acetic acid–water (4:1:5) (Partridge's mixture)
(3) 20 % KCl.

Detection was effected in UV light (2,537 nm) or by spraying with 0.1 N potassium permanganate.

In the first part of the experiments, i.e. before administering the extracts containing the C_2-factor, no spots were found on the chromatograms. After completing the diet with such an extract, the chromatograms showed spots corresponding, in the main, to phloroglucinol, 3,4,5-trihydroxyphenylacetic acid and to 4,5-dihydroxy-3-methoxyphenylacetic acid (Table I).

This last methoxylated compound originates from the trihydroxyl derivative which, like all compounds carrying an *ortho* diphenolic function, can be methylated according to the method described by Axelrod and Laroche [10]. This experiment establishes that this flavanol is hydroxylated in positions 5,7,3′,4′,5′. It is, therefore, 3′,4′,5′,5,7-pentahydroxyflavan-3-ol.

Sodium hydroxide or barium hydroxide are suitable agents to liberate this compound and obtain it in the free state; however, chromatographic analysis must be carried out immediately, or else polymerization begins.

Table I

R_f of the Compounds Excreted by the Urinary System

Compound	R_f		
	B.M.A.	Partridge's mixture	20% KCl
Phloroglucinol	0.19	0.77	0.46
3,4,5-Trihydroxyphenyl-acetic acid	0.12	0.69	0.81
4,5-Dihydroxy-3-methoxy-phenylacetic acid	0	0.65	0.44

The latter chromatographic analysis was made on cellulose layer, using Partridge's mixture as the developing solvent; detection was effected with a 1 % ethanolic *p*-toluenesulfonic acid solution and 2 % vanillin. After the plates had been heated at 110 °C for 5 minutes in a drying oven, a spot with a magenta colour (R_f = 0.47 ± 0.02) appeared, which, according to Haslam [11] corresponds to the monomeric form of 1-*epi*-3′,4′,5′,5,7-pentahydroxyflavan-3-ol.

REFERENCES

1. Gazave, J.-M., *J. Physiol., Paris, 58,* Suppl. I, 128 (1966).
2. Gazave, J.-M., Parrot, J.-L., Saindelle, A. and Canu, P., *J. Physiol., Paris, 60,* Suppl. I, 215 (1968).
3. Parrot, J.-L., Gazave, J.-M., Canu, P. and Palou, A.-M., *J. Physiol., Paris, 59,* n° I bis, 278 (1967).
4. Zaprometov, M. H., "Biochemistry of Catechins", Moscow, 1964, p. 198.
5. Tayeau, F. and Masquelier, J., *Bull. Soc. Chim. France, 15,* 1167 (1948).
6. Tayeau, F. and Masquelier, J., *Bull. Soc. Chim. Bol. France, 31,* 72 (1949).
7. Tayeau, F., Masquelier, J. and Lefevre, G., *Bull. Soc. Pharm. Bordeaux (France), 89,* 5 (1951).
8. Gazave, J.-M. and Parrot, J.-L., *J. Angéiol. langue française,* Expans. scient., 1973, p. 272.
9. Das, N. P. and Griffiths, L. A., *Biochem. J., 110,* 449 (1968).
10. Axelrod, J. and Laroche, M.-J., *Science, 130,* 800 (1959).
11. Haslam, E., "Chemistry of Vegetable Tannins", Academic Press, London and New York, 1966, p. 179.

THE ROLE OF THE INTESTINAL MICROFLORA IN FLAVONOID METABOLISM

by

L. A. GRIFFITHS

Department of Biochemistry, University of Birmingham
Birmingham, UK

Modern knowledge of flavonoid metabolism is largely based upon the researches of DeEds, Booth and their associates, who, during the fifties established the pattern of degradation of some key flavonoid compounds including quercetin. They were able to show that phenylacetic acids and phenylpropionic acids were formed from the B ring of flavonols and flavones, respectively (cf. [1]), although neither the fate of the A ring, nor the site of the enzyme systems responsible for the fission of the heterocyclic ring system of flavonoid molecules were established.

Subsequent studies in my laboratories on the metabolism of (+)-catechin [2] indicated that the *in vivo* metabolism of this hydroxyflavan in the rat was largely dependent upon the intestinal microflora, as oral co-administration with an aureomycin/phthalylsulfathiazole mixture led to a reduction in the amounts of catechin metabolites *viz.* phenolic acids and phenylvalerolactones excreted (Fig. 1). The metabolism of catechin in the guinea pig was similarly affected when these antibacterials were administered [3]. Subsequently it was shown [4] that when (+)-catechin was incubated with the intestinal microflora of the rat, δ-(3-hydroxyphenyl)-γ-valerolactone, δ-(3,4-dihydroxypheny!)-γ-valerolactone and *m*-hydroxyphenylpropionic acid (which had previously been identified as catechin metabolites in rat urine) were formed in the culture medium. Scheline [5] has shown that a suspension of micro-organisms derived from the rabbit intestine is able to metabolize (+)-catechin to the products, 5-(3,4-dihydroxyphenyl)-valeric acid and 5-(3-hydroxyphenyl)-valeric acid, indicating that variation in catechin metabolism by the intestinal microflora of different animal species may occur. Das [6] has shown that the major urinary metabolites of (+)-catechin in man include many metabolites previously detected in the rat and guinea pig, although the presence of other additional metabolites which were not identified indicates the possibility of further species variation.

Further studies by Nakagawa, Shetlar and Wender [7] on the metabolism of quercetin confirmed the earlier work of Booth *et al.* [8], and furnished evidence that the intestinal microflora was also implicated in quercetin metabolism, as a

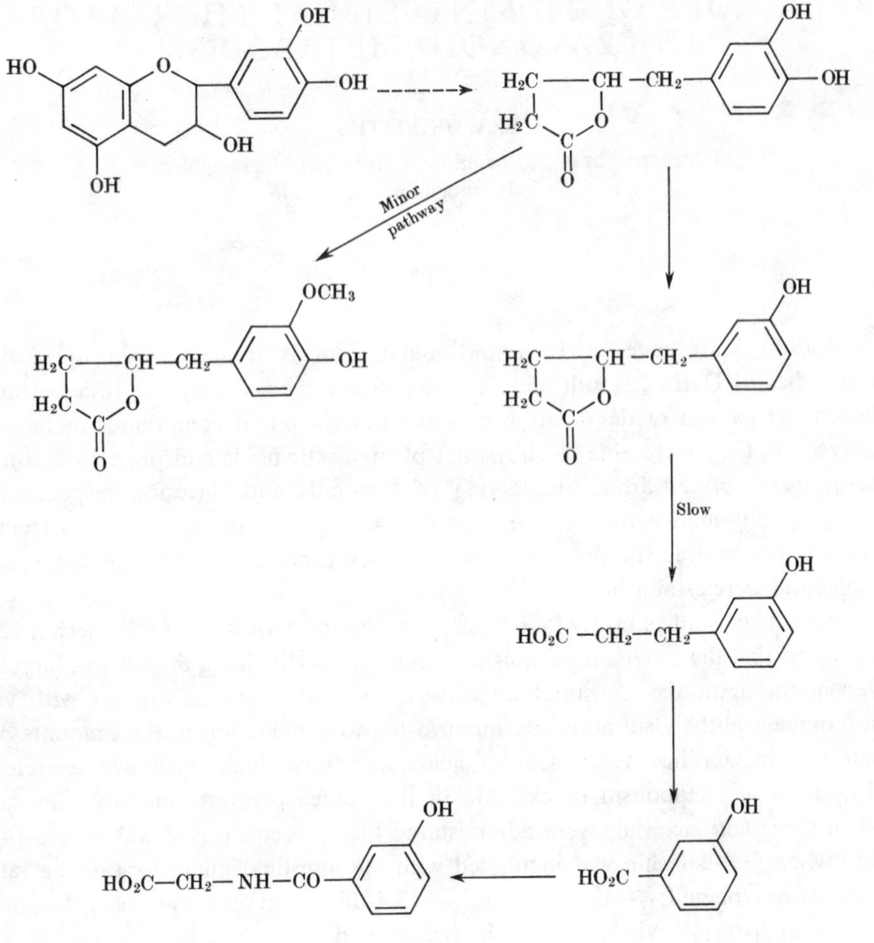

Fig. 1. **Metabolism of (+)-catechin in the guinea pig and rat (Das and Griffiths, 1969)**

marked reduction in the excretion of *m*-hydroxyphenyl metabolites of quercetin was observed in the presence of the antibiotic, neomycin. Scheline [9] has furthermore demonstrated the formation of *m*-hydroxyphenylacetic acid from rutin by the rat intestinal microflora *in vitro*.

These observations led us in the late sixties to begin an investigation upon a wide range of flavonoid compounds to determine the extent to which flavonoid catabolism *in vivo* could be correlated with the capacity of the intestinal microflora to degrade flavonoid molecules when incubated under anaerobic conditions *in vitro*. We attempted also to determine the effect of molecular structure upon the susceptibility of flavonoid molecules to ring fission *in vivo* and *in vitro*.

Since earlier investigations had centred mainly upon flavonoids possessing a 3′,4′-dihydroxylated B ring, our more recent investigations have been based largely upon two series of flavonoids, one possessing a 4′-mono- and the other a 3′,4′,5′-trihydroxylated B ring. Metabolite identification was based upon chromatographic and spectral methods which have been described [10, 11]. The results (Tables I and II) show that the intestinal microflora is able *in vitro* under conditions of anaerobic incubation to degrade flavonoids to essentially the same ring fission products as those detected in urine following oral administration of the specific flavonoid. It will also be noted that those compounds which are observed to be resistant to microbial degradation *in vitro* are likewise not metabolized *in vivo*. The ability of the microflora of the rat to metabolize other related flavonoids was explored and the results are shown in Table III.

It will be seen that all compounds possessing free 5- and 7-hydroxyl groups in the A ring and a free 4′-hydroxyl group in the B ring gave rise to ring fission products. The absence of a hydroxyl group or methoxylation in these positions reduced or abolished susceptibility to degradative attack. These observations are in accord with those of DeEds [1] who has noted that certain methyl ethers of quercetin including rhamnetin, tangeretin and nobiletin do not undergo ring fission in the rat.

The availability to us of a range of hydroxyethyl derivatives of rutin (constituents of the vascular therapeutic agent, Paroven®, Zyma S.A., Nyon, Switzerland) has permitted observations to be made upon the effect of progressive substitution of the free hydroxyls of rutin upon susceptibility of the molecule to degradative attack (Table IV).

It will be noted that in all cases incubation with the microflora resulted in metabolic hydrolysis of each glycoside to the corresponding aglycone, but that ring fission occurs only in respect of rutin and the mono-HR. It is evident therefore that increasing hydroxyethylation reduces the susceptibility of rutosides to ring fission, but does not prevent cleavage of the glycosidic linkage. The metabolites listed in Table IV are also known to be formed *in vivo* and have been isolated as the free compounds or as conjugates [12, 13, 14]. In addition, it is important to note that the hydroxyethylrutosides are also excreted unchanged and as glycoside and aglycone conjugates in bile after oral administration [12].

In a series of experiments in which hydroxyethylrutosides labelled with ^{14}C in the α and β positions of their hydroxyethyl side chains were orally administered to rats [14], the respired CO_2 was trapped, but no significant levels of radioactivity were found indicating that the intestinal microflora, at least *in situ*, is unable to degrade the side chains of the hydroxyethylrutosides to $^{14}CO_2$, although evidence has been presented by Ryan *et al.* [15] which suggests that

Table I

Metabolism of Apigenin and Related Compounds *In Vivo* and *In Vitro*

Compound administered	*Urinary metabolites*	*Microfloral metabolites in vitro*
Apigenin (R=OH, X_1=OH, X_2=OH, Y=H) Apiin (Apigenin 7–apiosylglucoside) Naringin (2,3–dihydroapigenin) Phlorrhizin (2′,4,4′,6′–tetrahydroxy-phloretin–2′–glucoside)	*p*–Hydroxyphenylpropionic acid, *p*–Hydroxycinnamic acid, *p*–Hydroxybenzoic acid, flavonoid aglycones and conjugates	*p*–Hydroxyphenylpropionic acid *p*–Hydroxycinnamic acid *p*–Hydroxybenzoic acid and flavonoid aglycones
Acacetin (R=OCH_3, X_1=OH, X_2=OH, Y=H)	Trace metabolites only	Trace metabolites only
7,4′–Dihydroxyflavone Chrysin (5,7–dihydroxyflavone) Tectochrysin (5–hydroxy–7–methoxyflavone)	No ring fission products	No fission products
Kaempferol (R=OH, X_1=OH, X_2=OH, Y=OH) Robinin (Kaempferol 7–rhamnoside–3–galactorhamnoside)	*p*–Hydroxyphenylacetic acid and kaempferol	*p*–Hydroxyphenylacetic acid and kaempferol

Table II

Metabolism of the Myricetin Group *In Vivo* and *In Vitro*

Compound administered	*Urinary metabolites*	*Microbial metabolites formed in vitro*
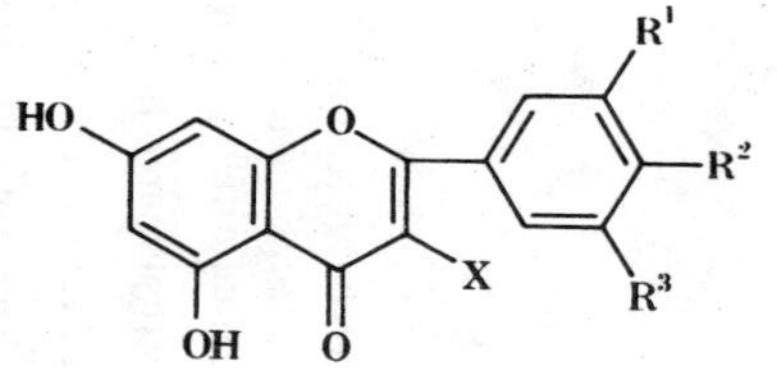		
Myricetin X=OH, R^1=OH, R^2=OH, R^3=OH Myricitrin X=O–rhamnose, R^1=OH, R^2=OH, R^3=OH	3,5–Dihydroxyphenylacetic acid, *m*–Hydroxyphenylacetic acid (trace) and myricetin	3,5–Dihydroxyphenylacetic acid *m*–Hydroxyphenylacetic acid 3,4,5–Trihydroxyphenylacetic acid, myricetin
Tricin X=H, R^1=OCH_3, R^2=OH, R^3=OCH_3 Tricetin X=H, R^1=OH, R^2=OH, R^3=OH	3,5–Dihydroxyphenylpropionic acid and the unchanged flavonoid	3,5–Dihydroxyphenylpropionic acid
5,7–Dihydroxy–3′,4′,5′ tri–methoxyflavone X=H, R^1=OCH_3, R^2=OCH_3, R^3=OCH_3	No phenolic acid metabolites	No metabolites detected
Robinetin (5–deoxytricetin)	No phenolic acid metabolites + robinetin	No metabolites
Delphinidin (3,5,7,3′,4′,5′–hexahydroxy–flavylium)	Neutral phenolic metabolite : Da	Metabolites Da and Db
Malvin (3,5,7,4′–tetrahydroxy–3′,5′–dimethoxyflavylium, 3,5 diglucoside)	3 Neutral metabolites M_2 M_3 and M_4	No phenolic metabolites

Table III

Degradation of Other Flavonoids by the Intestinal Microflora of the Rat *In Vitro*

Compound	*Microfloral products*
Genistein (X=OH, Y=OH)	*p*–Ethylphenol
Biochanin A (X=OH, Y=OCH_3)	Trace metabolite only
Daidzein (X=H, Y=OH)	Equol but no fission products
Formononetin (X=H, Y=OCH_3)	No fission products
Pelargonin (5,7,4′–trihydroxyflavylium 3,5–diglucoside)	*p*–Hydroxyphenyl-lactic acid
Epiafzelechin (5,7,4′–trihydroxyflavan–3–ol)	*p*–Hydroxyphenylpropionic acid and a neutral metabolite (δ–4–hydroxyphenyl–γ–valerolactone?)

Table IV

Metabolism of Rutin and Hydroxyethylrutosides by the Intestinal Microflora *In Vitro*

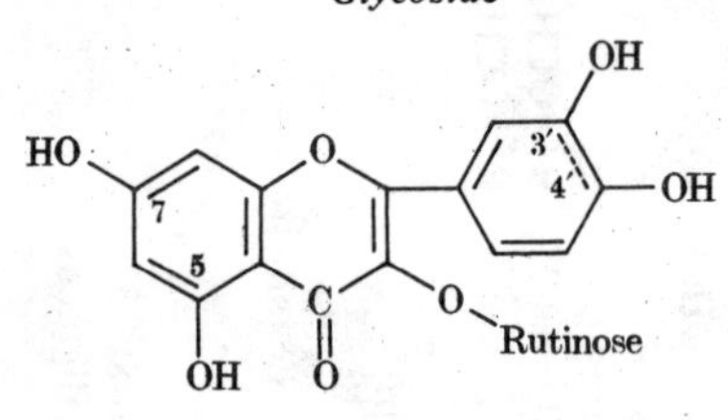

Glycoside	*In vitro Metabolites*
Rutin	Quercetin (tr.) *m*–Hydroxyphenylacetic acid
7-Mono-HR	7–Monohydroxyethylquercetin *m*–Hydroxyphenylacetic acid Metabolite A (Unidentified A ring fragment)
4′,7-Di-HR	4′,7–Dihydroxyethylquercetin
3′,4′,7-Tri-HR	3′,4′,7–Trihydroxyethylquercetin
3′,4′,5,7-Tetra-HR	3′,4′,5,7–Tetrahydroxyethylquercetin

the hydroxyethyl groups of [^{14}C]-hydroxyethylstarch are metabolized to $^{14}CO_2$ by the microflora of the rat. Recently we found that when 7-mono-HR labelled with ^{14}C in the hydroxyethyl side chain was incubated with the microflora *in vitro,* less than 1 % of the radioactivity was present in trapped CO_2, but radioactivity was found by radioscanning at the R_f values of both metabolite A and the aglycone [14].

A major problem in flavonoid research is the metabolic fate of the carbon atoms of the A ring. However, using isotopically labelled (+)-catechin selectively labelled with ^{14}C in ring A, we were able to show that radioactivity in the resulting metabolites was restricted to $^{14}CO_2$ and to certain phenylvalerolactones, which on theoretical grounds would be expected to contain the C atoms common to the A ring and the pyrone ring [16]. It appeared probable therefore that the $^{14}CO_2$ detected was derived from the remaining carbon atoms, i.e. those in positions 5, 6, 7 and 8 of the hydroxyflavan molecule. Although our data led us to suggest a possible formation of aliphatic intermediates as precursors of the $^{14}CO_2$, it was not until 1972 that evidence of the identity of such an aliphatic intermediate was obtained [17]. Using cell-free extracts of a *Pseudomonas* obtained from rat faeces, evidence for the metabolism of (+)-catechin *via* the pathway shown in Fig. 2 was presented [17, 18]. The formation of the four-carbon intermediate oxalacetic acid (which would be metabolized *in vivo* to $^{14}CO_2$) is in accord with our earlier findings, but it is important to note that none of the other intermediates reported by Jeffrey *et al.* [17, 18] have been detected by ourselves or by other workers as metabolites of (+)-catechin *in vivo* or *in vitro.* A further fact of importance is that the pseudomonad studied had been obtained from faeces rather than from intestinal contents *in situ,* so that the possibility of aerial contamination cannot be excluded; indeed the authors have also referred to their organism as a soil pseudomonad [18]. Thus, although the pathway proposed is of considerable interest, it remains in doubt whether it is of relevance to the *in vivo* metabolism of flavonoids by the intestinal microflora.

Further studies on the identity of the micro-organisms active in the ring fission of flavonoid molecules in the intestine of laboratory animals and man are thus clearly needed. In ruminant animals considerable progress has recently been made by Simpson and his colleagues [19, 20, 21]. Fifteen strains of *Butyrivibrio* sp. capable of degrading rutin have been obtained from the rumen of cattle, whilst three other organisms from a laboratory collection of rumen organisms have been implicated in rutin catabolism [19]. The rumen organism *Butyriovibrio* sp. C 3 has been shown to degrade rutin to products including phloroglucinol, 3,4-dihydroxyphenylacetic acid and CO_2 [20], and naringin to products which included phloroglucinol and *p*-hydroxyphenylpropionic acid

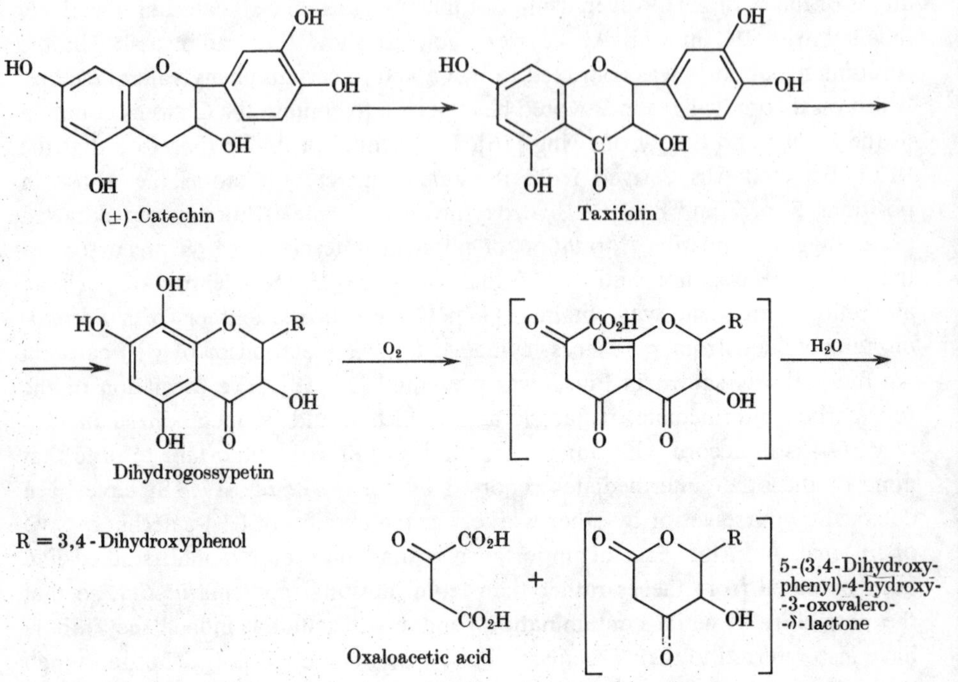

(±)-Catechin → Taxifolin → Dihydrogossypetin $\xrightarrow{O_2}$ [ ] $\xrightarrow{H_2O}$ Oxaloacetic acid + 5-(3,4-Dihydroxyphenyl)-4-hydroxy-3-oxovalero-δ-lactone

R = 3,4-Dihydroxyphenol

Fig. 2. Metabolism of (+)-catechin by a soil pseudomonad (Jeffrey *et al.*, 1972)

[21]. Their finding of phloroglucinol as a product of flavonoid catabolism is interesting, as this compound has not been found by Scheline or ourselves as a product of flavanol or flavone metabolism by the rat microflora *in vitro* or *in vivo* and suggests that flavonoid catabolism in the ruminant and non-ruminant may proceed by different pathways.

That the intestinal microflora may be of importance in the metabolism of flavonoid compounds not only following oral dosage, but also after parenteral administration is indicated by recent observations [12, 14, 22, 23] which show that a high proportion of certain flavonoids including trihydroxyethylrutoside

Table V

Excretion of ^{14}C Following Administration of [^{14}C]-Hydroxyethylrutosides to Rats with Cannulated Bile Ducts

Compound	Route	Excretion of ^{14}C in		
		Bile, % of dose (mean ± S.D.)	Urine, % of dose (mean ± S.D.)	Time (hr)
3′,4′,7-Tri-O-ß-hydroxyethyl-[hydroxyethyl-^{14}C]-rutoside	i.v.	66.5 ± 5.1^{3}	19.5 ± 4.9^{3}	24
	oral	10.6 ± 4.5^{3}	1.8 ± 1.3^{3}	48
3′,4′,5,7-Tetra-O-ß-hydroxyethyl-[5-hydroxyethyl-^{14}C]-rutoside	i.v.	29.2 ± 8.9^{6}	54.7 ± 10.9^{6}	24
	oral	0.4 ± 0.2^{2}	2.4 ± 0.8^{2}	48
7-Mono-O-ß-hydroxyethyl-[hydroxyethyl-^{14}C]-rutoside	i.v.	65.4 ± 1.1^{3}	17.3 ± 13.2^{3}	24
	oral	16.1 ± 8.3^{2}	3.3 ± 0.8^{2}	48

Table VI

Excretion of ^{14}C in Faeces After i.v. Administration of [^{14}C]-3′,4′,7-Tri-HR and [^{14}C]-7-Mono-HR

hr.	3′,4′,7-Tri-HR			7-Mono-HR
	Rat	Rabbit	Monkey	Rat
	% of dose in faeces *			*% of dose in faeces*
0- 24	17.0	29.5	27	12
24- 48	32.0	14.4	22.7	4.1
48- 72	6.0	4.5	5.3	1.2
72- 96	1.6	–	–	0.3
96-120	0.3	–	–	–
Total	56.9%	48.4%	55.0%	17.6%

* Mean values of three experiments

and catechin are excreted into the intestinal *via* the bile. Parenteral administration of [^{14}C]-hydroxyethylrutosides to rats in which biliary cannulae had been implanted permitted the demonstration of high levels of radioactivity in bile (Table V), which has been shown by radio-scanning to be associated with the unchanged rutosides and rutoside conjugates, whilst parenteral administration of these compounds to non-cannulated rats results in the excretion of considerable quantities of radioactivity as the aglycone in faeces (Table VI). That the aglycone arose from the rutoside by the action of the intestinal microflora was shown by the use of an oral antibiotic. The biliary route has been shown to be of importance also in the excretion of naturally occurring flavonoids [12], but in these cases, due to the absence of hydroxyethylation, the aglycones are usually degraded in the intestine to phenolic ring fission products.

Although the evidence we have considered indicates that the intestinal microflora is capable of effecting flavonoid ring fission *in vitro* and that it participates in the catabolism of flavonoids *in vivo,* it does not permit any conclusion to be drawn as to whether the ring fission products detected in the urine of orally dosed animals were derived in part from degradative processes occurring in the tissues of the animal or were of wholly microfloral origin. The use of germ-free animals has now permitted this problem to be resolved.

Rats bred under germ-free conditions were maintained in an isolator of the Trexler type on a powder diet sterilized by γ-irradiation. The flavonoid compounds also sterilized by γ-irradiation were administered orally at the level of 100 mg/rat. The urines and faeces were collected on the preceding day and for

Table VII

Major Urinary Metabolites of (+)-Catechin in Normal and Germ-free Rats

	Hydroxyphenylvalerolactones		*m*-Hydroxyphenylpropionic acid	*m*-Hydroxyhippuric acid	Hydroxyflavan conjugates
	3-Hydroxy-	3,4-Dihydroxy-			
Normal rats	+++	+	++++	++	++
Germ-free rats	−	−	−	−	+++
Conventionalized germ-free rats	++	+	++++	+	++

Table VIII

Major Urinary Metabolites of Bioflavonoids in Normal and Germ-free Rats

Bioflavonoid	*Normal rat*	*Germ-free rat*
Apigenin Naringin	*p*-Hydroxyphenylpropionic acid *p*-Hydroxycinnamic acid *p*-Hydroxybenzoic acid	Absent
Myricetin	3,5-Dihydroxyphenylacetic acid *m*-Hydroxyphenylacetic acid	Absent
Hesperidin	*m*-Hydroxyphenylpropionic acid *m*-Hydroxycinnamic acid	Absent

Table IX

Major Metabolites of Rutin and Hydroxyethylrutosides in Normal and Germ-free Rats

	In urine		*In faeces*	
	Normal	G.F.	Normal	G.F.
Rutin	*m*–Hydroxyphenylacetic acid Homovanillic acid Methylcatechol glucuronide	Absent	Corr. aglycone (tr.) *m*–Hydroxyphenylacetic acid	Rutin
7–Mono–O–ß–hydroxyethyl–rutoside	*m*–Hydroxyphenylacetic acid Metabolite A Methylcatechol glucuronide	Absent	Corr. aglycone	7–Mono–HR
3′,4′,7–Tri–O–ß–hydroxyethyl–rutoside	No ring fission products		Corr. aglycone	3′,4′,7–Tri–HR
4′,5,7–Tetra–O–ß–hydroxyethyl–rutoside	No ring fission products		Corr. aglycone	3′,4′,5,7–Tetra–HR

five days following administration of the test compound. Extracts of urine and faeces were obtained and examined by previously described methods [2, 23].

Following administration of (+)-catechin to germ-free animals, complete suppression of the normally found phenolic acid and phenylvalerolactone metabolites was observed, although metabolites arising by conjugation of the intact hydroxyflavan molecule were detectable in both urine and faeces (Table VII). These observations indicated that although heterocyclic ring fission was suppressed in these germ-free animals, absorption and metabolism of (+)-catechin by conjugation proceeded normally. Examination of other naturally occurring flavonoids including apigenin, naringin, myricetin, hesperidin and rutin under similar germ-free conditions showed similar total suppression of the normal ring fission products (Table VIII) [13]. Studies on the metabolism of the hydroxyethylrutosides have also revealed significant differences in the metabolism of these compounds in the normal and germ-free animal (Table IX) of which the most important appears to be suppression of aglycone formation in the intestine [14].

These results demonstrate that orally ingested flavonoids do not in the rat give rise to ring fission products in the absence of the normal intestinal microflora and support the view that the phenolic acids detected in the urines of normal animals after oral ingestion of flavonoid compounds are of wholly microfloral origin.

REFERENCES

1. DeEds, F., "Flavonoid Metabolism in Comprehensive Biochemistry", Vol. 28, Elsevier, London, 1968, p. 127.
2. Griffiths, L. A., *Biochem. J., 92,* 173 (1964).
3. Das, N. P. and Griffiths, L. A., *Biochem. J., 110,* 449 (1968).
4. Das, N. P., *Biochim. Biophys. Acta, 177,* 668 (1969).
5. Scheline, R. R., *Biochim. Biophys. Acta, 222,* 228 (1970).
6. Das, N. P., *Biochem. Pharmac., 20,* 3455 (1971).
7. Nakagawa, Y., Shetlar, M. R. and Wender, S. H., *Biochim. Biophys. Acta, 97,* 233 (1965).
8. Booth, A. N., Murray, C. W., Jones, F. T. and DeEds, F., *J. Biol. Chem., 223,* 251 (1956).
9. Scheilne, R. R., *Acta Pharmacol. Toxicol., 26,* 332 (1968).
10. Griffiths, L. A. and Smith, G. E., *Biochem. J., 128,* 901 (1972).
11. Griffiths, L. A. and Smith, G. E., *Biochem. J., 130,* 141 (1972).
12. Barrow, A. and Griffiths, L. A., *Biochem. J., 125,* 24P (1971).
13. Barrow, A. and Griffiths, L. A., *Xenobiotica, 2,* 575 (1972).
14. Barrow, A. and Griffiths, L. A., Unpublished observations (1973).
15. Ryan, A. J., Holder, G. M., Mate, C. and Adkins, G. K., *Xenobiotica, 2,* 141 (1972).

16. Das, N. P. and Griffiths, L. A., *Biochem. J.*, *115*, 831 (1969).
17. Jeffrey, A. M., Jerina, D. M., Self, R. and Evans, W. C., *Biochem. J.*, *130*, 383 (1972).
18. Jeffrey, A. M., Knight, M. and Evans, W. C., *Biochem. J.*, *130*, 373 (1972).
19. Cheng, K. J., Jones, G. A., Simpson, F. J. and Bryant, M. P., *Canad. J. Microbiol.*, *15*, 1365 (1969).
20. Krishnamurty, H. G., Cheng, K. J., Jones, G. A., Simpson, F. J. and Watkin, J. E., *Canad. J. Microbiol.*, *16*, 759 (1970).
21. Cheng, K. J., Krishnamurty, H. G., Jones, G. A. and Simpson, F. J., *Canad. J. Microbiol.*, *17*, 129 (1971).
22. Das, N. P. and Sothy, S. P., *Biochem. J.*, *125*, 417 (1971).
23. Griffiths, L. A. and Barrow, A., Symposia Angiologica Santoriana. 4th Int. Symp., Fribourg-Nyon Angiologica, *9*, *162* (30) (1972).

THE FATE OF TYROSINE IN GERMINATING BARLEY

by
G. SCHULTZ, D. KEIL and H. KRÜGER
Institut für Tierernährung, Tierärztliche Hochschule
Hannover, FRG

Tyrosine is the substrate for the biosynthesis of protein as well as for the formation of several compounds of the so-called secondary metabolism in the plant. This is especially true of the compounds of cinnamoyl (or phenylpropane) metabolism in Gramineae, to which belong the cinnamic acids, benzoic acids, flavonoids, stilbenes, lignins and others [1, 2]. Other groups derived from tyrosine are the respective amines, hydroxynitriles (or cyanogenic glycosides) (*cf.* Conn and Butler [3] and others).

In barley the interesting ones are the compounds of cinnamoyl metabolism and the amines, especially the *N*-methyl-tyramines (*cf.* [4]).

The aim of this work was to investigate the mode and extent of biosynthetic pathways of the endospermal tyrosine in different parts of the plant during germination.

Raoul [5, 6], James and Butt [7] and Rabitzsch [8] found that *N*-methyltyramines occur only in a restricted interval of time in the root of barley, namely, in the first 20 days after germination. The biogenesis of the amines derived from tyrosine has been elucidated by Kirkwood and Marion [9], Leete, Kirkwood and Marion [10], Leete and Marion [11] with the help of isotope experiments. Mudd [12], Mann and Mudd [13], and Mann, Steinhart and Mudd [14] demonstrated that the enzymes for methylation of these tyramines also exist only during the same interval of time and, therefore, the limited occurrence of *N*-methyl-tyramines is easily explained. According to Frank and Marion [15] the tyramines are later polymerized. Probably the amino nitrogen is also included in this reaction. This would be inferred from the earlier work of Phillips and Goss [16].

Application of labelled substances to the root as done by Marion *et al.* [9, 10, 11, 15] localizes the activity mainly in the root: in the case of tyrosine or tyramine it was up to 85 % [11]; in that of hordenine up to 95 % [15]. On the other hand, Seikel and Geissman [17], Seikel and Bushnell [18], McClure and Wilson [19], Carlin and McClure [20] have detected in the earliest stages of development of the leaf 6-*C*-glucosyl-flavones, namely, saponarin (6-*C*-glucosyl-

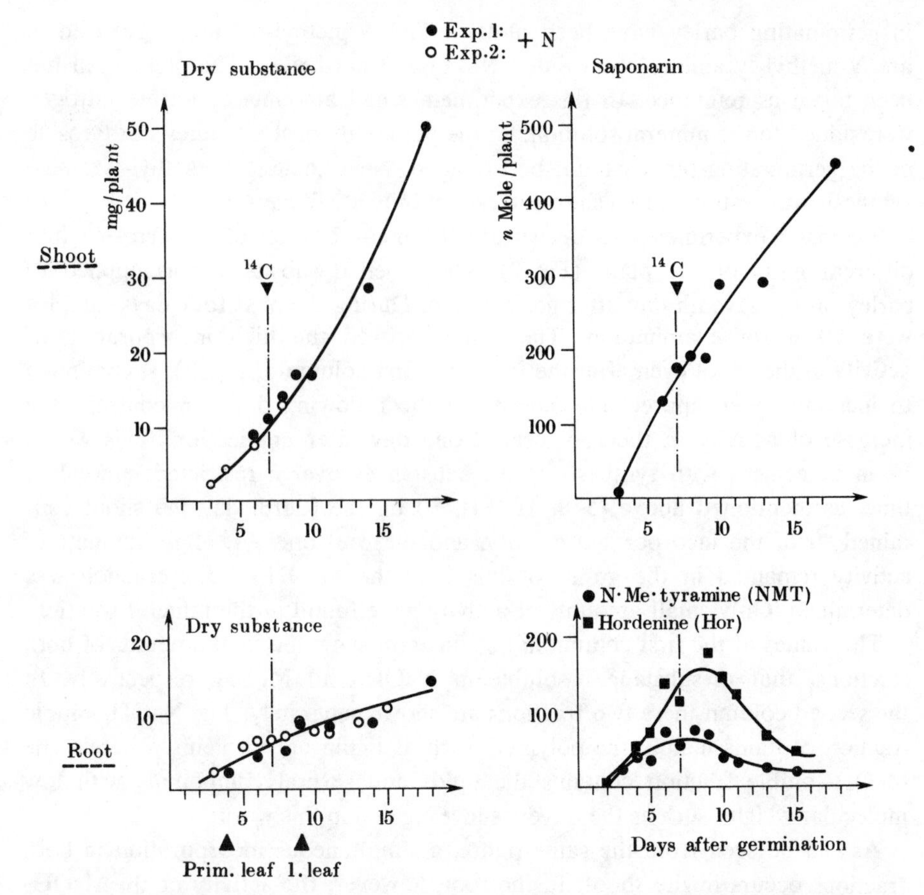

Fig. 1. **Dry substances of the shoot (top left) and of the root (bottom left) (in mg/plant), saponarin of the shoot (top right) and *N*-methyl-tyramine and hordenine of the root (bottom right) (in nanomole/plant) in the germinating plant**

5,7,4′-trihydroxyflavone 7-glucoside), and small amounts of iso-orientin (6-*C*-glucosyl-5,7,3,4′-tetrahydroxyflavone 7-glucoside) and its 3-methyl ether. But the question still remained if a relationship exists between the occurrence of tyrosine in the endosperm of the grain and the synthesis of protein and compounds of cinnamoyl metabolism.

So we proposed to study these substances under conditions approaching the natural ones. This could be achieved by the method of μl-injection into the soft, milky endosperm.

In Fig. 1 the values for dry substances of the shoot and those of the root, the nanomoles of flavonoids of the shoot and of *N*-methyl-tyramines of the root in germinating barley have been plotted. The *N*-methyl-tyramines referred to are *N*-methyl-tyramine (abbreviated NMT) and hordenine. The plant itself has been taken as reference. In this experiment small amounts of nitrate nitrogen were added to the mineral solution. So the possibility of the synthesis of tyrosine in the germinating plant cannot be excluded. Nevertheless, *N*-methyl-tyramine synthesis was restricted to a short interval of only a few days.

We made experiments to investigate the incorporation of ^{14}C-tyrosine into different parts of the plant (Fig. 2). The labelled substance was applied to barley on the seventh day after germination. During the next four days samples were taken for examination. The values proved the high incorporation of activity in the shoot even after the first day (first column of Fig. 2). It continued to increase to an appreciable amount in the following days. In contrast, the increase of activity in the root ceased one day after application. This would be in agreement with synthesis of the substances over a restricted interval of time, as mentioned above [5–8, 12–14]. So on the fourth day the shoot contained $^{2}/_{3}$ of the incorporated activity, and the root only $^{1}/_{3}$. High amounts of activity remained in the grain; of this only the MeOH-soluble fraction was determined. Only small amounts of activity were found in the mineral solution.

The values of the first column in this diagram show the total amounts of both fractions, that of substances soluble in NaOH and MeOH, respectively. In the second column these two fractions are shown separately. The NaOH-soluble fraction contains mainly the polymers such as lignin and proteins, whereas the MeOH-soluble fraction contains the lipids and several compounds with low molecular weight, such as the flavones investigated in this work.

As can be seen from the same figure, a simultaneous incorporation in both fractions occurs in the shoot. In the root, however, the activity of the MeOH-soluble fraction decreased, whereas that of the NaOH-soluble fraction increased to the same extent. The latter is an indication for a transformation of monomers to polymers.

Concerning the activity/plant, the substances isolated from the MeOH-soluble fraction behave similarly: the values of the flavone, saponarin, in the shoot increased, whereas those of the *N*-methyl-tyramines in the root decreased.

Figure 3 shows these values together with those of nanomoles/plant and specific activity for the same experiment.

Though the activity/plant and the specific activity of saponarin in the shoot increased as mentioned above, the amount in nanomoles/plant decreased a few days after germination in this experiment conducted without additional nitrogen. Probably there are two reasons for it: firstly, the tyrosine pool in the endosperm is depleted; secondly, the flavone has a short biological half-life period. The

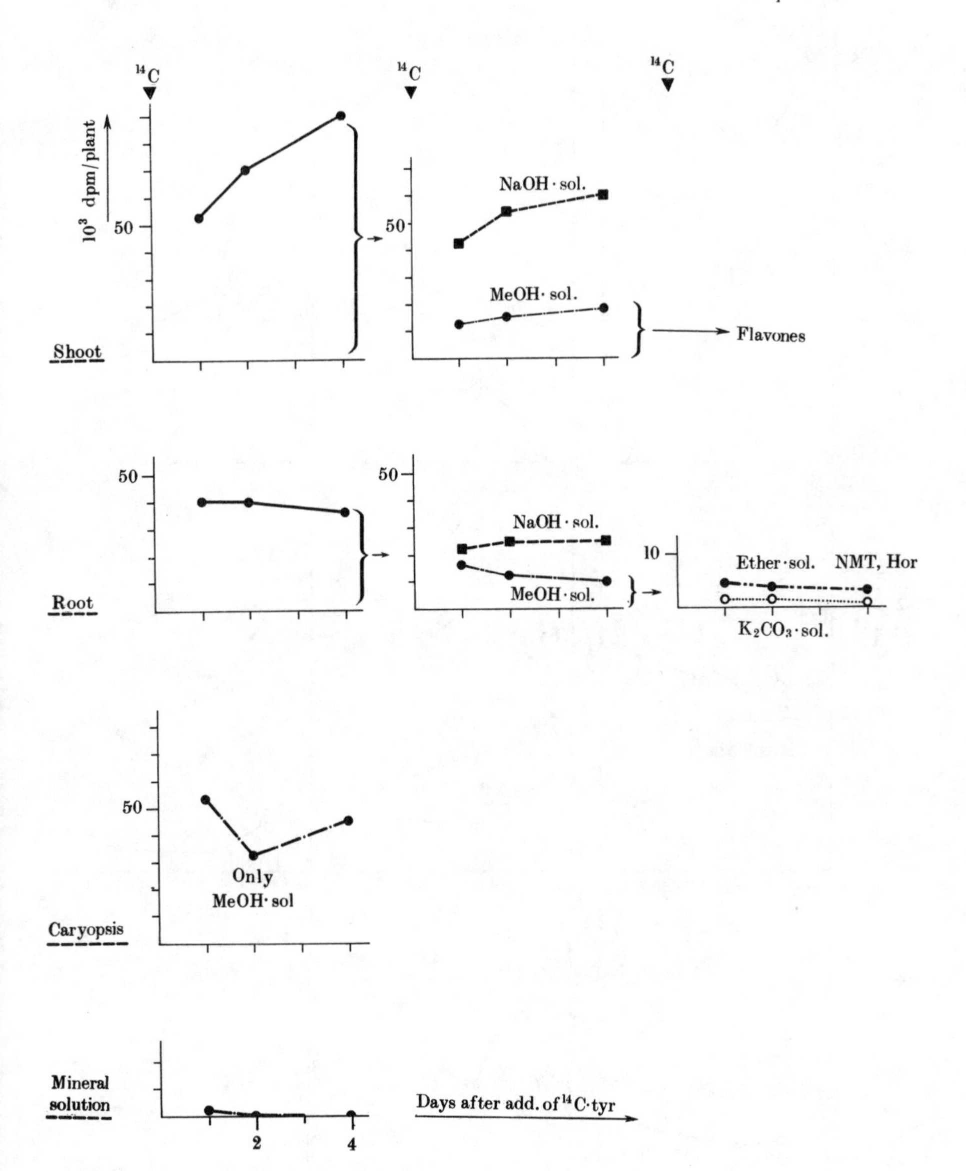

Fig. 2. Incorporation of ^{14}C-tyrosine in different parts of germinating barley (see text)

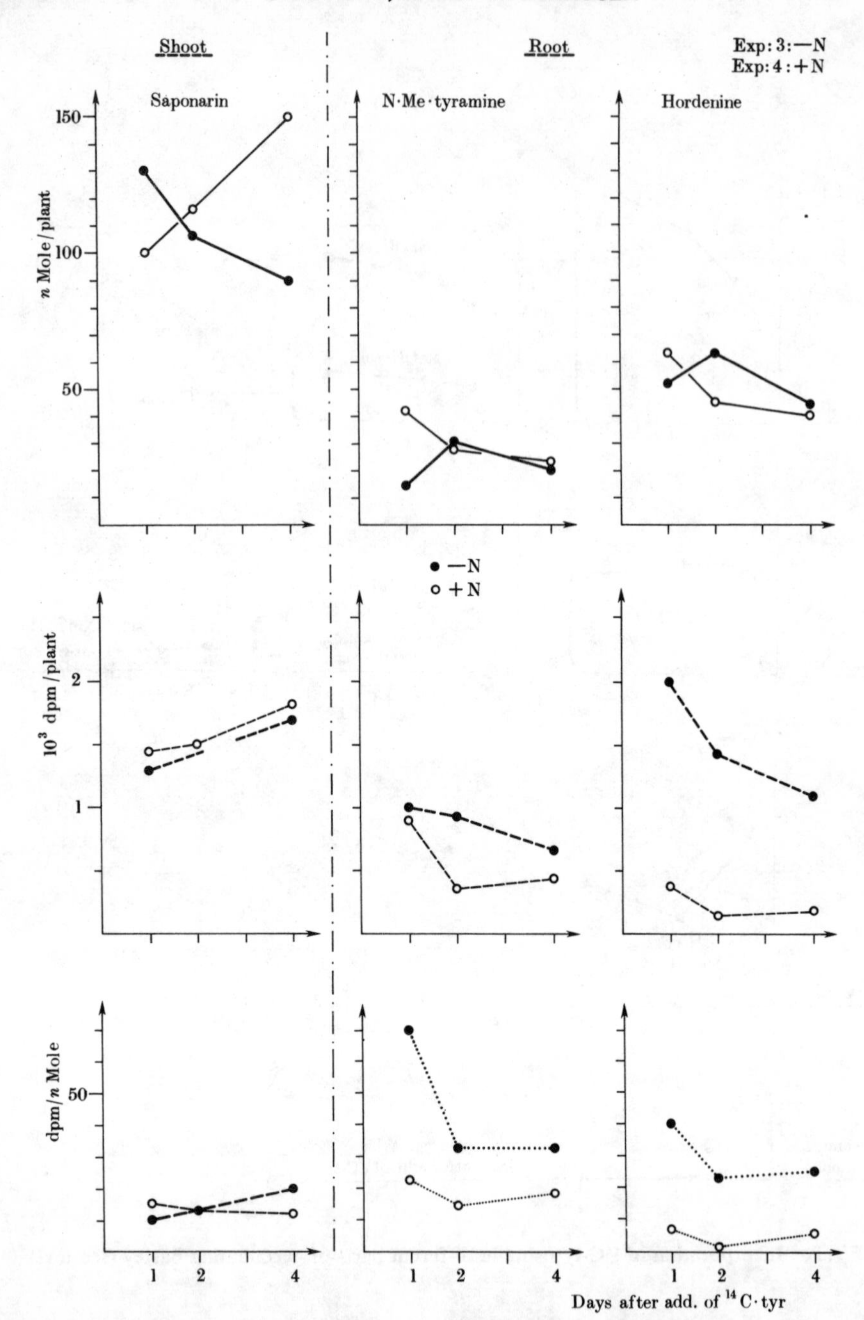

Fig. 3. Nanomole/plant (top), ^{14}C activity in 10^3 dpm/plant (middle) and specific activity in dmp/nanomole (bottom) of saponarin, *N*-methyl-tyramine and hordenine in germinating barley (see text)

increase in activity/plant and specific activity indicates that the reaction is in the steady state.

On the other hand, the *N*-methyl-tyramines of the root behave differently. The amount of NMT and hordenine in nanomoles/plant remained fairly constant, while the activity/plant and the specific activity decreased appreciably. It is not yet possible to explain the latter observation on the basis of our experiments.

In the same diagram the results of another experiment with additional nitrogen are plotted. These data are shown as thin lines. As was to be expected, the specific activity of saponarin in the shoot decreased slowly. That is due to the synthesis of unlabelled tyrosine in the germinating plant. It is of interest that the *N*-methyl-tyramines in the root showed a similar behaviour; but the decrease is much more.

In all cases the specific activity of hordenine is lower than that of *N*-methyl-tyramine. This would be in agreement with the fact of hordenine being a sequential product of *N*-methyl-tyramine [13, 14]. Likewise, the specific activity of the flavone, saponarin, is much lower than those of the *N*-methyl-tyramines. Regarding the experimental conditions, namely, starting from tyrosine, this can be explained on the basis of the formation of flavones being a multi-step reaction (*cf.* [1, 2]), whereas that of *N*-methyl-tyramines being an oligo-step reaction (*cf.* [9–11, 13, 14]).

From the analytical point of view, among the substances soluble in MeOH, flavones and *N*-methyl-tyramines are characterized by high specific activity. The distribution of the activity on TLC showed extreme high values in the respective zones (Fig. 4). The relatively low background seems to indicate the absence of labelled substances.

Figure 5 should demonstrate the effect of different methods of application on transport and metabolism. In the case of germinating barley it should be assumed that the substances applied to the root were used up by decarboxylating systems and/or there was a barrier of permeability between the root and shoot. On the other hand, if the substances are applied into the endosperm, they may penetrate through a large area of the scutellum into the root and shoot, respectively. Therefore, the importance of tyrosine occurring in the endosperm for the first stages of development of barley is presented in a new light. The main function may not only be the formation of *N*-methyl-tyramines including their polymers and that of proteins in the root, but also the synthesis on a larger scale of proteins and cinnamoyl compounds including their polymers, especially lignins, in the shoot. It is well known that there is — besides other ways of metabolism — a strong requirement of enzymes and substrates of cinnamoyl

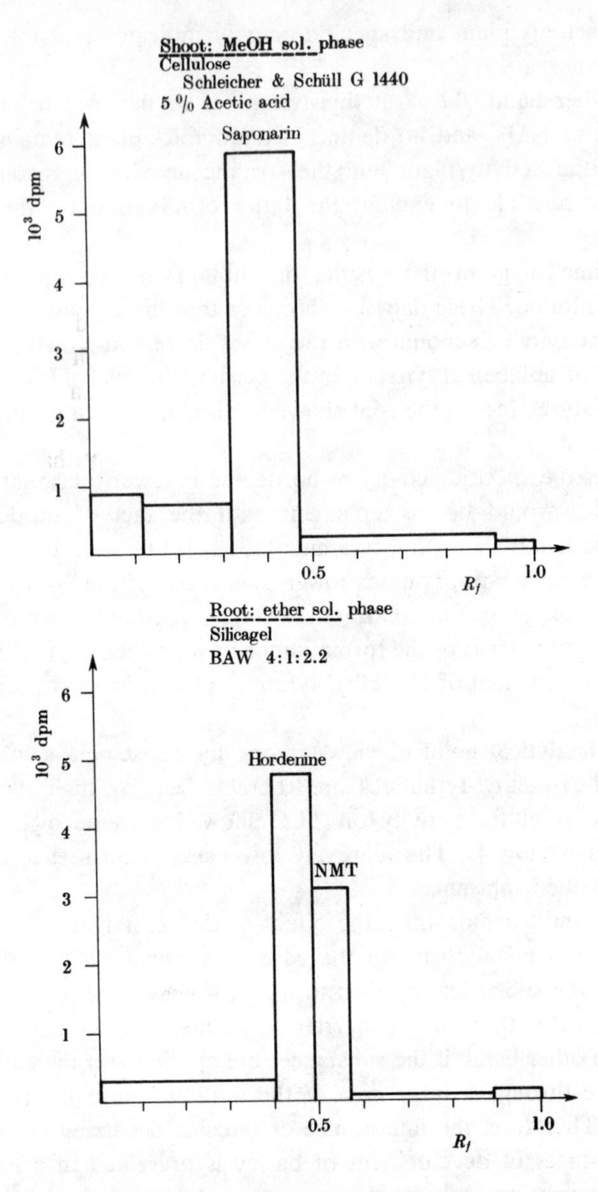

Fig. 4. ^{14}C-activity in the zones of TLC

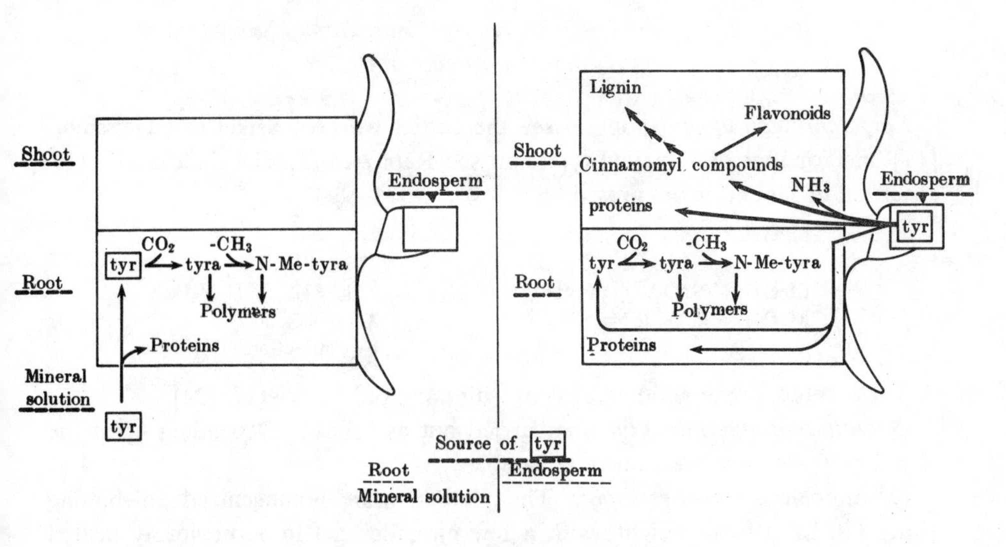

Fig. 5

metabolism [21, 22] for lignification during the rapid elongation of the axis to the stalk (Freudenberg and Neish [23]). This represents a very important stage in the development of the Gramineae.

EXPERIMENTAL

The Cultivation of the Seedlings

For swelling the grains of barley (variety "Villa"), they were submerged for 1 day in tap water which was aerated with the help of a membrane pump. The swollen grains were put in a tray (20 × 30 cm) in rows on filter-paper, covered with tap water in a thin layer. For preventing putrefaction, the filter paper under the grains was lifted by glass stripes. The ratio of day/night period was 12:12 hours. The illumination was conducted by neon tubes of "white light" quality with an intensity of 3,000–4,000 lux (5 cm over the seedlings). The temperature during the day was 22–24 °C, which was reduced to 15 °C during the night. In the experiments with nitrate, the water was replaced by a mineral solution containing (in g/l) 0.80 KNO_3; 0.50 $CaCl_2$; 0.25 $MgSO_4 \cdot 7\ H_2O$; 0.18 $Na_2HPO_4 \cdot 2\ H_2O$; 0.45 $KH_2PO_4 \cdot 2\ H_2O$; 0.01 $FeSO_4 \cdot 7\ H_2O$; 0.0013 $ZnSO_4 \cdot 7\ H_2O$; 0.0002 $MnCl_2 \cdot 4\ H_2O$.

Identification and Quantitative Determination of Saponarin and N-methyl-tyramine in Barley

For identification of saponarin see the earlier work of Seikel and Geissman [17] and for that of *N*-methyl-tyramine see Rabitzsch [8]. The data of the UV spectrum of saponarin are presented below:

MeOH	271, 332
MeOH + 0.1 % $AlCl_3$ · aqua	276, 301, 345
MeOH + NaOAc (saturated)	258_i, 271, 350, 394
MeOH + 0.002 M NaOEt	271, 306_i, 389
	i = inflection

These values are in good agreement with those of Mabry *et al.* [24].

Quantitative determination was carried out as follows. Depending upon the size, 10–30 seedlings were taken as samples.

(a) Saponarin in the shoot: The shoots were homogenized in boiling water (10 ml/g fresh weight) with a few mg silica gel in a previously heated mortar immediately after harvesting, in order to make the enzymes inactive. The extraction was performed quantitatively by refluxing the suspension with water twice, each for 10 min. with ml/g fr.w. The aqueous extracts were then poured through a polyamide column (min: 7 × 1 cm/g fr.w.; polyamide Woelm) for selective adsorption of the flavone. After washing the column with 2 vol. of distilled water, the flavone was eluted with 2 vol. of MeOH. The eluate was evaporated cautiously in a rotating evaporator ("Rotavapor") at 40 °C almost to dryness. An adequate purification for quantitative determination was achieved with the help of preparative TLC on cellulose (cellulose layers, Schleicher and Schüll G 1440, ca. 0.25 mm) with 5 % acetic acid as solvent (loading of the layer: max = 0.5 g fr.w. per 20 × 20 cm). After marking in UV 360 nm, the zone of saponarin (for R_f values see [17]) was scraped out and extracted once with 4 ml, and then twice each with 3 ml of MeOH. This was done by stirring vigorously the suspension and centrifuging the adsorbent. The long-wave maximum in MeOH, $\lambda = 332$ nm, and in MeOH + 0.05 N NaOEt, $\lambda = 389$ nm, respectively, gave reliable results. Because saponarin was not crystalline, no molar extinction coefficient was obtainable. So the values of only the sugar-free compound, namely apigenin (Roth, Karlsruhe), were determined ($\varepsilon^{336\,nm}_{MeOH} = 19.2 \cdot 10^3\ cm^{-1}\ mole^{-1}$; $\varepsilon^{395\,nm}_{NaOEt} = 24.8 \cdot 10^3\ cm^{-1}\ mole^{-1}$) and were used for approximate calculation.

(b) N-methyl-tyramines in the root: The roots were dipped into boiling MeOH immediately after harvestings (10 ml/g fr.w.; 10 min) and subsequently extracted in a Soxhlet apparatus with MeOH (min 50 ml/g fr.w.; 24 hours) (= MeOH soluble fraction).

After evaporation of the extract to dryness, the resulting residue was subjected to partitition between ether and 5 % K_2CO_3 20 ml in a liquid–liquid extractor for about 15 hours. After evaporation, the ether phase was fractionated with the help of preparative TLC on silica gel (silica gel F 254 layers, Merck) with BAW 4:1:2,2 as solvent (max ≙ 0.5 g fr.w./20 × 20 cm). The detection of the tyramines was achieved by spraying the margins of the layers (1/4 + 1/4 in rel. area) with a fresh mixture of ferric chloride and potassium hexacyanoferrate(III) (1 vol. 5 % K_3 [Fe(CN$_6$)] + 2 vol. 10 % $FeCl_3$ + 8 vol. H_2O) (Prussian Blue reaction, *cf.* [25]). The zones of the tyramines between the margins of the layer were scraped out and extracted with 0.01 N HCl, once with 2 ml, and then twice, each with 1 ml. The substances were determined with Folin–Ciocalteua reagent for phenols (5 ml of the respective tyramine solution in 0.01 N HCl + 0.5 ml reagent (Merck) + 4.5 ml solution of sodium carbonate (30 g $Na_2CO_3 \cdot 10\ H_2O$ in 100 ml). Reading at 750 nm was done after standing for 30 min (*cf.* [26]). The calibration was conducted with commercial reference substances (tyramine hydrochloride and hordenine sulfate, Merck).

Preparation of the MeOH-soluble and NaOH-soluble Fractions

The MeOH-soluble fraction was prepared as described above.

For obtaining the NaOH-soluble fraction [15], the residue of the MeOH-soluble fraction was refluxed with 0.5 N NaOH twice, each for 8 hours with 3 ml/≙ g fr.w.

Application of ^{14}C-tyrosine to the Seedlings and ^{14}C-determination

7 days after submerging into the water (here equalized as the time of germination), 2 μl of ^{14}C-tyrosine (Radiochemical Centre, Amersham, England) solution of pH 10 (containing 0.1 μCi/0.01 μmole/μl) was injected with a micro-doser (Draeger, Heidelberg) in the soft, milky endosperm of the grain.

Aliquots (max. vol. 200 μl) of the fractions and isolated substances, respectively, described above were mixed with 10 ml scintillation fluid (MeOH/toluol 1:1, containing 65 mg PPO + 0.3 g POPOP/l). ^{14}C-Determinations were performed in a Packard 2320 Tri-Carb-scintillation spectrometer.

*

The author thanks Dr. K. Viswanathan for advice in translating the manuscript.

REFERENCES

1. Grisebach, H., "Biosynthetic Patterns in Microorganisms and Higher Plants." Wiley and Sons, New York, London, Sidney, 1967.
2. Grisebach, H. and Barz, W., *Naturwissenschaften, 56,* 538 (1969).
3. Conn, E. E. and Butler, G. W., "The biosynthesis of cyanogenic glycosides and other simple nitrogen compounds" in "Perspectives in Phytochemistry" (ed. J. B. Harborne and T. Swain), Academic Press, London, 1969, p. 47.
4. Hegnauer, R., "Chemotaxonomie der Pflanzen." Vol. 1. Birkhäuser, Basel, Stuttgart, 1962.
5. Raoul, Y., "Etude biochimique de l'hordenine". Thesis, Univ. Paris, 1936.
6. Raoul, Y., *Ann. ferment., Paris, 3,* 129, 193, 385 (1937).
7. James, W. O. and Butt, V. S., *Abh. Dtsch. Akad. Wiss., Berlin, Kl. Chem., Geol., Biol., 7,* 182 (1957).
8. Rabitzsch, G., *Plant med., 7,* 268 (1959).
9. Kirkwood, S. and Marion, L., *Can. J. Chem., 29,* 30 (1951).
10. Leete, E., Kirkwood, S. and Marion, L., *Can. J. Chem., 30,* 749 (1952).
11. Leete, E. and Marion, L., *Can. J. Chem., 31,* 126 (1953).
12. Mudd, S. H., *Biochim., Biophys. Acta, 38,* 354 (1960)
13. Mann, J. D. and Mudd, S. H., *J. Biol. Chem., 238,* 381 (1963).
14. Mann, J. D., Steinhart, C. E. and Mudd, S. H., *J. Biol. Chem., 238,* 676 (1963).
15. Frank, A. W. and Marion, L., *Can. J. Chem., 34,* 1641 (1956).
16. Phillips, M. and Goss, M. J., *J. Agric. Research, 51,* 301 (1935).
17. Seikel, M. K. and Geissman, T. A., *Arch. Biochem. Biophys., 71,* 17 (1957).
18. Seikel, M. K. and Bushnell, A. J., *J. Org. Chem., 24,* 1995 (1959).
19. McClure, J. W. and Wilson, K. G., *Phytochemistry, 9,* 763 (1970).
20. Carlin, R. M. and McClure, J. W., *Phytochemistry, 12,* 1009 (1973).
21. Ebel, J. and Grisebach, H., *Z. physiol. Chem., 354,* 1183 (1973).
22. Gross, G. G., *Z. physiol. Chem., 354,* 1195 (1973).
23. Freudenberg, K. and Neish, A. C., "Constituents and Biosynthesis of Lignin". Springer, Berlin, 1968.
24. Mabry, T. J., Markham, K. R. and Thomas, M. B., "The Systematic Identification of Flavonoids." Springer, Berlin, 1970.
25. Randerath, K., "Dünnschicht-Chromatographie," 2. ed. Verlag Chemie, Weinheim/Bergstrasse, 1965, p. 215.
26. Klinisches Labor, 11. ed. (ed. E. Merck) Darmstadt, 1970.

THE EFFECT OF FLAVONOLS ON THE *IN VITRO* RESPONSE OF CHICKEN LYMPHOCYTES TO PHYTOHEMAGGLUTININ

by

M. BAKAY, R. PUSZTAI and I. BÉLÁDI

Institute of Microbiology, University Medical School
Szeged, Hungary

In earlier publications we described the antiviral activity of flavonols on herpesviruses [1–3]. It has also been observed that cells treated with quercetin exhibited decreased sensitivity to *Herpesvirus hominis* suggesting changes due to quercetin on the cell surface. This prompted us to study the effect of flavonols on the behaviour of other cells, and lymphocytes were used for this purpose. Lymphocytes cultured *in vitro* can be induced to enlarge, initiate DNA synthesis by the addition of phytohemagglutinin (PHA), a nonspecific antigen. This stimulation can be followed by detecting either the morphological alterations or the increased synthesis of RNA, DNA, or protein.

The effect of quercetin and rutin was tested on the response of chicken lymphocytes to PHA. Lymphocytes were obtained from the spleens of 2 months old chickens. In some experiments peripheral blood lymphocytes of chickens were also used. The lymphocytes were cultured at 38 °C for 2–4 days in a synthetic medium without serum. PHA M (Difco) was given at the beginning of the culture period; the controls received no PHA. To determine DNA synthesis in the presence and absence of PHA, 1 μCi ^{3}H-thymidine (Amersham, England, spec. act. 5 Ci/mole) was added at the end of the culture period to each culture for a 6-hour period. The incorporation of tritiated thymidine into the acid-insoluble material was determined by liquid scintillation counting.

Table I summarizes the results of experiments in which quercetin and rutin were given to chicken spleen lymphocytes in the presence and absence of PHA.

It can be seen that 200 μg of rutin was toxic, reducing ^{3}H-thymidine incorporation compared with the control lymphocytes. The ^{3}H-thymidine incorporation was not altered by 20 μg of rutin, however, the response to PHA was inhibited, and 2 μg of rutin had no effect. A slight influence on the response of spleen lymphocytes to PHA could be observed with quercetin.

Similar results were found with rutin and quercetin in testing peripheral blood lymphocytes of chickens (Table II).

Rutin impaired the response of peripheral lymphocytes and quercetin had no effect.

Table I

^{3}H-Thymidine Incorporation into Chicken Spleen Lymphocytes

Flavonols		Counts/min/culture	
		Without PHA	With PHA
	—	1,620	9,486
	200/μg	334	1,490
Rutin	20/μg	1,462	2,356
	2/μg	1,702	9,666
	—	618	2,710
Quercetin	30/μg	674	2,052

Table II

^{3}H-Thymidine Incorporation into Chicken Peripheral Blood Lymphocytes

Flavonols		Counts/min/culture	
		Without PHA	With PHA
	—	598	2,497
Rutin	20/μg	888	834
Quercetin	30/μg	472	2,422

Lymphocytes taking part in the immune response are important cells of vertebrates. To study their response to PHA offers a method by which the efficiency of cellular immune response could be detected. Our results have shown that rutin, like immunosuppressive drugs, alters the *in vitro* response of chicken lymphocytes to PHA. We assume that rutin inhibits the binding of PHA to the carbohydrate receptor sites on lymphocytes, as it has been observed by Borberg *et al.* [4] using N-acetyl-D-galactosamine. Further experiments are in progress to support this suggestion and to study the possible *in vivo* immunosuppressive effect of rutin.

*

We wish to thank Professor G. Ivánovics for his interest and support.

REFERENCES

1. BÉLÁDI, I., PUSZTAI, R. and BAKAY, M., *Naturwissenschaften, 52,* 402 (1965).
2. PUSZTAI, R., BÉLÁDI, I., BAKAY, M., MUCSI, I. and KUKÁN, E., *Acta Microbiol. Acad. Sci. Hung., 12,* 327 (1965/66).
3. BAKAY, M., MUCSI, I., BÉLÁDI, I. and GÁBOR, M., *Acta Microbiol. Acad. Sci. Hung., 15,* 223 (1968).
4. BORBERG, H., YESNER, I., GESNER, B. and SILBER, R., *Blood, 31,* 747 (1968).

NEW DATA ON THE METABOLISM OF FLAVONOIDS

by
Ö. TAKÁCS* and M. GÁBOR**
*Department of Biochemistry and **Department of Pharmacodynamics,
Medical University
Szeged, Hungary

As it is well known, Szent-Györgyi and Rusznyák [1] recognized the role of citrin ("vitamin P") in 1936. Many reports have been published about the metabolism on oral administration of different flavonoids to experimental animals [2–9], but few metabolic studies have been reported on isolated organs [10].

In previous investigations [11] it was found that the isolated liver of rat maintained on a normal diet is able to metabolize rutin. In the present paper we shall demonstrate the results of the continuation of the previous experiments. Our aim was to study the metabolism of rutin and structurally related flavonoids, to establish whether the decompositions proceed by the same or different routes.

Flavonoids with a 4′-monohydroxylated B ring are very widespread in plants, and thus their dietary importance is well recognized. Special mention should be made of the metabolism of aglycones (apigenin) and flavone glycosides (naringin) in *Citrus* fruits. In our earlier experiments it was found that in the course of the metabolism of rutin, aglycones are first formed in the isolated liver, followed by ring cleavage or oxidative splitting-off of the B ring, giving rise to various phenolic acid derivatives.

After oral administration to experimental animals and to man of various flavonoids, Booth *et al.* [2] and Griffiths [5] detected essentially the same phenolic acid derivatives in the urine and faeces as those which we described in 1972 in the case of the isolated, perfused rat liver. For this reason, it seemed practical in the subsequent experimental series to study the metabolism of the individual aglycones (apigenin, luteolin) under similar experimental conditions. The methods used were briefly as follows.

The isolated liver was perfused with an oxygenated Tyrode solution. The medium contained 20 μg per ml rutin, or the same amount of apigenin. At the end of the experiment 50–100 ml of perfusate was lyophilized, the dry material was extracted with methanol, the solvent evaporated in a Rotavapor apparatus at 30 °C, and the residue was dissolved in 5–10 ml methanol. For analysis, gel filtration was performed on a Sephadex LH-20 column as described by Johnston *et al.* [12], the elution being carried out with methanol. The fractions were further analyzed by thin-layer chromatography and UV spectroscopy. The examinations were made simultaneously by several methods, the systems used primarily being those of Mabry *et al.* [13] and Griffiths [5]. An effort was made to identify the metabolites on the basis of the R_f values given by these latter authors, and also by UV spectrophotometric analysis. The spots were developed with 5 % aluminium chloride or with Barton's reagent.

Results

On a Sephadex LH-20 column the rutin perfusate gave six fractions. These fractions were examined by thin-layer chromatography, leading to the following picture (Fig. 1). The chromatogram shows the spots of the overall perfusate, the six fractions obtained by gel filtration, and apigenin, luteolin, quercetin and rutin, which were used as standards. The perfusate and its gel-filtered fractions give R_f values identical with those of the standard spots of apigenin, luteolin and rutin; a few other spots can also be seen.

As it has been reported [11], two-dimensional thin-layer chromatography revealed the presence of twelve spots in the perfusate (Fig. 2). Of these, spots *1, 2, 3* and *4* exhibit an intense fluorescence in UV light on the action of aluminium chloride. Spots *1–4* were isolated by Sephadex gel filtration and analyzed spectrophotometrically. Table I lists the metabolites identified on the basis of the R_f values and the spectrophotometric analysis. The fluorescent spots could be identified with apigenin, luteolin, and di- and trihydroxyflavone.

Table I

Spot No.	R_f I	R_f II	Spectrophot. analysis	Colour with Barton's r	Compound
1.	0.83	0.0	328, 312sh, 253sh,	++	4′,7–Dihydroxyflavone
2.	0.67	0.0	336, 269,	+++	Apigenin
3.	0.53	0.0	349, 291sh, 267, 253	+++	Luteolin
4.	0.41	0.0	343, 309, 250sh, 235	++++	3′,4′,7–Trihydroxyflavone
5.	0.51	0.15	–	+	Unidentified
6.	0.50	0.22	359, 259,	++	Rutin
7.	0.50	0.33	–	+	(3,4 Dihydroxyphenyl)–γ–valerolactone
8.	0.52	0.37	310	++	Isoferulic acid
9.	0.59	0.50	278	++	*m*–Hydroxyphenylpropionic acid
10.	0.615	0.56		+++	3–Methoxy–4–hydroxyphenylacetic acid
11.	0.40	0.64	254	++	*m*–Hydroxybenzoic acid
12.	0.26	0.69	279	+++	*m*–Hydroxyphenylacetic acid

Solvent I: *t*–BuOH : AcOH : H_2O (3:1:1)
Solvent II: 20% KCl

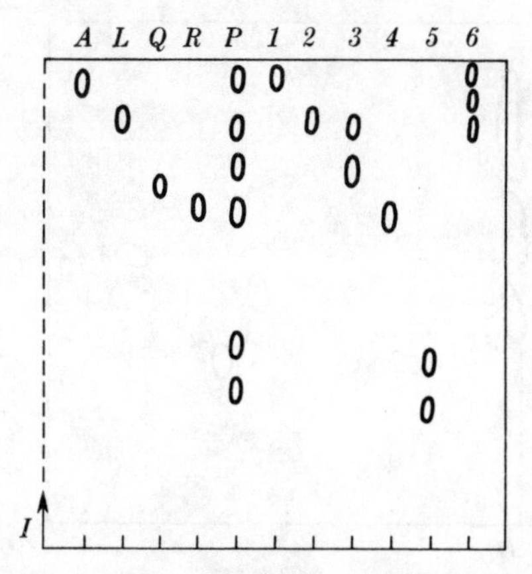

Fig. 1. **Chromatogram of the perfusate (*P*) of liver. Fractions obtained from gel filtration (*1, 2, 3, 4, 5, 6*) and standards: *A* — apigenin; *L* — luteolin; *Q* — quercetin; *R*— rutin**

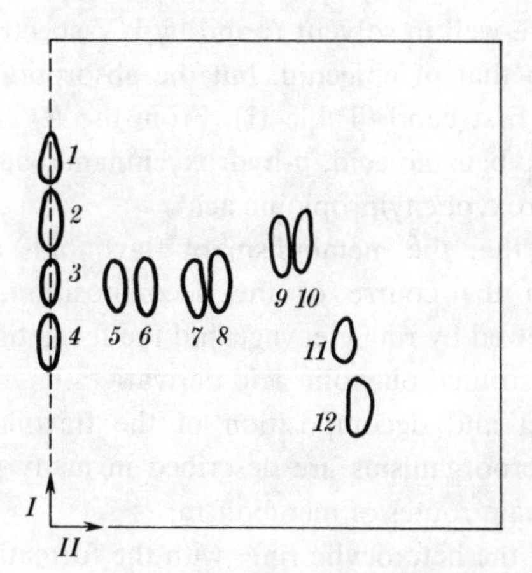

Fig. 2. **Chromatogram of the perfusate of liver with rutin. Solvents: *I* — t-BuOH : AcOH : H_2O (3:1:1); *II* — 20 % KCl**

Spot *5* could not be identified in this way, while spot *6* corresponded to undecomposed rutin, spot *7* to 3,4-dihydroxyphenyl-γ-valerolactone, and spots *8–12* to various phenolic acid derivatives. Quantitative determinations were not performed, and thus the intensities of the spots are indicated with crosses.

The assumed intermediates had to be examined further individually. As the first step, perfusion experiments were carried out with apigenin. Analysis of the perfusate gave the following result. Two-dimensional thin-layer chromatography led to three fluorescent spots (Fig. 3). Of these, spot *1* was identified as apigenin.

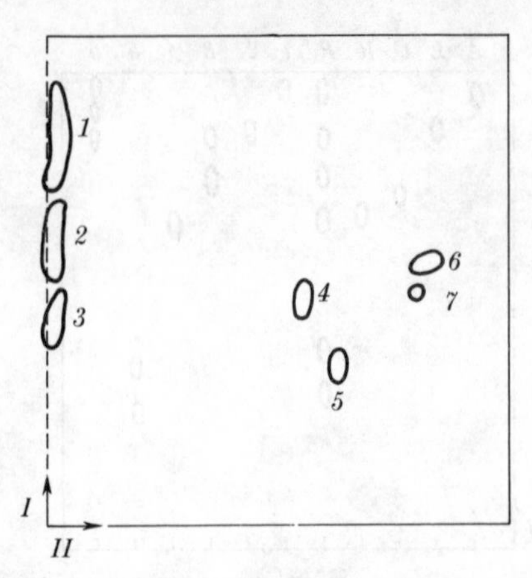

Fig. 3. Chromatogram of the perfusate of liver with apigenin. Solvents: as in Fig. 2

Spots *2* and *3* separate well in solvent *I*, and by UV spectrophotometry a curve is obtained similar to that of apigenin, but the absorption maximum is somewhat different in the first band (Table II). From the R_f values spots *4–7* were identified as *p*-hydroxybenzoic acid, *p*-hydroxycinnamic acid, *p*-hydroxyphenylacetic acid and *p*-hydroxyphenylpropionic acid.

Our results prove that the metabolism of flavonoids also proceeds in the isolated rat liver. In the course of the decomposition, aglycones are first produced; this is followed by ring cleavage and the formation of the corresponding *para-* or *meta-*substituted phenolic acid derivatives.

The transformation and decomposition of the flavonoids in the mammal organisms and in microorganisms are described in many papers. According to these, there are two main routes of metabolism:

1. The splitting of the heterocylic ring, with the formation of carbon dioxide and two substituted phenolic acids, which represent rings A and B.

Table II

Spot No.	R_f I	R_f II	Abs. maxima (nm) in methanol	Colour with Barton's r	Compound
1.	0.84	0.02	336, 269	++++	Apigenin
2.	0.60	0.02	330, 269	+++	Unknown
3.	0.45	0.02	325, 269	++	Unknown
4.	0.47	0.55	254	+++	*p*–Hydroxybenzoic acid
5.	0.34	0.61	310	+++	*p*–Hydroxycinnamic acid
6.	0.54	0.82	279	++	*p*–Hydroxyphenylacetic acid
7.	0.47	0.78	278	+	*p*–Hydroxyphenylpropionic acid

Solvents I and II: as in Table I

2. The oxidative cleavage of ring B, leading to phenyl-substituted propionic or acetic acid.

Our experiments indicate that in the course of the decomposition aglycones are first formed. These in part remain unchanged, and in part decompose to phenolic acid derivatives.

On the basis of investigations in recent years, Griffiths and Barrow [8] attribute great importance to the intestinal microflora in flavonoid metabolism. Our experiments on isolated, perfused rat liver show that it can be taken as proved that in addition to the intestinal microflora the liver also plays a part in the break-down of flavonoids. Our investigations further prove that certain intermediate flavonoid compounds are formed prior to ring cleavage; these are primarily luteolin, apigenin and two unknown flavonoids.

The results can be summarized as follows. In experiments on isolated, perfused rat liver it has been established that the tissue of the liver is capable of metabolizing rutin and apigenin. In the case of rutin the metabolites are produced *via* certain intermediate compounds. In the break-down process of rutin eleven metabolites are formed; of these, four compounds are intermediate flavonoid derivatives, while the others are variously substituted phenolic acid derivatives. In the case of apigenin, two intermediate aglycones and four phenolic acid derivatives could be detected.

REFERENCES

1. Rusznyák, St. and Szent-Györgyi, A., *Nature, 138,* 27 (1936).
2. Booth, A. N., Murray, C.W., Jones, F. T. and DeEds, F., *J. Biol. Chem., 223,* 661 (1956).
3. Watanabe, H., *Bull. Agric. Chem. Soc. Japan, 23,* 263 (1959).
4. Watanabe, H., *Bull. Agric. Chem. Soc. Japan, 23,* 164 (1959).
5. Griffiths, L. A., *J. Biol. Chem., 92,* 173 (1964).
6. DeEds, F., "Flavonoid Metabolism," in "Comparative Biochemistry. Metabolism of Cyclic Compounds," Vol. 20 (eds Florkin, M. and Stolz, E. H.), Elsevier, Amsterdam, 1968, p. 127.
7. Das, N. P. and Griffiths, L. A., *Biochem. J., 110,* 449 (1968); Das, N. P., *Biochem. Pharmacol., 20,* 3435 (1971).
8. Griffiths, L. A. and Barrow, A., *Angiologica, 9,* 162 (1972).
9. Gábor, M., "The Anti-Inflammatory Action of Flavonoids". Akadémiai Kiadó, Budapest, 1972.
10. Förster, H., Bruhn, U. and Hoos, I., *Arzneimittel-Forsch. (Drug Res.), 22,* 1312 (1972).
11. Takács, Ö., Benkő, S., Varga, L., Antal, A and Gábor, M., *Angiologica, 9,* 175 (1972).
12. Johnston, K. M., Stern, D. J. and Waiss, A. C. Jr., *J. Chromatogr., 33,* 539 (1968).
13. Mabry, T. J., Markham, K. R. and Thomas, M. B., "The Systematic Identification of Flavonoids." Springer-Verlag, Berlin-Heidelberg-New York, 1970.

SOME OBSERVATIONS ON THE MECHANISM OF THE ANTIOXIDANT EFFECTS OF CERTAIN FLAVONOID COMPOUNDS

by

E. SZ.-GÁBOR

Department of Chemistry, College of Food Technology
Szeged, Hungary

There are a great number of publications about the antioxidant activity of flavonoid compounds [1–13], but not many of them deal with the anthocyanins [14–18].

Our examinations were extended to this field. Some anthocyanins are present as natural antioxidants in different fruits and they are very important compounds for canned fruit enriched in vitamin C, as the inhibitors of the harmful oxidative transformation of the vitamin to biologically inactive materials.

Our previous experiments have shown that the antioxidant effect of anthocyanin compounds on L-ascorbic acid depends not only on a number of external factors such as pH, the L-ascorbic acid : oxygen ratio, reaction time, and the redox potential of the system, but also on the nature of organic acids present.

In this paper we wish to give an account of studies on the special effects of the various organic acids.

L-Ascorbic acid was oxidized with atmospheric oxygen at room temperature in solutions stirred electromagnetically at 120 r.p.m.

The extent of oxidation was measured by determining the amount of L-ascorbic acid remaining at the end of reaction, which was expressed in percentage of the initial quantitiy. L-Ascorbic acid was determined iodometrically, using the dead-stop end-point technique.

The solutions used in the experiments were as follows:

Bell I 96	L-ascorbic acid (L-AS), 0.005 N,
„	iodine, 0.0025 N,
„	L-arginine, 0.25 % (for adjustment of the pH),
„	citric acid, 0.03 M,
„	tartaric acid, 0.03 M,
„	acetic acid, 0.03 M,
„	anthocyanin extract (AC), prepared from plum skin with methanol containing 0.1 % HCl; the concentration of anthocyanin in the reaction mixture was 1 mg/10 ml.

EXPERIMENTAL

A number of reference measurements were first made in the absence of the anthocyanin extract, to establish changes of the potentials and of the L-ascorbic acid content under the given conditions.

Figure 1 shows the changes of the potentials of distilled water, citric acid, acetic acid and tartaric acid solutions as a function of the reaction time. It can readily be seen that the more favourable low potentials develop in the presence of citric acid and tartaric acid. With tartaric acid the potential is smaller with the advance of time.

Figure 2 is a diagram of the variations of the L-ascorbic acid contents of solutions of L-ascorbic acid in the presence of various organic acids. The potential values are also indicated.

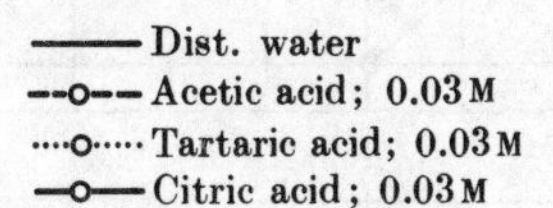

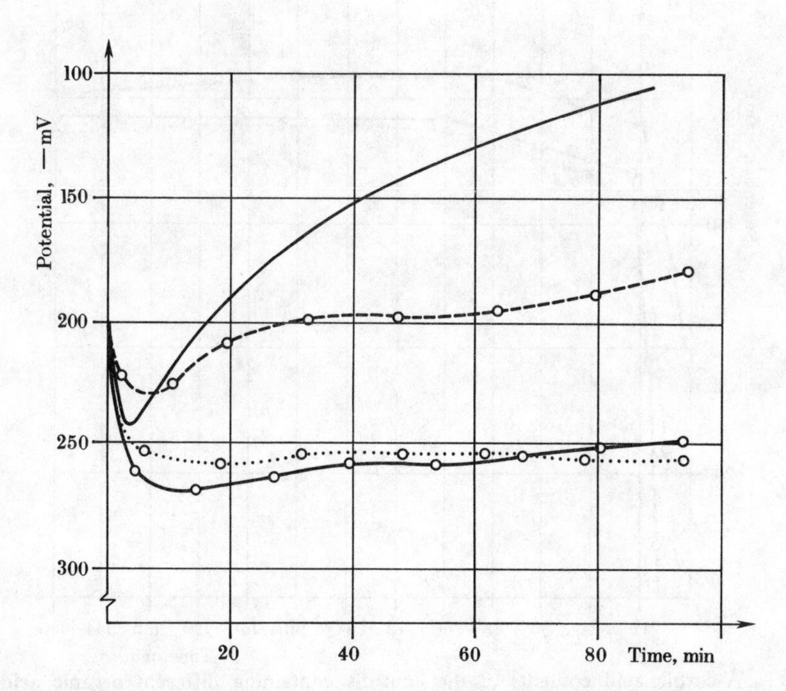

Fig. 1. Potential values of different solutions (Stirred electromagnetically at room temperature)

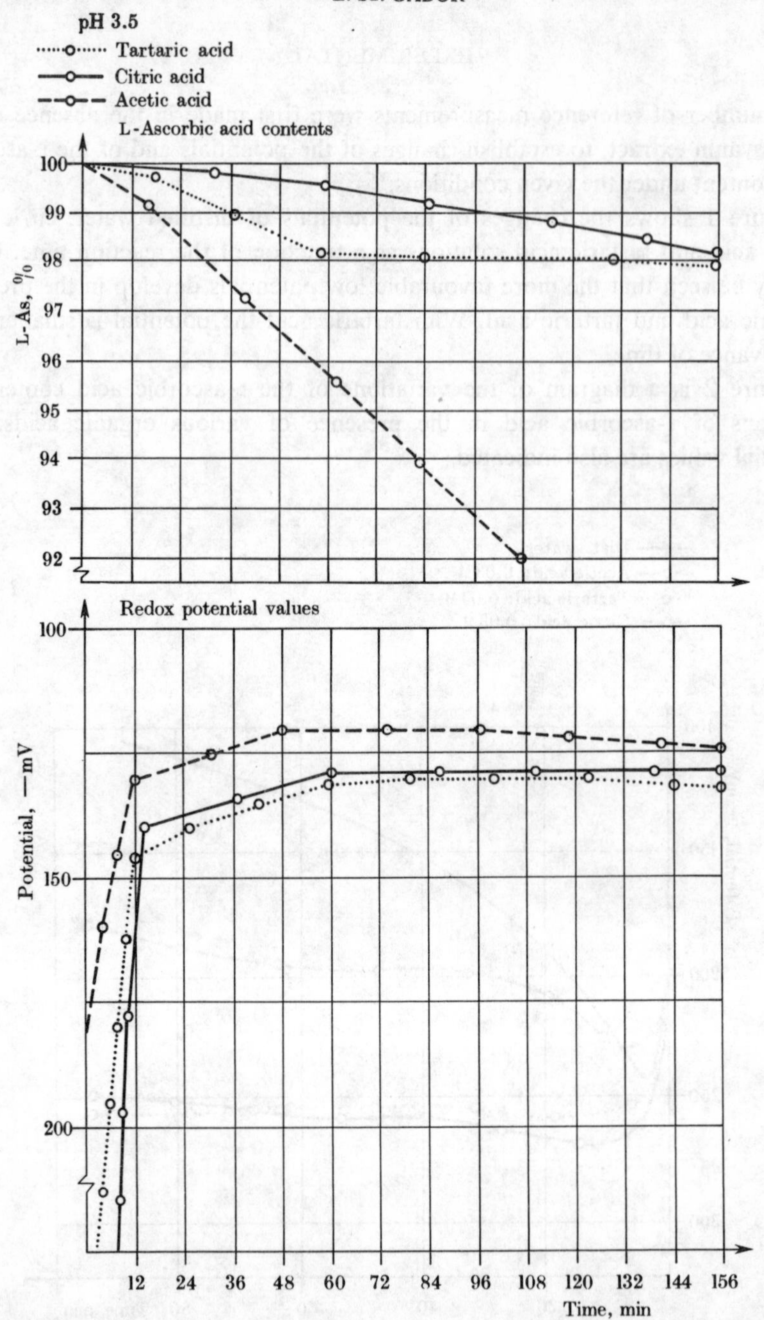

Fig. 2. L-Ascorbic acid contents of the controls containing different organic acids and redox potential values of the mixtures at pH 3.5. (Oxidized with atmospheric oxygen using an electromagnetic stirrer at room temperature)

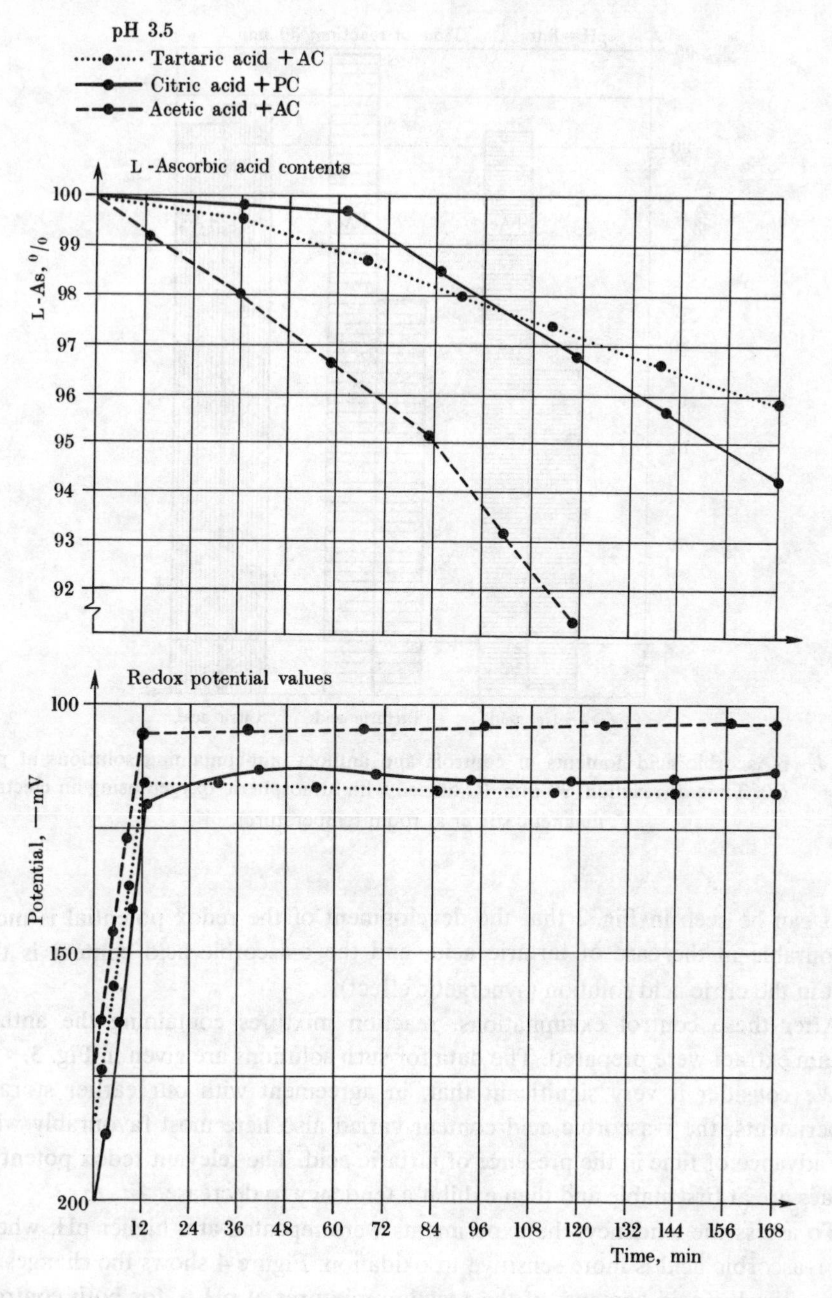

Fig. 3. L-Ascorbic acid contents in the presence of anthocyanin compounds extracted from plum skin and the redox potential values of the mixtures at pH 3.5. (Oxidized with atmospheric oxygen using an electromagnetic stirrer at room temperature)

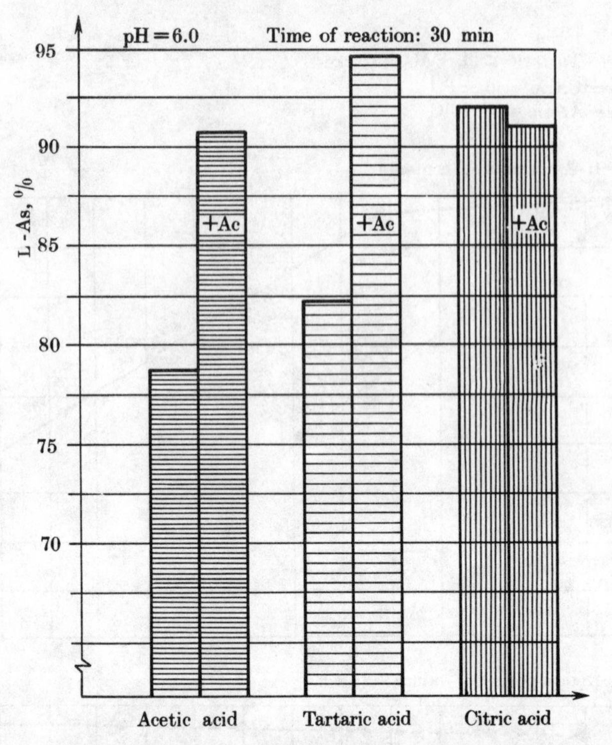

Fig. 4. L-Ascorbic acid contents of controls and anthocyanin-containing solutions at pH 6.0. Time of reaction: 30 min. (Oxidized with atmospheric oxygen using an electromagnetic stirrer at room temperature)

It can be seen in Fig. 2 that the development of the redox potential is most favourable in the case of tartaric acid, and the L-ascorbic acid content is the best in the citric acid solution (synergetic effect).

After these control examinations, reaction mixtures containing the anthocyanin extract were prepared. The data for such solutions are given in Fig. 3.

We consider it very significant that, in agreement with our earlier storage experiments, the L-ascorbic acid content varied also here most favourably with the advance of time in the presence of tartaric acid. The relevant redox potential values are at first stable and then exhibit a tendency to decrease.

To assess the tendency, the experiments were repeated at a higher pH, where the L-ascorbic acid is more sensitive to oxidation. Figure 4 shows the changes of the L-ascorbic acid contents of the reaction mixtures at pH 6, for both controls and anthocyanin-containing solutions, after a reaction time of altogether 30 minutes.

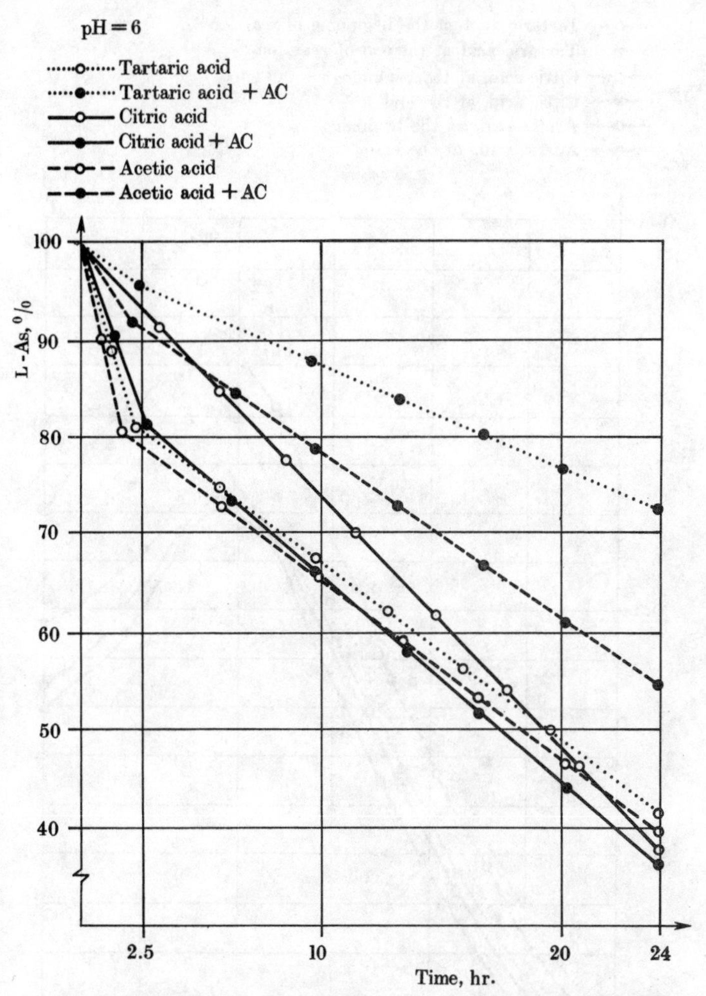

Fig. 5. L-Ascorbic acid contents of different solutions at pH 6.0. (Oxidized by spontaneous diffusion of atmospheric oxygen at room temperature)

The favourable effect of tartaric acid is even more clearly expressed here.

To draw nearer to the problems of food technology, a study was made of the variation of the L-ascorbic acid in solutions of similar composition, in the course of oxidation resulting from the spontaneous diffusion of oxygen. The data given in Fig. 5 indicate that the most favourable results are obtained in the tartaric acid solution.

These good results are not attributed solely to the low redox potential of the system. A study was further made of the visible spectrum of the anthocyanin in

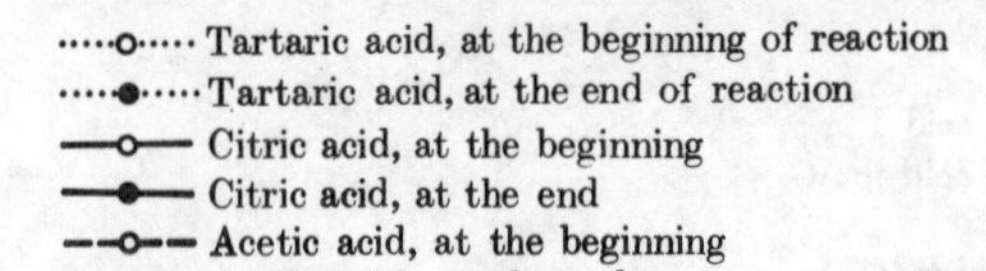

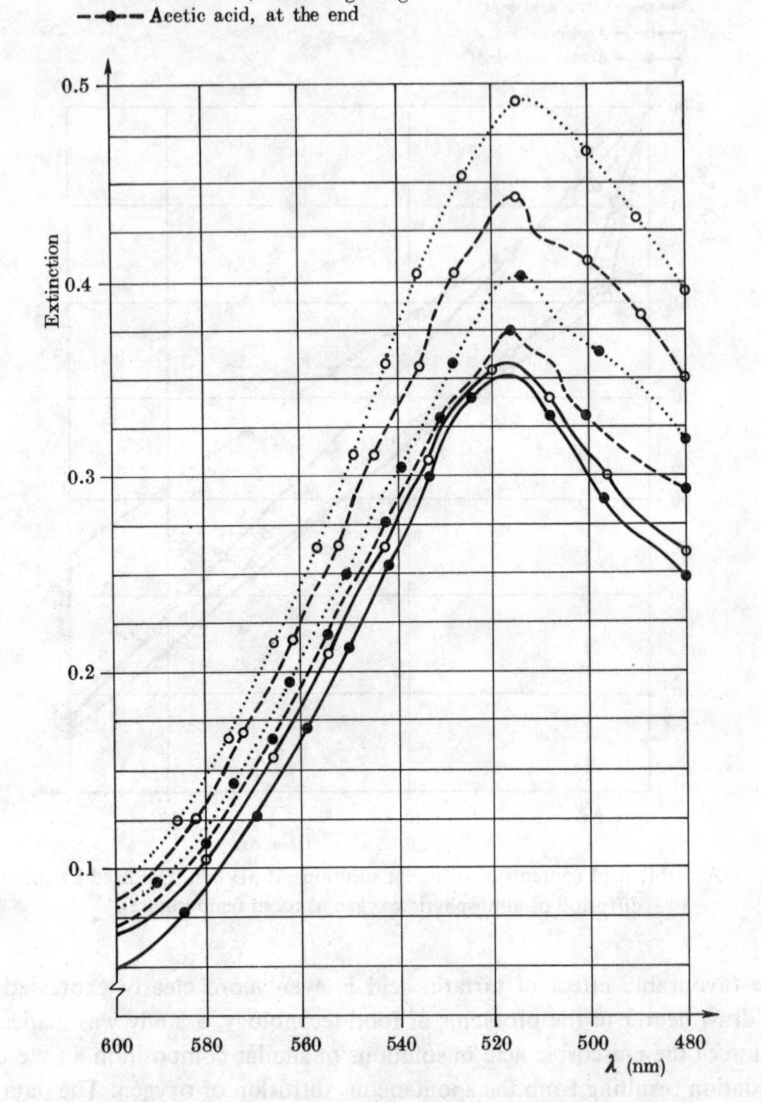

Fig. 6. **Spectra of anthocyanin compounds in solutions containing different organic acids. (Oxidized by spontaneous diffusion of atmospheric oxygen)**

the reaction mixtures containing the three organic acids, at zero reaction time and after oxidation following spontaneous diffusion of oxygen for 2.5 hours.

The spectra in Fig. 6 show the difference, which can be evaluated by examining the maximum at 515 nm; the extinction is the highest in the tartaric acid system, and barely falls by the end of the reaction. The lowest values were found in the citric acid solution.

This permits the conclusion that the anthocyanin compound may be modified in the various systems, and this may affect their antioxidant effects.

To sum up the results of the examinations with the solutions containing anthocyanins, tartaric acid has a favourable effect in preventing the oxidative transformation of L-ascorbic acid. The effect is presumably due to the anthocyanin compounds. The assumption is substantiated by the photometric spectra. However, this must be confirmed in future experiments mainly by molecular structural examinations.

Storage experiments are planned with preserved plum preparations enriched in the nutrition-physiologically important vitamin C; this may finally decide the significance of tartaric acid as an accessory substance promoting the antioxidant effects of the anthocyanins.

REFERENCES

1. Medovar, B. Ya., *Voprosi Pitaniya, 1967* (No. 2), p. 53.
2. Ki Soon Rhee and Watts, B. M., *J. Food Sci., 31,* 669 (1966).
3. Pratt, Dan. E., *J. Food, Sci., 30,* 737 (1965).
4. Letan, A., *J. Food Sci., 31,* 518 (1966).
5. Letan, A., *J. Food Sci., 31,* 395 (1966).
6. Gábor, E.-Sz., *Élelmezési Iparok, 19,* 309 (1965).
7. Hilditch, T. P., "The Chemical Constitution of Natural Fats", Chapman and Hall Ltd., London, 1947.
8. Clementson, C. A. B. and Andersen, L., *Ann. New York Acad. Sci., 136,* Art 14, 339 (1961).
9. Davidek, L., *Biokhimiya, 25,* 1105 (1961).
10. Davidek, L., *Biokhimiya, 26,* 93 (1961).
11. Szigyártó, E. G., *Nahrung, 13,* 355 (1969).
12. Gábor, E.-Sz. and Vámos, É. K., *Élelmiszervizsgálati Közlemények, 16,* 75 (1970).
13. Fragner, J., "Vitamine",Vol. I, VEB Gustav Fischer Verlag, Jena, 1964, p. 487.
14. Geissman, T. A., "The Chemistry of Flavonoid Compounds", Pergamon Press, Oxford, London, New York, Paris, 1962.
15. Harborne, J. B., "Anthocyanins and their Sugar Components", Springer Verlag, Wien, 1962.
16. Ribéreau-Gayon, P., "Les Composés Phénoliques des Végétaux", Dunod, Paris, 1968.
17. Gábor, E.-Sz., *Élelmiszeripari Főiskola Tudományos Közlemények, Szeged, 1,* 29 (1971).
18. Gábor, E.-Sz., *Élelmiszeripari Főiskola Tudományos Közlemények, Szeged, 2,* 49 (1973).

RELATIONSHIP BETWEEN THE ELECTRONIC STRUCTURE AND BIOLOGICAL ACTION OF SOME FLAVONOIDS

by

Z. DINYA* and E. HETÉNYI**

* Institute of Organic Chemistry, Kossuth Lajos University, Debrecen

** Research Institute for Organic Chemical Industry, Budapest, Hungary+

In recent years the number of biologically active γ-pyrone derivatives has increased [1, 2]. Examples of such compounds introduced as drugs are shown in Fig. 1.

In our earlier papers [3, 4] interest was centred on the important biological actions of various flavone derivatives; observations made by Da Re *et al.* [5, 6] called our attention particularly to the cardiovascular effect of these compounds. Later our studies have been extended to the isoflavones. In the literature of these compounds only their oestrogenic action and the action on the nervous system have been emphasized [7]; we have also given an account of the relationships between the cardiac effects and chemical structure [3]. To elucidate the exact relationships between constitution and the biological action, our studies have been continued by applying quantum chemical methods. This work is reported in this paper.

The compounds first investigated included the flavone derivative (No. 1) and isoflavone derivatives (Nos. 2–7) containing various substituents (H, CH_3, COOH, $CH(CH_3)_2$, C_2H_5, C_6H_5) at the C-2 atom, and an $-O-CH_2-COO^-$ function at C-7 (Table I). We have established that by changing the C-2 substituents in the above order, the positive inotropic effect of these compounds on the frog heart decreased. In fact, an explicit cardiodepressive (toxic) effect has been observed in the case of the compound (No. 7) containing a phenyl substituent. Thus the 2,3-diphenylchromone derivative which can be regarded as 2-phenylisoflavone or 3-phenylflavone, has lost the properties characteristic of both the flavones and isoflavones.

The interaction between a pharmacon and receptor — and therefore the biological effect — is characteristically influenced by structural changes in the pharmacon molecule [8–13]. For a given pharmacon series Cammarata [14] gave an equation of the following type to describe the correlation between the

+Present address: BIOGAL Pharmaceutical Works, 4041 Debrecen, Hungary

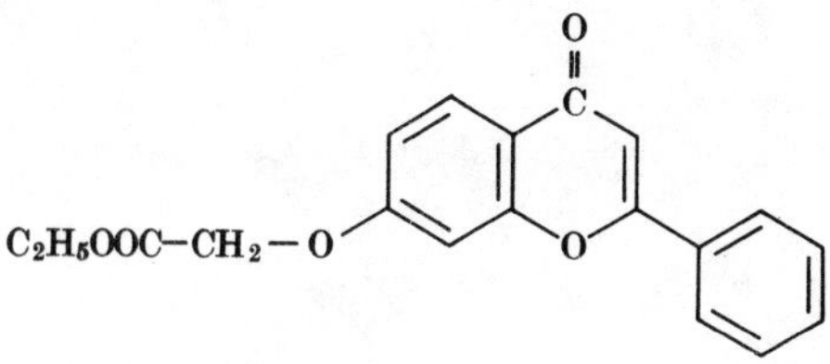

O
C
$C_2H_5OOC-CH_2-O$
O

Recordyl
(7-Flavonoxyacetic acid ethyl ester)

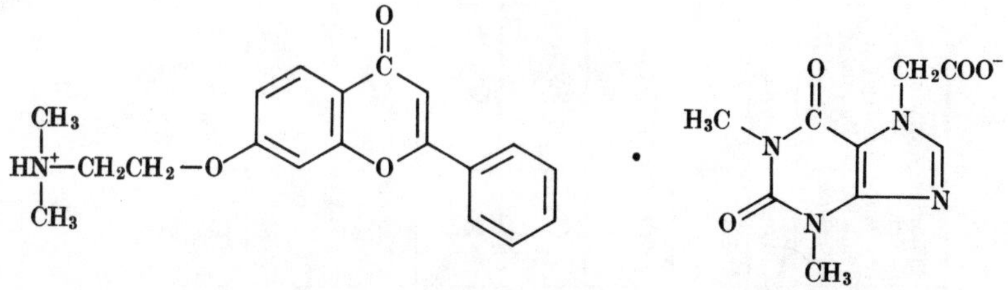

O
C
CH_3
$HN^+-CH_2CH_2-O$
CH_3
O
•
O
CH_2COO^-
H_3C
C
N
N
N
O
N
CH_3

Perflavon (Theophylline-7-acetic acid 7-[2-(dimethylamino)ethoxy]-flavone salt)

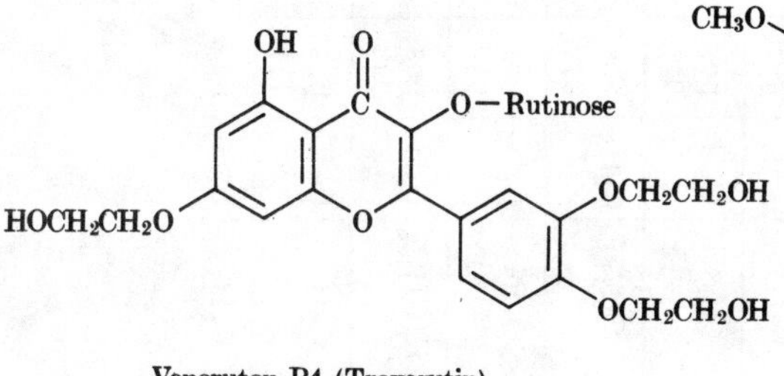

OH
O
C
O—Rutinose
$HOCH_2CH_2O$
O
OCH_2CH_2OH
OCH_2CH_2OH

OCH_3
CH_3O
O
C
O
O
O
CH_2
C
CH_3

Venoruton P4 (Troxerutin)
(7, 3′,4′-Tris[O-(2-hydroxyethyl)]rutin)

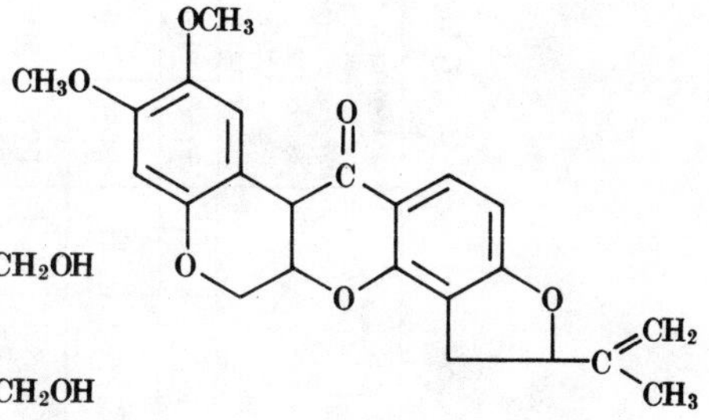

Rotenone

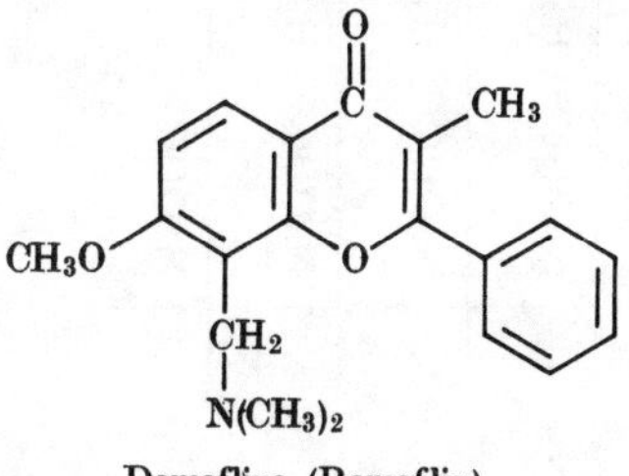

O
C
CH_3
CH_3O
O
CH_2
$N(CH_3)_2$

Demefline (Remeflin)
(8-Dimethylaminomethyl-7-methoxy-3-methylflavone)

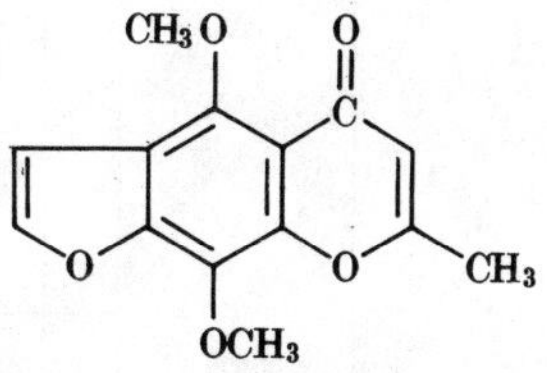

CH_3O
O
C
O
O
CH_3
OCH_3

Khellin
(5,8-Dimethoxy-2-methyl-6,7-furanochromone)

Fig. 1

Table I

(chromone structure: R_1, R_2, R_3)				Inotropic activity Percent changes of the contractile amplitude		Change in coronary resistance	Inhibition of fibrillation 20mM KCl *m.Sartorius*	Antiarrhythmic activity 2 ml/kg $CaCl_2$ solution, i.v. (10% $CaCl_2$)	Coronary vasodilatator effect Vasopressin 2 I.U./kg, i.v.	Number of experiments
No.	R_1	R_2	R_3	Frog	Rat	Rat	Frog	Rat	Rat	
1.	NaOOC–CH_2–O–	C_6H_5	H	pos.	pos.	0.80	–	–	±	10
2.	NaOOC–CH_2–O–	H	C_6H_5	pos. 94%	pos.	0.70	Ø	100%	Ø	10
3.	NaOOC–CH_2–O–	CH_3	C_6H_5	pos. 92%	pos.	0.75	Ø	95%	Ø	10
4.	NaOOC–CH_2–O–	COONa	C_6H_5	pos. 33%	pos.	0.80	Ø	80%	Ø	10
5.	NaOOC–CH_2–O–	$CH(CH_3)_2$	C_6H_5	pos. 29%	pos.	0.80	Ø	–	Ø	10
6.	NaOOC–CH_2–O–	C_2H_5	C_6H_5	pos. 11%	pos.	0.95	Ø	–	–	10
7.	NaOOC–CH_2–O–	C_6H_5	C_6H_5	neg.	neg.	1.2	Ø	Ø	Ø	10

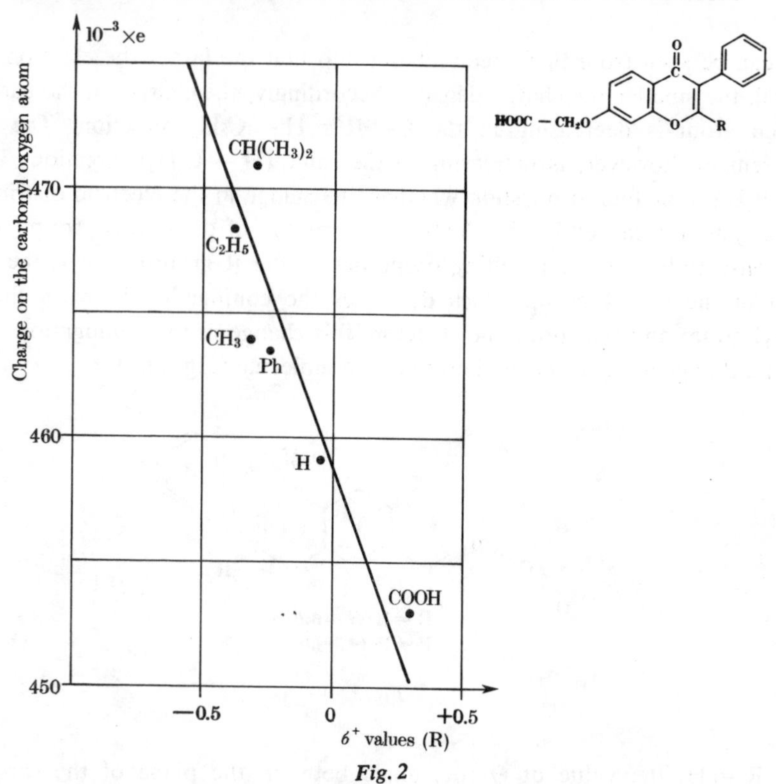

Fig. 2

electronic, hydrophobic and steric effects of a substituent and the change in the biological effect (δA_n):

$$\delta A_n = \bar{a}\sigma + \bar{b}\pi + \bar{c}E^S + \bar{d}$$

where $\bar{a}$, $\bar{b}$, $\bar{c}$ and $\bar{d}$ are constants, and σ, π and E^S represent the Hammett [15], Hansch [16] and the Taft [17] constants, respectively.

The next problem is the application of this equation to the group of compounds discussed above. It can be stated with certainty that a change in the substituent at C-2 does not result in a conformational change.. We have calculated the quantum chemical parameters [8, 9] of the derivatives having various substituents at the C-2 carbon atom. The data [8, 9] characteristic of the electron distribution of the system (charge density, superdelocalizability, etc.) change characteristically, first of all, in the γ-pyrone ring. The relationship shown in Fig. 2 has been obtained between the charge on the oxygen atom of the carbonyl group and the σ* values [15] expressing the electronegativity of the substituent on the C-2 carbon atom.[+]

[+] The charge is given in electron charge (e) units.

It can be seen from this linear relationship that the larger the electron withdrawal, the smaller the charge density. Accordingly, the charge on the carbonyl oxygen atom is decreasing in the C-2-R = H → C_6H_5 direction. The steric requirement, however, is increasing in the same (H → C_5H_5) direction. Therefore, it is reasonable to question whether the change in the electron distribution of the system is caused by the electronic properties of the C-2-R group (i.e. the electron-withdrawing or repelling properties of the R group) or it is the steric effect of the C-2-R group which decreases the conjugation between the C-3 phenyl group and the chromone skeleton, this change in the conjugation resulting in a change in the electron density of the molecule (Fig. 3).

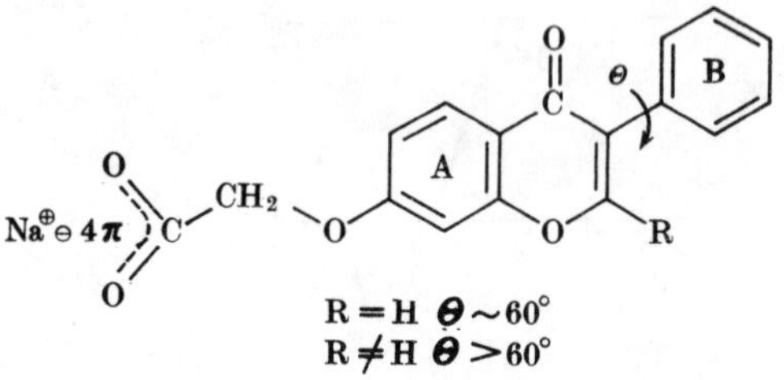

R = H $\Theta \sim 60°$
R ≠ H $\Theta > 60°$

Fig. 3

If R = H, the value of Θ (the angle between the plane of the chromone skeleton and ring B) is approx. 60 ° [18]. If R ≠ H ring B "turns out" to a greater extent, therefore the conjugation is decreased. It can be postulated that the effect of the change in the conjugation is the more important factor. The effect of C-2 substitution manifests itself also in the charge of the ethereal oxygen atom of the pyrone ring. The value of the charge is increasing in the H → C_6H_5 direction, but, on the whole, this atom has a net positive charge (— 0.2 e). According to our calculations a change of the substituent at the C-2 carbon atom has no effect on the charge of the ethereal oxygen atom bound to the C-7 carbon atom of the flavone skeleton. The electron density, however, of this oxygen atom is greater than that of the pyrone oxygen in the ring (although also this has a positive charge (— 0.084 e)). The charge of the carbon atoms in ring A changes only slightly by changing the C-2-R structure (Fig. 4).

It can be seen that the carbon atoms neighbouring the C-7 atom (C-6 and C-8, according to the original numbering of the flavonoid skeleton) are negatively charged, and C-8 has a greater charge.

It can be stated moreover that the electrophilic superdelocalizability indices (S_i^E) [8, 9] of the C-6 and C-8 atoms are influenced characteristically by the C-2-R function (Table II).

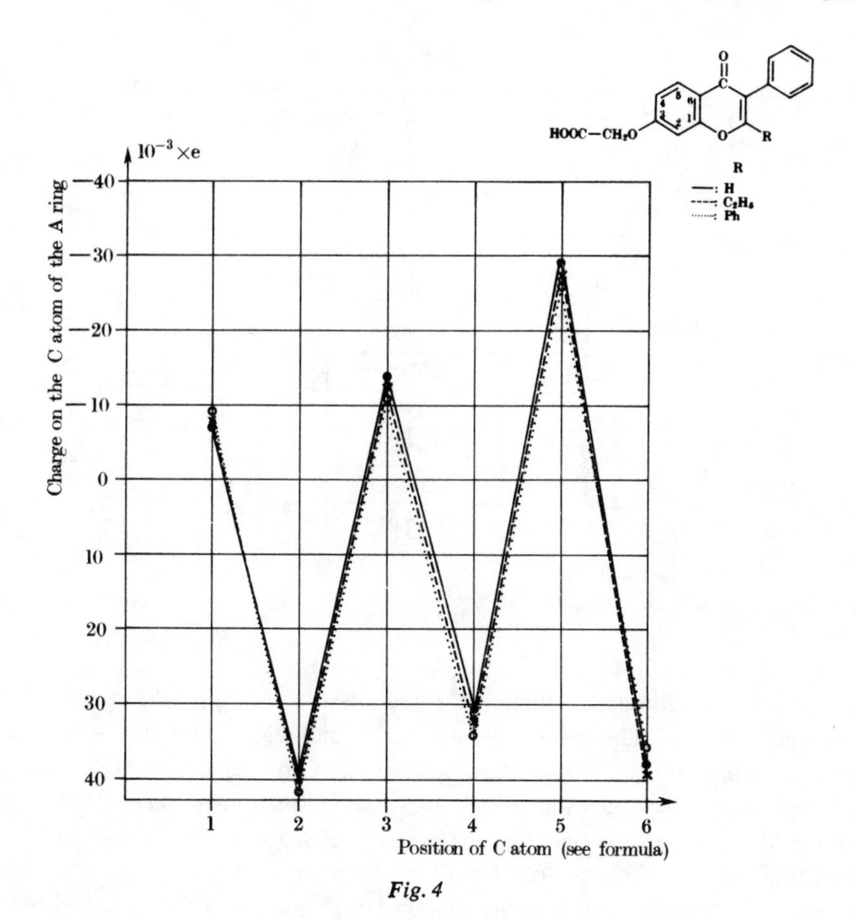

Fig. 4

It can be seen that both S_6^E and S_8^E decrease in the $H \rightarrow C_6H_5$ direction. There is, moreover, a very characteristic change in the values of E^{HOMO} (Energy of the Highest Occupied Molecular Orbital). This value of energy is characteristic of the electron donor character of a given molecule [8, 9]; the smaller this value, the better electron donor the given system, i.e. it will function as a more suitable donor partner in a charge transfer interaction. Accordingly, the best electron donor is the C-2-H compound. (It should be noted that the electron structures of the isoflavone and the flavone skeletons are not influenced by saturated functions, e.g. CH_2–COOH and CH_2–CH_2–$N(CH_3)_2$, attached to the C-7 oxygen atom, and the results given here are valid for both types of compounds.)

Table II

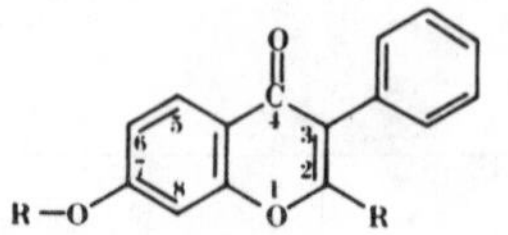

R	S_6^E * $[\beta_0^{-1}]$	S_8^E * $[\beta_0^{-1}]$	E^{HOMO} * $[\beta_0]$	P **
H	1.0124	0.9600	0.6334	1,986
CH_3	1.0110	0.9595	0.6465	1,535
C_2H_5	1.0101	0.9585	0.6861	1,550
$CH(CH_3)_2$	1.0094	0.9583	0.6989	1,409
COOH	1.0124	0.9599	0.6925	1,399
C_6H_5	1.0099	0.9579	0.7299	36,535

* β_0 = quantum chemical unit
** values calculated according to [9]

The results of the calculations have been used to interpret the changes in the biological effects. Unfortunately, no exact biological data are available at present, which would permit the application of the regression technique [10, 11, 14]. Therefore, our statements are mainly qualitative ones. It was seen above that the positive inotropic effect is decreasing in the R = H → COOH direction, and in the case of R = C_6H_5 an expressed cardiodepressive effect prevails. It was also seen that the charge density values of the carbonyl oxygen and the oxygen atom of the pyrone ring change characteristically in the same direction. The changes in the values of S_6^E and S_8^E and the electron donor character (E^{HOMO}) of the compounds are similar. These data might prove that the oxygen atoms of the carbonyl group and the pyrone ring, and the aromatic A ring have an important role in the biological effect, i.e. in the pharmacon–receptor interaction. The change of these parameters, however, is caused by a decrease in the conjugation which is the result of the steric effect of the C-2-R group (*cf.* Fig. 3).

This fact indicates that also the steric effects should be taken into account in the study of the relationship between the biological action and the chemical structure. Various models have been used to study the positions of C-2 and C-6 phenyl groups in the diphenylchromone derivative (Fig. 5). The computations were performed by using various Θ and ψ angle values; the data in Table II relate to $\Theta = 60°$ and $\psi = 80°$.

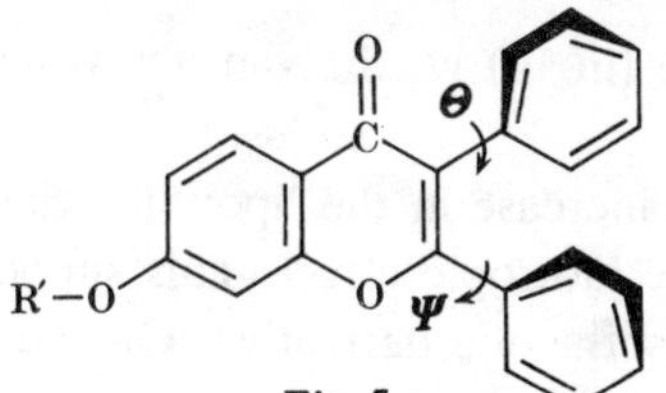

Fig. 5

The two groups cannot assume a coplanar position ($\psi = \Theta = 0°$). It was found, however, that the computed data relating to the various Θ and ψ values do not differ from the data of other derivatives so considerably which, on the basis of the electron structural parameters, might account for the strikingly different biological effects. The most remarkable change was shown by the hydrophobic parameter P (Table II) which differs by an order of magnitude from that of the parent compound. On this basis it might be postulated that the very different lipophilic character of the compounds has a very important role in the opposite biological action of the 2-phenylisoflavone derivative. The role of the steric factors is supported to a certain extent by Fig. 6, which shows the

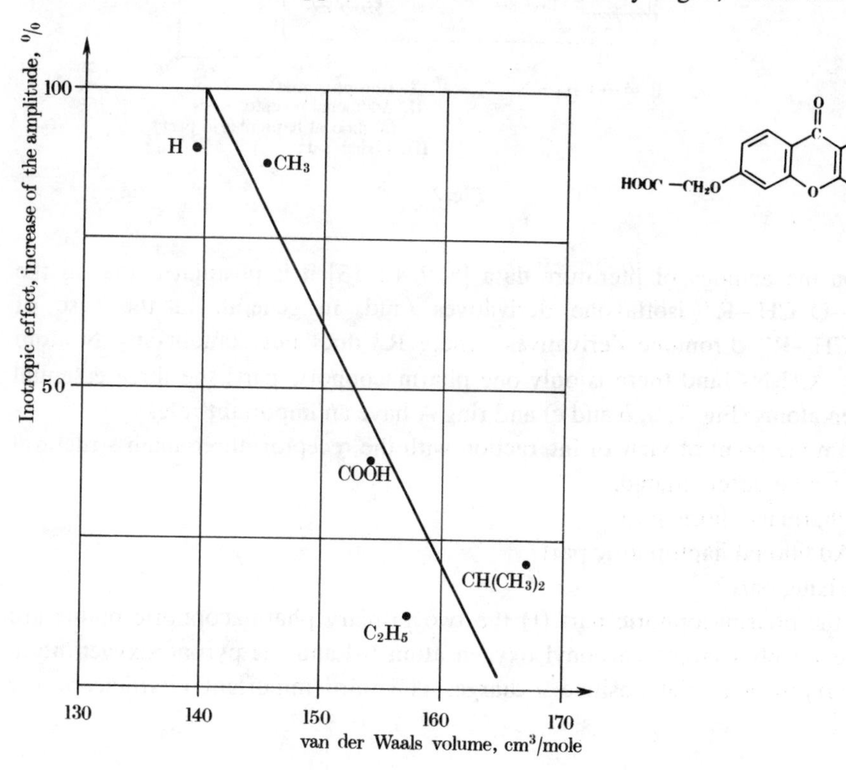

Fig. 6

increase in the amplitude (in %) *vs.* the van der Waals volumes determined by Bondi's method [19].

It can be seen that an increase in the lipophilic character is accompanied by a decrease in the positive inotropic effect. This supports our above statements relating to the 2-phenylisoflavone derivative. On the basis of our calculations

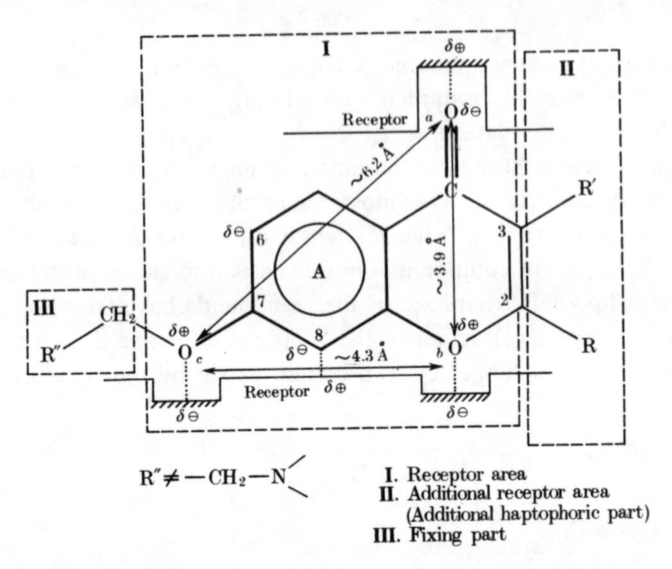

Fig. 7

and on the analogy of literature data [8, 9, 12, 13] it is postulated that in the 2-R-7-O–CH_2–R″ isoflavone derivatives (and, in general, in the case of 7-O–CH_2–R″ chromone derivatives, where R″ does not contain any N atom (R″ ≠ CH_2N<) and there is only one pharmacophoric part) the three essential oxygen atoms (Fig. 7: *a, b* and *c*) and ring A have an important role.

From the point of view of interaction with the receptor, three main structural units can be differentiated:

I. Pharmacophoric part
II. Additional haptophoric part
III. Fixing part

In the pharmacophoric part (I) the two primary pharmacophoric points are the negatively charged carbonyl oxygen atom *(a)* and the pyrone oxygen atom *(b)* carrying a partial positive π-charge. The third important pharmacophoric point is the oxygen atom ($\delta+\pi$) *(c)* attached to the C-7 atom. The condition for the Ӓriens-type "three-point linkage" is therefore met. The distances of these points from each other (data computed on the basis of the standard geometry [18]) are also shown in Fig. 7. The $\delta\ominus$ charges on the C-6 and C-8 atoms of ring A ensure the possibility for ionic bonding [9, 20–22]. By its π-electron system, ring A can interact with the pharmacophoric part by a charge transfer mechanism. The groups (in area II) on the C-2 and C-3 atoms can effect secondary additional bonding interaction with the receptor. This interaction is predominantly hydrophobic in character and has a charge transfer nature in the case of the phenyl substituent. The group on the C-7-O atom (e.g. CH_2–COOH) (III) is fixing the molecular complex formed with the receptor. On the basis of our receptor model, the biological properties outlined above can be explained (Fig. 7; R = X, R′ = C_6H_5, R″ = COO^-).

Substitution on C-2 causes an important change, first of all, on the haptophoric part (II). If R′ = C_6H_5, i.e. in the case of an isoflavone derivative, it can be postulated that the C-3 phenyl group "shields" the carbonyl oxygen atom as compared with the 2-phenyl analogue (flavone derivative), that is, in the case of the isoflavone derivative the binding of the carbonyl oxygen atom to the receptor is more hindered. For the 2,3-diphenyl derivative it should be postulated that the role of part II is modified at the expense of part I, from the point of view of interaction with the receptor (the hydrophobic character is increased). Further, the binding tendencies of points *a* and *b* are changed (the conjugation is decreased by steric effects). Finally, the diphenylchromone derivative has stronger electron acceptor character than either the flavone or the isoflavone compounds. To sum it up, in this group of compounds our experimental results outlined hitherto can be explained by these three factors.

The second group of compounds examined by us consisted of isoflavone derivatives carrying dimethylaminoethyl group at the C-7-O atom (Table III).

In comparison with the 7-O–CH_2–COOH derivatives (see above) a very important structural alteration has taken place which is also reflected by the change of the characteristic biological effect. Namely, these compounds have negative inotropic activity. This change can be explained by the different structure of the side chain attached to the C-7 carbon atom.

In these compounds a nicotine pharmacophore is present, i.e. a phenylcholine ether unit having nicotine-like action (Fig. 8, II).

The pharmacological properties of the phenylcholine ethers have been studied by several workers [8, 9] from the point of view of quantum pharmacology. According to Hey [20, 21] a partial positive charge ($\delta\oplus$) is necessary for the activity at a certain distance from the quaternary nitrogen atom. According to him, the $\delta\oplus$ charge is located on the ethereal oxygen

Table III

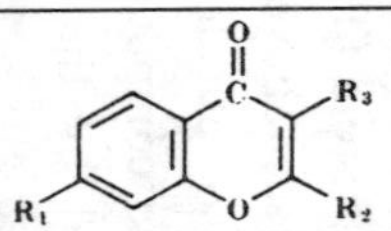

No.	R_1	R_2	R_3	Inotropic activity Percent changes of the contractile amplitude		Change in coronary resistance	Inhibition of fibrillation 20 mM KCl *m. Sartorius*	Antiarrhythmic activity 2 ml/kg $CaCl_2$ solution, i.v. (10% $CaCl_2$)	Coronary vasodilatator effect Vasopressin 2 I.U./kg, i.v.	Number of experiments
				Frog	Rat	Rat	Frog	Rat	Rat	
1.	$HCl \cdot N(CH_3)_2-(CH_2)_2-O-$	H	C_6H_5	neg.	–	1.10	+++	100%	Ø	10
2.	$HCl \cdot N(CH_3)_2-(CH_2)_2-O-$	CH_3	C_6H_5	neg.	–	1.12	+++	100%	Ø	10
3.	$HCl \cdot N(CH_3)_2-(CH_2)_2-O-$	C_6H_5	H	neg.–pos. (biphase effect)	pos.	–	+++	100%	Ø	20
4.	$HCl \cdot N(CH_3)_2-(CH_2)_2-O-$	C_6H_5	C_6H_5	neg.	neg.	1.20	Ø	Ø	Ø	10

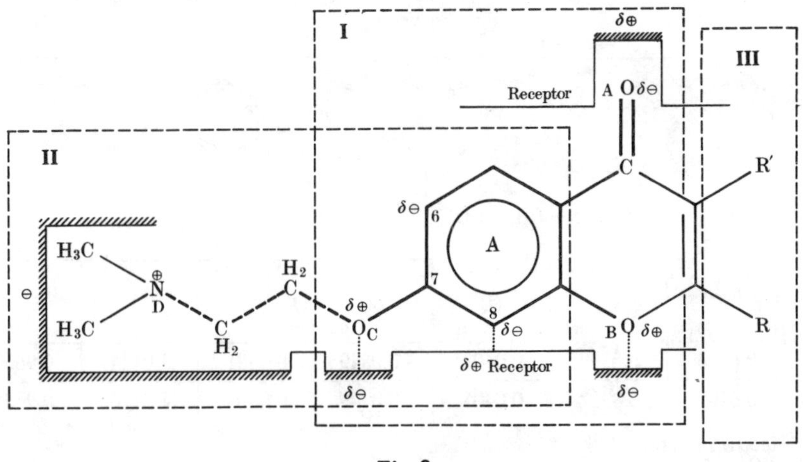

Fig. 8

atom. It is presumed by Ormerod [22] and Sekul and Holland [23] that the presence of a partial negative charge (δ⊖) which is analogous to the negative charge on the carbonyl oxygen atom in acetylcholine, at a certain distance from the "onium head", is a condition for the nicotine-like action. Fukui *et al.* [24] did not find any correlation between the charge density of the ethereal oxygen atom and the relative nicotine-like activity. They found, however, relationships between the frontal electron density of the ethereal oxygen atom [9] and the superdelocalizabilities of the *ortho* positions of the aromatic ring and the biological action. A detailed study by Kier [9, 25] has shown that the δ⊖ charge (which is on the so-called secondary bonding site) should be at a distance of 4.85 ± 0.1 Å from the quaternary nitrogen atom in the case of nicotine-like molecules. It has been stated by Crow *et al.* [26] that the aromatic ring of phenylcholine ethers as an electron donor, provides the secondary bonding site in the charge transfer mechanism. On the basis of computations Coubeils *et al.* [27] have recently concluded that the role of the ethereal oxygen atom, first of all, is to ensure the proper position of the N^+ cationic group in relation to the plane of the aromatic ring. For this position geometrical conditions have been provided by these authors [27]. According to this, the N^+ group should be at a distance of 5.42 Å from the centre of the aromatic ring.

Our results are in good agreement with the data in the literature. Firstly, it can be postulated that for the structure of the dimethylaminoethyl side chain the conditions given for the distance between the charge centres ($N^+ \rightarrow$ δ⊖) are met also in case of our compounds. The negative centre is provided by the δ⊖ charges on the C-6 and C-8 carbon atoms (Table IV).

Table IV

$(CH_3)_2N—CH_2CH_2O$ (9) — structure: 3-phenyl-chromone with R at C-2, ether O at C-7; positions 6, 7, 8 marked

R	E^{HOMO} [ß₀]	Q_6 * [e]	Q_8 * [e]	Q_9 * [e]	FE_9^{HOMO} **	S_6^E [$ß_0^{-1}$]	S_8^E [$ß_0^{-1}$]
H	0.6334	0.0309	0.0402	−0.0849	0.0001	1.0124	0.9600
CH_3	0.6465	0.0312	0.0405	−0.0349	0.0003	1.0110	0.9595

* e = electron charge
** FE^{HOMO} = frontal electron density of the highest occupied molecular orbital [28]

It can be seen in Table IV that the electron density of the C-7-O ethereal oxygen atom is not influenced by introducing a methyl substituent into the molecule at the C-2 carbon atom. The value of FE^{HOMO}, however, is slightly changed, which is in agreement with the result obtained by Fukui *et al.* (*cf.* [9, 26]). The changes in the electron donor character of the molecules are indicated by the E^{HOMO} values. For the interpretation of the biological effects, it is postulated that there are two pharmacophoric parts in these compounds (Fig. 8) instead of only one (*cf.* Fig. 7). Part II in Fig. 8 corresponds to the phenylcholine ethers, while part I corresponds to the analgiphoric grouping characteristic of the C-7-O-substituted flavones or chromones, discussed above. The role of the haptophoric part (III) is the same as in the earlier compounds. Consequently, for these molecules it can be suggested that two receptor parts exist simultaneously, of which the predominant is part II in the case of the isoflavone derivatives.

Further work dealt with a comparison of the properties of flavone and isoflavone derivatives having identical substituents at the C-7 oxygen atom. If this function is the $(CH_3)_2N–CH_2–CH_2$ group, differences are observed between the flavone and isoflavone derivatives.

The *flavone derivative* causes a biphase effect on frog heart (a short initial negative effect followed by an extending positive one). It prevents the KCl fibrillation on isolated frog Sartorius muscle as well as the potassium ion transport in other tests. Furthermore, it has a definite antifibrillatory effect on the heart.

On the contrary, the *isoflavone derivative* has a definite negative inotropic effect. It has no influence upon the fibrillation by KCl on the Sartorius muscle and on the potassium ion transport in other tests. Its antifibrillatory effect is the same as that of the flavone preparation.

Neither compound can be shown to have any β-blocking activity. At the same time, as was seen above, if the isoflavone skeleton contains an $O–CH_2–COOH$ function at the C-7 atom, a definitely positive inotropic effect appears. The differences can be interpreted on the basis of the receptor models described above (Figs 7 and 8). The flavone → isoflavone transition results in an important change, first of all, in the haptophoric addition receptor part (Fig. 7: II; Fig. 8: III). The electron structure of the molecule is influenced by the position of the phenyl group attached to C-2 (flavone) or C-3 (isoflavone). Some of the calculated data are shown in Table V. It can be seen that the carbonyl oxygen atom of the flavone derivative has a greater negative charge ($Q_{=O}$) and the π-electron structure ($\frac{ED\pi}{n}$) is more stable. The isoflavone compound is a better electron donor (E^{HOMO}) than the flavone derivative. The oxygen atom of the pyrone ring has a greater negative charge in the case of the flavone compound than in the isoflavone derivative. The carbon atoms of the aromatic ring A have a resulting negative charge, and this sum of charge is a bit larger for the flavone compound. The character of the aromatic ring B of the flavone derivative is different from that in the isoflavone [18, 29], as it is an electron donor in the flavone, whereas electron acceptor in the isoflavone.

The different biological data can be interpreted on the basis of these properties. We suggested earlier that, depending on the side chain on the C-7 oxygen atom, one or two receptor parts can exist (Figs 7 and 8). The positive inotropic effect can be assigned, in all probability, to the C-7-O-substituted chromone skeleton (Fig. 7: I). If the side chain is a dimethylaminoethyl group, the receptor part II, which can be assigned to the phenylcholine ether grouping, is the decisive structural unit (Fig. 8: II). The importance of the receptor parts I and II (Fig. 8) is not influenced by the haptophoric part III, i.e. by the flavone or the isoflavone character. The flavone derivative ($7\text{-}C–O–CH_2–CH_2–N(CH_3)_2$) behaves in a manner as if it were two compounds at the same time. In the isoflavone derivative the C-3 phenyl group shields the carbonyl group, i.e. it prevents the interaction between the carbonyl oxygen and the receptor part. At the same time, as a result of the electron acceptor properties of the phenyl group, a charge transfer interaction between part III and the receptor is out of question. The result is that in this case the receptor part II (Fig. 8) (phenylcholine ether) becomes predominant and this causes a negative inotropic effect. On the contrary, in the flavone derivative this greater charge density on the carbonyl oxygen atom, the more negative character of ring A

Table V

$R = (CH_3)_2N{-}CH_2CH_2{-}$

		(isoflavone structure, R–O at position 7)	(flavone structure, R–O at position 7)
Q_1	[e]	−0.2148	−0.1998
Q_2	[e]	−0.0849	−0.0849
Q_3	[e]	0.4593	0.4685
$\frac{ED\pi}{n}$	[ß₀]	0.3187	0.3256
InP		7.5941	7.4929
E^{HOMO}	[ß₀]	0.6334	0.6682
E^{LEMO}	[ß₀]	−0.5914	0.4928
$\sum_{i=4}^{9} Q_{\pi}^{A}$	[e]	0.0627	0.0650

and the elimination of the shielding effect of the phenyl group result in an increased importance of the receptor part I (Fig. 8), and in this case the roles of parts I and II appear simultaneously in an increased degree. This derivative, as a result of the increasing role of part I, behaves as if it were two compounds. On this basis it can be said that the flavone skeleton is more favourable from the point of view of the receptor activity. Clearer effects can, however, be expected in the case of the isoflavone derivatives.

Summing up, it can be stated that the data obtained from quantum chemical calculations provide a basis for the study of the relationship between the biological action and the chemical structure of flavonoid compounds. We do not consider our study to be complete. It is just a beginning of the application of the quantum chemical and quantum pharmacological methods to flavonoids. In the course of our further work we want to prepare and test new model compounds with special attention to the study of steric properties as these may have decisive role in this group.

Computation. The quantum chemical calculations were performed by the application of the HMO method generally used for similar purposes [9, 30], applying the earlier parameters [31, 32]. The effect of the methyl group was computed on the basis of the inductive model [30]. In the calculation of the interaction between the phenyl groups and the chromone skeleton the resonance integral parameter was taken into account in the usual manner [30]:

$$\beta_{ij} = \beta^{\circ}_{ij} \cdot \cos \Theta$$

where $\Theta = 20°$ and $60°$ were used for the flavone (C-2-Ph) and the isoflavone (C-3-Ph) ($\beta^{\circ}_{ij} = 1$) compounds, respectively. The effect of the saturated side chain on the C-7-O atom was taken into account by applying the —0.1 factor [30] in the Coulomb integral parameter of the –O– atom. The computations were performed on an ODRA-1204 computer.

REFERENCES

1. BINON, F., *Chim. Ther., 7,* 156 (1972).
2. KLOSA, J., *J. prakt. Chem., 22,* 259 (1963).
3. HETÉNYI, E., SZABÓ, V., LÉVAI, A. and BOGNÁR, R., Soc. Pharmacol. Hung. V. Conferentia Budapest, Akadémiai Kiadó, Budapest, 1971.
4. HETÉNYI, E., LÉVAI, A. and BOGNÁR, R., *Acta Pharm. Hung., 44,* 1 (1974).
5. DA RE, P., COLLEONI, A. and SETNIKAR, J., *Il Farmaco, Ed. Sci. (Milano), 13,* 561 (1958).
6. DA RE, P. and COLLEONI, A., *Ann. Chim. (Rome), 49,* 1632 (1959).
7. DA RE, P. and VERLICCHI, L., *Ann. Chim. (Rome), 50,* 1273 (1960).
8. KOROLKOVAS, A., "Essentials of Molecular Pharmacology. Background for Drug Design". Wiley-Interscience, New York, 1970.
9. KIER, L. B., "Molecular Orbital Theory in Drug Research". Academic Press, New York, 1971.
10. CAMMARATA, A., "Molecular Orbital Studies in Chemical Pharmacology" (ed. KIER, L. B.), Springer-Verlag, Berlin, 1970.
11. HANSCH, C. and FUJITA, T., *J. Am. Chem. Soc., 86,* 1616 (1964).
12. ARIENS, E. J., "Molecular Pharmacology. The Mode of Action of Biologically Active Compounds". Academic Press, New York, 1964.
13. ARIENS, E. J. (ed.), "Drug Design," Med. Chem. Series, Vol. 11, Academic Press, New York, 1971.
14. CAMMARATA, A., *J. Med. Chem., 12,* 314 (1969).

15. BRAUN, H. C. and OKAMOTO, Y., *J. Am. Chem. Soc., 80,* 4979 (1958).
16. FUJITA, T., IVASA, I. and HANSCH, C., *J. Am. Chem. Soc., 86,* 5175 (1964).
17. TAFT, R. W., "Steric Effects in Organic Chemistry" (ed. NEWMAN, M.), Wiley and Sons Inc., New York, 1956.
18. DINYA, Z., KISS, A., LITKEI, GY. and LÉVAI, A., In the press.
19. BONDI, A., *J. Phys. Chem., 68,* 441 (1964).
20. HEY, P., *J. Physiol. (London), 110,* 28 (1949).
21. HEY, P., *Brit. J. Pharmacol., 7,* 117 (1952).
22. ORMEROD, W. E., *Brit. J. Pharmacol., 11,* 267 (1956).
23. SEKUL, A. A. and HOLLAND, W. C., *J. Pharmacol., Exptl. Therap., 132,* 171 (1961).
24. FUKUI, K., NAGATA, C. and IMAMURA, A., *Science, 132,* 87 (1960).
25. KIER, L. B., *Mol. Pharmacol., 4,* 70 (1968).
26. CROW, J., WASSERMANN, O. and HOLLAND, W. C., *J. Med. Chem., 12,* 764 (1969).
27. COUBEILS, J. L., COURRIÈRE, PH. and PULLMAN, B., *J.Med. Chem., 15,* 453 (1972).
28. FUKUI, K., YONEZAWA, T. and NAGATA, C., *Bull. Chem. Soc. Japan, 27,* 423 (1954).
29. EFIMOV, A. A. and KOMAROW, V. M., *Opt. i. Spektr., 30,* 19 (1971).
30. STREITWIESU, A., "Molecular Orbital Theory for Organic Chemists". Wiley, New York, 1961.
31. DINYA, Z. and SZABÓ, S., *Acta Chim. (Budapest), 72,* 65 (1972).
32. DINYA, Z. and BOGNÁR, R., *Acta Chim. (Budapest), 73,* 453 (1972).

ABSORPTION AND METABOLISM OF FLAVONOIDS

by
H. FÖRSTER

Zentrum der Biologischen Chemie der Universität
Frankfurt, FRG

Bioflavonoids are widely used for therapeutical purposes and they are administered by both oral and parenteral routes [1]. In former studies the oral absorption of bioflavonoids has been questioned [1, 15]. As the authors did not detect the administered flavonoids in the urine, they concluded that these substances were not significantly absorbed [15]. On the other hand, Demole and Guerne [5] discovered in test animals an excretion of rutosides in the urine after oral administration. Mirkovitch *et al.* [13] administered ^{14}C-labelled rutosides to dogs and detected the radioactivity in both the blood plasma and the urine of the animals.

In our studies conducted in rats we detected the rutosides in the *vena portae* as well as in the arterial blood of the living animal during perfusion of the small intestine with solutions containing either 3′,4′,7-trihydroxyethyl- (tri-HR) or tetrahydroxyethylrutoside (tetra-HR) [7]. In the perfusion experiment with tri-HR we discovered an additional fluorescent substance in the blood of the animals [7]. Further investigations on the metabolism of various rutosides and their biliary excretion were carried out in the perfused rat liver [6].

The aim of this paper is to summarize our experimental results with various hydroxyethyl-substituted rutosides [6, 7, 9]. The absorption experiments were performed in male Sprague–Dawley rats according to a modified perfusion method [8]. After narcotizing the animals, vascular catheters were inserted in the *arteria carotis* and the *vena portae*. Blood coagulation was inhibited by administration of heparin. The entire small intestine of the animals was continuously perfused with solutions containing isotonic sodium chloride and the respective rutoside, the perfusion rate being 1 ml/min. The experiment lasted for 3 hours; blood samples were taken every full hour.

The liver perfusion studies were carried out with Wistar rats in a humid chamber according to a modified method of Hems *et al.* [12]. The perfusion solution used was composed of Krebs–Ringer-bicarbonate and 7 % of Haemaccel®, mixed with heparinized fresh rat blood (1:1). This experiment, too, lasted

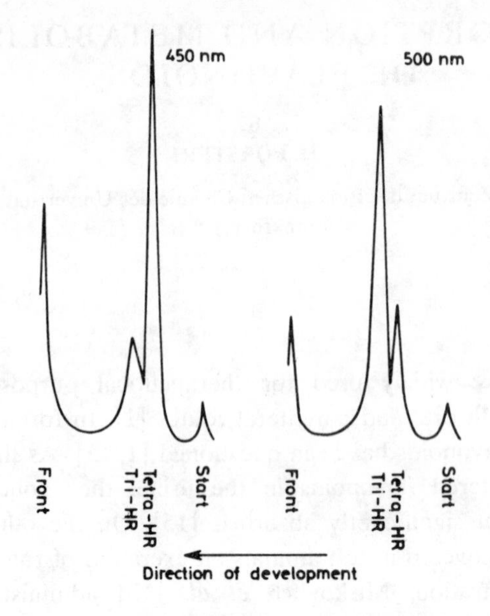

Fig. 1. Fluorescence of 3′,4′,7-trihydroxyethylrutoside (97.5 %) and tetrahydroxyethylrutoside (2.5 %) using different wavelengths

for 3 hours; samples were taken every full hour. The 110 ml perfusion solution was recirculated in a closed system.

Analyses of the rutosides were made by thin-layer chromatography. The blood was deproteinized by acetone. Samples of 10–40 μl were applied to Kieselgel plates (Merck 5/21). The solvent used was water : acetic acid : n-butanol (1:1:3). The incubation period was 6 hours. After drying, the plates were sprayed with methanolic aluminium chloride solution. Under these conditions the R_f-value of 3′,4′,7-trihydroxyethylrutoside is 0.48, the R_f-value of tetrahydroxyethylrutoside being 0.35 [6, 7]. The plates were photographed by means of a polaroid camera during fluorescence excitation at 365 nm. With excitation again at 365 nm, the plates were then measured with a fluorescence spectrophotometer (Fig. 1). The emission maximum of tetrahydroxyethylrutoside and of 3′,4′,7-trihydroxyethylrutoside was according to the blue fluorescence approx. 450 nm (Fig. 2), whereas according to the yellow fluorescence the emission maximum of 4′,7-dihydroxyethylrutoside, 3′,4′,7-trihydroxyethylrutoside and the 4′,7-dihydroxyethylrutoside metabolite was found to be about 500 nm [6, 7, 9] (see also Fig. 2). As shown in Fig. 1, the fluorescence of tetrahydroxyethylrutoside is much more intense than the fluorescence of 3′,4′,7-trihydroxyethylrutoside. Therefore, the fluorescence-chromatographic detection of tetrahydroxyethyl-

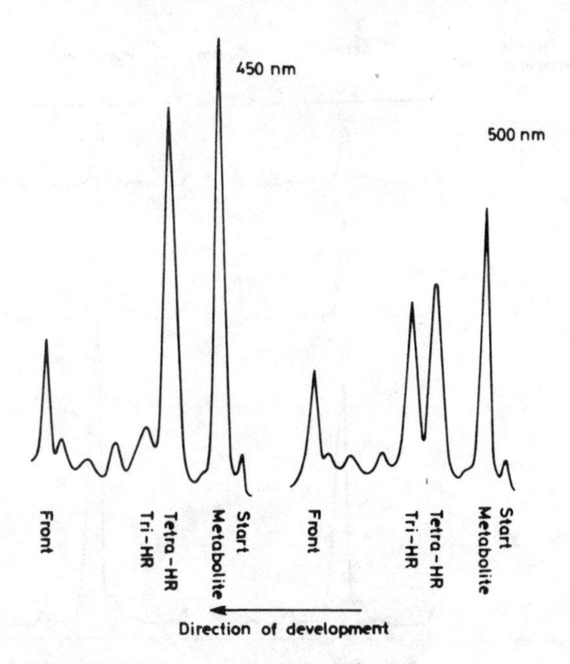

Fig. 2. Fluorescence of 3′,4′,7-trihydroxyethylrutoside, tetrahydroxyethylrutoside and the metabolites in the portal vein blood of the animals using different wavelengths

rutoside is easier than that of 3′,4′,7-trihydroxyethylrutoside or 4′,7-dihydroxyethylrutoside. While the substance was administered, tetrahydroxyethylrutoside was detected in both the portal and the arterial blood of the intestinal preparation of the living rat. The concentration of the substance in the portal blood was found to be above the concentration found in the arterial blood, the difference being 1–5 mg/100 ml. No additional fluorescent substance was detected in any of our experiments with tetrahydroxyethylrutoside [7].

The quantitative analyses made of the fluorescence in the blood during a 5% tetrahydroxyethylrutoside perfusion of the intestinal lumen are presented in Fig. 3. In the course of the experiment the fluorescence was found to increase. This experiment demonstrates that the fluorescence in portal blood is stronger than in arterial blood. Before the experiment was started, practically no fluorescence could be detected in the blood of the animals. According to the calibration value, small amounts of 3′,4′,7-trihydroxyethylrutoside were found in the substance tetrahydroxyethylrutoside. These small amounts of 3′,4′,7-trihydroxyethylrutoside were also observed in the blood during intestinal perfusion. In some perfusion experiments with tetrahydroxyethylrutoside we

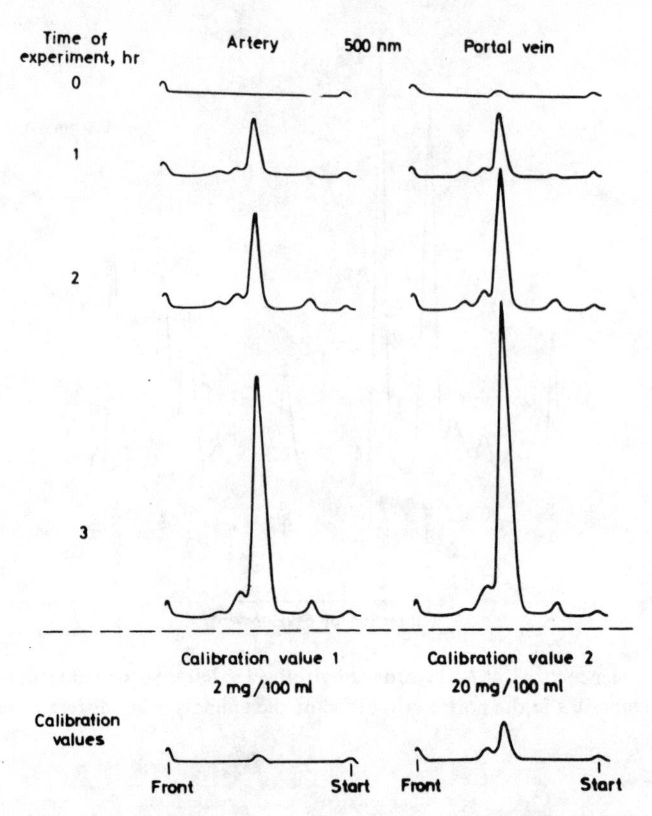

Fig. 3. Thin-layer chromatograms of arterial and portal vein blood during intestinal perfusion with 5 % tetrahydroxyethylrutoside

furthermore discovered some fluorescence in the same spot in which a new metabolite was detected during perfusion with 3′,4′,7-trihydroxyethylrutoside.

In our investigations with 3′,4′,7-trihydroxyethylrutoside, this metabolite was found in rather high concentrations.

The next experiment was performed with the intestinal perfusion using 3′,4′,7-trihydroxyethylrutoside. We unfortunately discovered that the 3′,4′,7-trihydroxyethylrutoside compound contained also significant amounts of tetrahydroxyethylrutoside (*cf.* Figs 2 and 4). From the chromatogram one may see that one hour after the beginning of the experiment two spots with blue and one spot with yellow fluorescence were present, and it is to be assumed that the intensity may have gradually increased. This is especially true of that spot which is absent from the calibration values and which represents a new substance. This substance was obviously formed in the animal body following absorption of the rutosides.

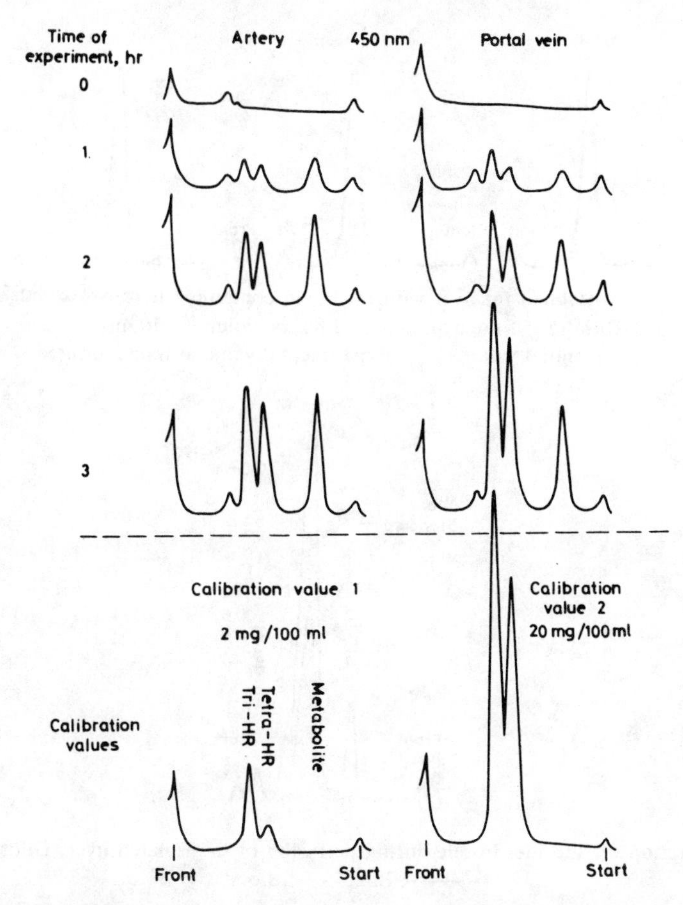

Fig. 4. Thin-layer chromatogram of arterial and portal vein blood during intestinal perfusion with 2.5 % 3′,4′,7-trihydroxyethylrutoside (95 %) contaminated with tetrahydroxyethylrutoside (2.5 %).

Figure 4 represents the quantitative analysis by means of fluorescence chromatography of this experiment. With the exception of the blank value, three fluorescent peaks were noted, one of them representing the newly synthesized metabolite. Whereas the concentrations of 3′,4′,7-trihydroxyethylrutoside and tetrahydroxyethylrutoside were, in general, found to be higher in the portal blood than in the arterial blood, the contrary was true of the metabolite. This result indicates that the metabolite is formed in the liver of the animals.

Furthermore, in the liver perfusion studies we found that tetrahydroxyethylrutoside was excreted in the bile to a considerable amount. The uptake of tetrahydroxyethylrutoside by this organ was not very good and — as can be

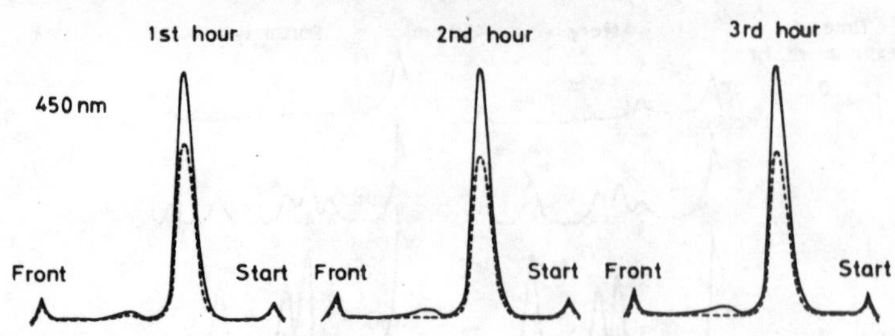

Fig. 5. Perfusion of isolated rat liver with rat blood containing tetrahydroxyethylrutoside. Duration of the experiment: 3 hours; volume: 110 ml.
———: Initial value; - - - -: Experimental value at time indicated

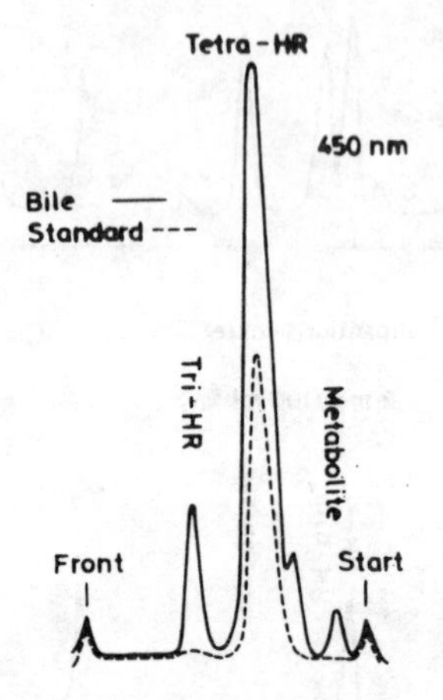

Fig. 6. Excretion of rutosides in bile during perfusion of the isolated liver (Dilution 1:10)
———: Bile; - - - -: Blood

seen in Fig. 5 — no further fluorescent substances were detected in the blood. In spite of this, additional fluorescent substances were found by us in the bile (Fig. 6). This might be a result of the contamination with 3′,4′,7-trihydroxyethylrutoside.

Figure 5 demonstrates the quantitative analysis of the thin-layer chromatogram obtained by means of a fluorescence spectrophotometer. The decrease in the concentration of tetrahydroxyethylrutoside observed during the first hour of the experiment is partly due to the extension of the partition space. This extension resulted from an increase in the fluid volume caused by the inclusion of the liver. Figure 6 shows the results obtained by chromatography of the bile. One may see that in addition to tetrahydroxyethylrutoside, 3′,4′,7-trihydroxyethylrutoside and the metabolite were excreted. However, the concentration of the metabolite was found to be rather low.

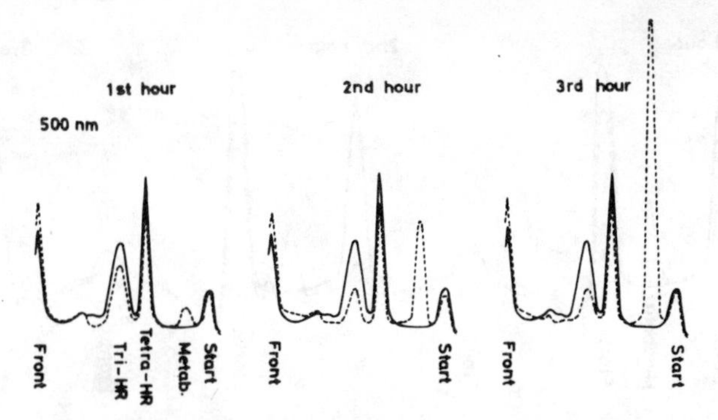

Fig. 7. Perfusion of isolated rat liver with rat blood containing 40 mg/100 ml 3′,4′,7-trihydroxyethylrutoside. Duration of the experiment: 3 hours; volume: 110 ml. ———: Initial value; - - - -: Experimental value at time indicated

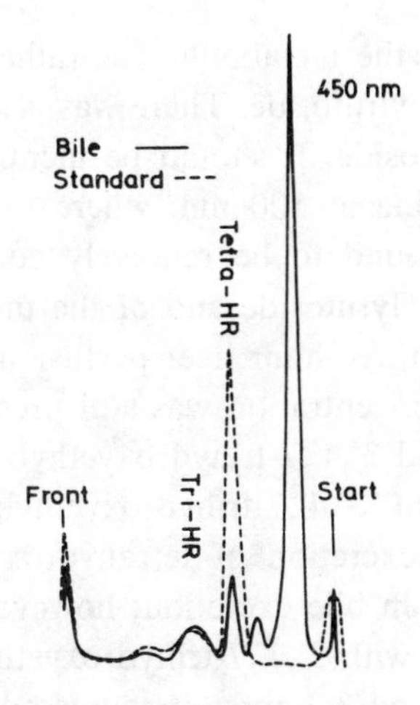

Fig. 8. Excretion of rutosides in the bile during perfusion of the isolated liver with trihydroxyethylrutoside (Dilution 1:25) ———: Bile; - - - -: Blood

In similar experiments with 3′,4′,7-trihydroxyethylrutoside the metabolite was detected in the blood as soon as one hour after the beginning of the experiment (Fig. 7). As can be seen, the metabolite was also excreted in the bile (Fig. 8). In comparison with tetrahydroxyethylrutoside, the concentration of 3′,4′,7-trihydroxyethylrutoside decreased rather rapidly and a corresponding increase in the metabolite concentration was evident. However, a decrease in the concentration of tetrahydroxyethylrutoside being present as contamination was hardly to be noted throughout the whole experiment. According to the chromatogram

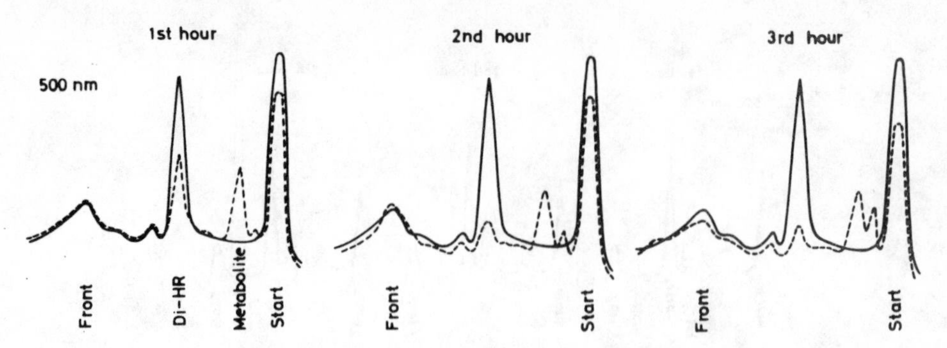

Fig. 9. Perfusion of isolated rat liver with rat blood containing 30 mg/100 ml 4′,7-dihydroxyethylrutoside. Duration of the experiment: 3 hours; volume: 110 ml. ————: Initial value; - - - -: Experimental value at time indicated

of the bile, the excretion of the metabolite was rather high, especially in comparison with tetrahydroxyethylrutoside. There was also a considerable excretion of 3′,4′,7-trihydroxyethylrutoside. It should be mentioned that the registration of the fluorescence was made at 500 nm, where the sensitivity of 3′,4′,7-trihydroxyethylrutoside was found to be relatively good as compared with the sensitivity of tetrahydroxyethlyrutoside and of the metabolite (see also Fig. 1).

In summary, following a three-hour liver perfusion, 70–80 % of the original tetrahydroxyethylrutoside concentration was still present in the blood, whereas only 25–50 % of the original 3′,4′,7-trihydroxyethylrutoside concentration was recovered. The excretion of 3′,4′,7-trihydroxyethylrutoside in the bile was higher compared with the excretion of tetrahydroxyethylrutoside. Both substances effected an increase in bile excretion; however, the increase was found to be higher in experiments with 3′,4′,7-trihydroxyethylrutoside. The basal bile excretion amounted to 0.7 ml/3 hours; it increased to 1.0 ml/3 hours after addition of tetrahydroxyethylrutoside and to 1.2 ml/3 hours after addition of 3′,4′,7-trihydroxyethylrutoside. The concentration of 3′,4′,7-trihydroxyethylrutoside in the bile was found to be about 25 times higher than that in the blood, whereas this ratio was found to be only between 8:1 and 20:1 with tetrahydroxyethylrutoside. It was also observed that the metabolite concentration was 10 times higher in the bile than in the blood.

In further experiments the metabolism of 4′,7-dihydroxyethylrutoside was investigated in the perfused liver. The problem with this substance is its low solubility in water. We therefore used a concentration of only 30 mg/100 ml instead of 40 mg/100 ml as in the other experiments. The uptake of 4′,7-dihydroxyethylrutoside by the liver is rather high, it is comparable with the uptake of 3′,4′,7-trihydroxyethylrutoside (Fig. 9).

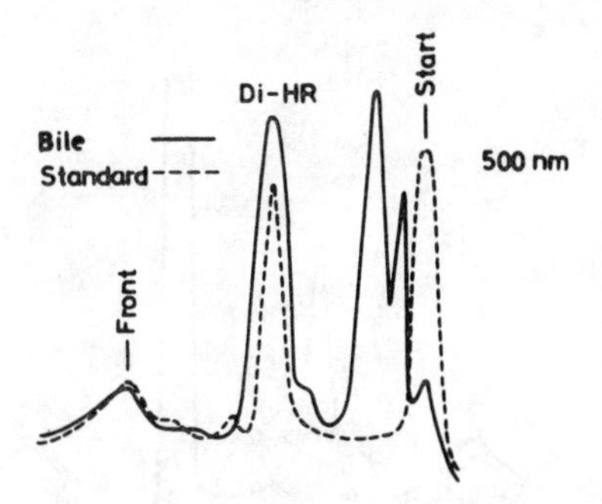

Fig. 10. Excretion of rutosides in the bile during perfusion of the isolated liver with 4′,7-dihydroxyethylrutoside (Dilution 1:10)
———: Bile; - - - -: Blood

After a three-hour liver perfusion, the concentration of 4′,7-dihydroxyethylrutoside in the blood was found to be 20–50 % of the initial concentration. 4′,7-Dihydroxyethylrutoside was contamined by 3′,4′,7-trihydroxyethylrutoside. For this reason, both 3′,4′,7-trihydroxyethylrutoside and the blue fluorescent metabolite were found in the bile in low concentration. At the end of the experiment we also discovered this blue fluorescent metabolite in the blood. Another metabolite with a yellow fluorescence was discovered in the course of the same experiment. The fluorescence of this metabolite was similar to that of 3′,4′,7-trihydroxyethylrutoside or 4′,7-dihydroxyethylrutoside. This metabolite was found to be excreted in high concentrations in the bile (Fig. 10). In experiments in which 4′,7-dihydroxyethylrutoside was added, the bile excretion was found to be reduced as compared with the controls. The bile excretion, however, was increased after addition of tetrahydroxyethylrutoside and especially of 3′,4′,7-trihydroxyethylrutoside.

In further experiments the metabolism of 7-monohydroxyethylrutoside was investigated. This substance, too, is not readily soluble in water. A rapid decrease in the concentration of 7-monohydroxyethylrutoside in the blood was noted (Fig. 11). In a part of the experiments a yellow fluorescent metabolite was detected during the liver perfusion experiment. As shown in Fig. 11, the concentration of this substance did not increase in the course of the experiment. Additionally, this metabolite was not seen in the bile in most experiments (Fig. 12). We assume that this substance was 7-monohydroxyethylrutoside substituted in the 4′- or 3′-position. On the other hand, the decrease in fluorescent substances was so high that conversion to a non-fluorescent substance is supposed. This should be due to ring fission.

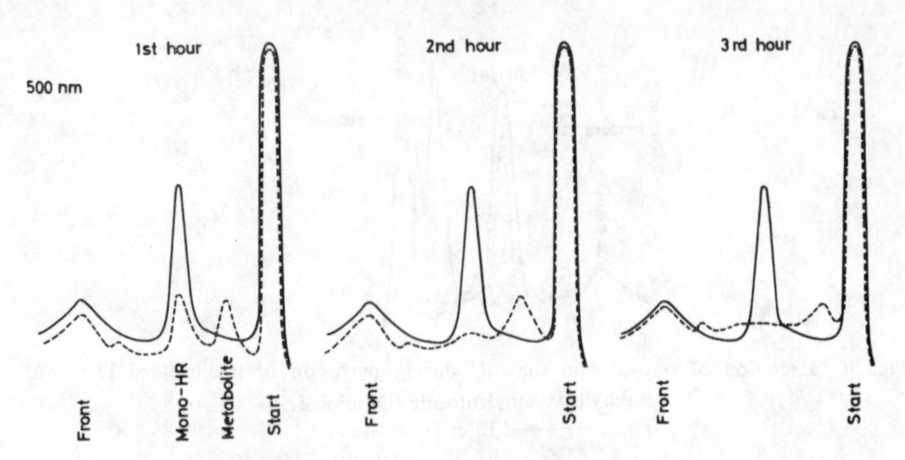

Fig. 11. Perfusion of isolated rat liver with rat blood containing 40 mg/100 ml 7-monohydroxyethylrutoside. Duration of the experiment: 3 hours; volume: 110 ml.
———: Initial value; - - - -: Experimental value at time indicated

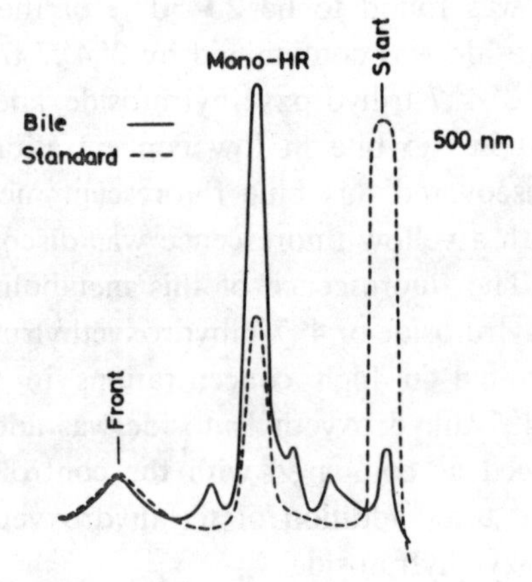

Fig. 12. Excretion of rutosides in the bile during perfusion of the isolated liver with 7-monohydroxyethylrutoside (Dilution 1:10).
———: Bile; - - - -: Blood

We assume that the metabolites found in our experiments represent the glucuronides of the original compounds. These metabolites were destroyed by incubation with glucuronidase. Furthermore, the metabolites were more easily soluble in water than 3′,4′,7-trihydroxyethylrutoside and 4′,7-dihydroxyethylrutoside, i.e. the R_f-value was considerably decreased by the additional group. This may be interpreted as incorporation of a larger hydrophilic residue into the molecule. On the other hand, 7-monohydroxyethylrutoside was rapidly taken up by the liver and only partly excreted in the bile. Since little fluorescent substances were detected in these experiments after the perfusion lasting for three hours, one may speculate that 7-monohydroxyethylrutoside partly might be decomposed by ring fission.

These results are in agreement with those obtained by Griffiths *et al.,* who performed studies in rats with cannulated bile ducts [10, 11]. After intravenous administration of 3′,4′,7-trihydroxyethylrutoside, they detected two blue fluorescent compounds in the bile which were believed to be glucuronide conjugates of the intact 3′,4′,7-trihydroxyethylrutoside. After intravenous administration, the excretion of 3′,4′,7-trihydroxyethylrutoside in the bile was found to be higher than that of tetrahydroxyethylrutoside. This is in accordance with our experimental results. According to Griffiths *et al.* the glucuronidation of flavonoids seems to occur by a common mechanism [10, 11]. Das *et al.* [2–4] observed a glucuronidation of (+)-catechin and also fission products in the intact rat. However, according to studies conducted by Takács *et al.* [14] in the perfused rat liver, rutin was degraded mainly by ring fission, and 11 different compounds were isolated.

In contrast to Griffiths and Barrow [10], we did not detect in our experiments large quantities of compounds with a yellow fluorescence after addition of 3′,4′,7-trihydroxyethylrutoside to the blood or to the intestinal perfusion medium. The formation of these substances might be a consequence of bacterial degradation of 3′,4′,7-trihydroxyethylrutoside in the large intestine, as Griffiths suggested in his lecture. Contrary to the studies performed by the aforementioned authors, 3′,4′,7-trihydroxyethylrutoside did not come into contact with the large intestine in our experiments, and furthermore our experiments were finished after 3 hours.

The results of the present work can be summarized as follows. In animal experiments evidence was obtained that absorption of rutosides occurs in the small intestine. The substances absorbed by the intestinal lumen are partly metabolized by the liver. Ring fission is likely to occur if 7-monohydroxyethylrutoside is used. On the other hand, the free hydroxyl groups in 4′,7-dihydroxyethylrutosides and 3′,4′,7-trihydroxyethylrutosides were found to have undergone substitution, probably mainly by glucuronidation. According to the findings of Griffiths in his studies on microorganisms, we assume that the substituents in the position 3′ or 4′ might be a protection against ring fission in liver metabolism. There is evidence that substitution occurs mainly in the side ring, because when using di-HR the substituted rutosides have mainly a yellow fluorescence. A compound with blue fluorescence was detected only when 3′,4′,7-trihydroxyethylrutoside was used for liver perfusion.

REFERENCES

1. Boehm, K., "Die Flavonoide", Editio Cantor, Aulendorf, 1967.
2. Das, N. P., *Biochem. Pharmacol., 20,* 3435 (1971).
3. Das, N. P. and Griffiths, L. A., *Biochem. J., 110,* 449 (1968).
4. Das, N. P. and Sothy, S. P., *Biochem. J., 125,* 417 (1971).
5. Demole, V. and Guerne, R., *Helv. Physiol. Pharmacol. Acta, 18,* C18 (1960).
6. Förster, H., Bruhn, U. and Hoos, I., *Arzneimittel-Forsch., 22,* 1312 (1972).
7. Förster, H. and Ziege, M., *Fortschr. Medizin, 89,* 627 (1971).
8. Förster, H., Meyer, E. and Ziege, M., *Rev. Europ. Clin. Biol., 17,* 958 (1972).
9. Förster, H. and Harth, P. (In preparation).
10. Griffiths, L. A. and Barrow, A., *Angiologica, 8,* 162 (1971).
11. Griffiths, L. A. and Smith, G. E., *Biochem. J., 130,* 141 (1972).
12. Hems, R., Ross, B., Berry, M. N. and Krebs, H. A., *Biochem. J., 101,* 284 (1966).
13. Mirkovitch, V., Robison, J. W. L., Bel, F. and Gumma, A., *Arzneimittel-Forsch., 23,* 967 (1973).
14. Takács, Ö., Benkő, S., Varga, L., Antal, A. and Gábor, M., *Angiologica, 8,* 175 (1971).
15. Wienert, V. and Gahlen, W., *Hautarzt, 21,* 278 (1970).

CLOSING OF THE SYMPOSIUM

by

PROF. L. FARKAS

Ladies and Gentlemen,

My closing words can only be those of thanks expressed first of all to our lecturers and all participants who honoured this Symposium by their presence and contributed to its success by their activity. Thank you all for coming to us.

If the hosts say that the Symposium has been a success, it might be considered self-praise; but it was your contribution which made it successful. It appears that this meeting has achieved its purpose in refreshing personal contacts and offering many interesting lectures on different topics of flavonoid chemistry and biochemistry as well as illustrations of the most up-to-date techniques of research in this field. I hope there are many who have the same feeling.

If this is so, you will come again when you receive the invitation to the next, the Fifth Hungarian Bioflavonoid Symposium.

In the good prospect of this, the Fourth Symposium is now closed.

AUTHOR INDEX

Italic page numbers indicate a citation by reference number only.

SUBJECT INDEX